Geomechanics of Coal Seams

SCIENCES

Mechanics, Field Director – Gilles Pijaudier-Cabot

Geomechanics, Subject Head – Gioacchino Viggiani

Geomechanics of Coal Seams

Coordinated by

Matthieu Vandamme
Jean-Michel Pereira

iSTE

WILEY

First published 2025 in Great Britain and the United States by ISTE Ltd and John Wiley & Sons, Inc.

ISTE Ltd
27-37 St George's Road
London SW19 4EU
UK

www.iste.co.uk

John Wiley & Sons, Inc.
111 River Street
Hoboken, NJ 07030
USA

www.wiley.com

Library of Congress Control Number: 2025932187

British Library Cataloguing-in-Publication Data
A CIP record for this book is available from the British Library
ISBN 978-1-78945-201-3

ERC code:
PE10 Earth System Science
PE10_5 Geology, tectonics, volcanology
PE10_10 Mineralogy, petrology, igneous petrology, metamorphic petrology

Contents

Introduction

Why a Book on the "Geomechanics of Coal Seams"?

Jean-Michel PEREIRA and Matthieu VANDAMME
Navier, ENPC, CNRS, Institut Polytechnique de Paris, École nationale des ponts et chaussées, Université Gustave Eiffel, Marne-la-Vallée, France

In the collective consciousness, coal is a material closely associated with the industrial revolution of the 19th century. Today, in an era of environmental urgency, it is rightly criticized as a particularly polluting energy source that emits significant amounts of greenhouse gases. So why a book on the "geomechanics of coal seams?". Because, by focusing on the geomechanical aspect, we consider coal as a reservoir rock rather than as an energy source. The targeted applications are natural gas production (recently designated as energy transition by the European Union[1]; a *coal gas* extraction project in Moselle has just been granted by the French government[2]) and, potentially, underground carbon dioxide storage, although pilot projects in recent years have underscored the challenges of using coal seams for this purpose. In both applications, the reservoir undergoes evolving mechanical stresses during its operation. These stresses must be managed to ensure optimal and safe exploitation. However, coal seams possess specific characteristics that make their hydromechanical behavior significantly

1 Commission Delegated Regulation (EU) 2022/1214 [Online]. Available at: https://eur-lex.europa.eu/legal-content/EN/TXT/?uri=CELEX:32022R1214.

2 Decree of November 20, 2023, granting the concession for conventional liquid or gaseous hydrocarbon mines, known as the "Concession de Bleue Lorraine" (Moselle), to the company La Française de l'Énergie SAS:
https://www.legifrance.gouv.fr/jorf/id/JORFTEXT000048443696.

different from that of other reservoir rocks. Controlling these stress variations therefore necessitates the application of geomechanics, with "geomechanics" referring to the study of the hydromechanical behavior of geological materials and reservoirs.

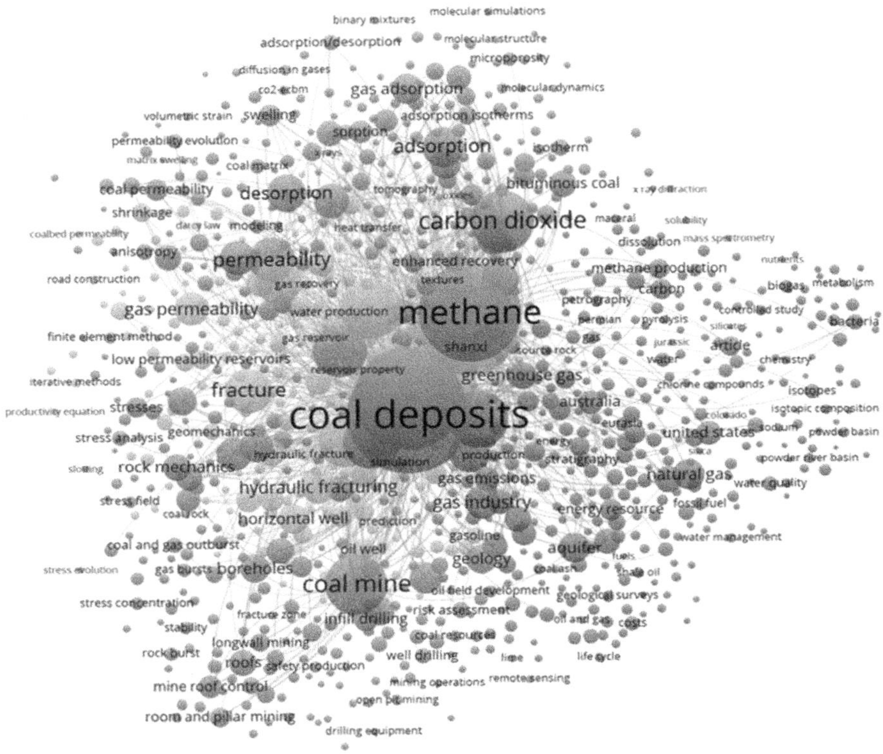

Figure I.1. Keyword co-occurrence network constructed[3] from a corpus based on the presence of the terms "coal seam" or "coalbed" along with "production" or "storage". The network reveals four groups (clusters) of keywords. The green cluster pertains to topics related to methane and carbon dioxide, highlighting the presence of keywords associated with adsorption/desorption and swelling. For a color version of this figure, see www.iste.co.uk/vandamme/coal.zip.

3 The SCOPUS queries, covering all indexed documents and conducted in July 2022, are as follows:
TITLE-ABS-KEY ((coalbed OR "coal seam") AND (storage OR sequestration OR production OR recovery)), labeled "production or storage," returned 8,030 documents.
TITLE-ABS-KEY ((coalbed OR "coal seam") AND (production OR recovery)), labeled "production", returned 7,244 documents.
TITLE-ABS-KEY ((coalbed OR "coal seam") AND (storage OR sequestration)), labeled "storage", returned 1,810 documents.

These two applications are reflected in the keyword co-occurrence network shown in Figure I.1, which indicates that the keywords most frequently associated with scientific studies addressing production or storage issues in *coal seams* include "*methane*" and "*carbon dioxide*" as prominent terms.

The increasing interest in the study of coal as a reservoir rock is clearly demonstrated in Figure I.2, which illustrates the annual number of scientific publications focusing on *coal seams*. This figure also reveals that the scientific community is more focused on coal seams for production purposes (including both coal and natural gas) than for storage (carbon dioxide).

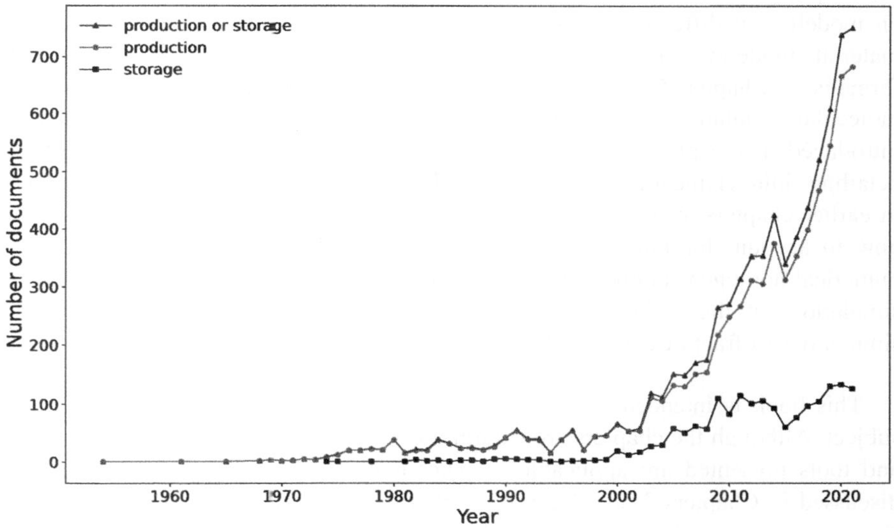

Figure I.2. *Number of scientific articles published per year mentioning the term "coal seam" or "coalbed". For a color version of this figure, see www.iste.co.uk/vandamme/coal.zip*

Beyond its practical applications, the study of coal seam geomechanics holds significant scientific interest. Like other reservoir rocks (such as carbonates) or natural materials (like soils or wood), coal is a complex material due to its genesis, heterogeneity and multi-scale nature, requiring the use of advanced tools for detailed study and modeling. However, coal also exhibits unique characteristics: one being the adsorption–deformation coupling due to its microporous nature; another being that it is a naturally fractured material.

This book, dedicated to the geomechanics of coal seams, is divided into seven chapters. Chapter 1 provides a detailed introduction to coal, including its genesis, characteristics, resources, uses and its role in the context of climate change. Chapters 2 and 3 are predominantly experimental, as experiments are essential for characterizing the geomechanical behavior of the material. Chapter 2 focuses primarily on the hydraulic properties of coal, as these transport properties within the reservoir are key during injection or production processes, emphasizing the fact that coal is a fractured material. Chapter 3 is dedicated to the sorption and poromechanical properties of coal. These sorption properties allow coal to contain large volumes of gas, while the poromechanical properties enable the expression of the coupling between fluid pressures, stresses and deformations. Chapters 4–7 focus on modeling at different scales. Chapter 4, the first modeling chapter, introduces material modeling at the continuous medium scale and the poromechanical formalism. Chapter 5 delves into the finest scale of the material, presenting molecular simulation methods in connection with the continuous medium models introduced in the previous chapter. In contrast, Chapter 6 explores larger scales, detailing finite element methods that, based on the constitutive equations introduced in earlier chapters, enable the simulation of reservoir exploitation. It also explains how to account for fault presence in calculations. Finally, Chapter 7 presents a numerical homogenization method (a “squared” finite element method) that enables simulations at the scale of naturally fractured material, building on finer scale simulations of fractures and the coal matrix.

This book is intended for students, engineers and researchers interested in the subject. Although the chapters are naturally centered on coal, the concepts, methods and tools presented are applicable to other materials. The experimental methods discussed in Chapters 2 and 3 can be applied to other rocks and porous materials. The poromechanical approach introduced in Chapters 4 and 5 is relevant for studying the mechanical behavior of any porous material in the presence of fluids. Moreover, the molecular simulation methods (Chapter 5), finite element methods (Chapter 6) and “squared” finite element methods (Chapter 7) are generic techniques in materials science and solid mechanics. We therefore hope these chapters will spark interest beyond the specific field of coal geomechanics.

We sincerely thank all the contributing authors for accepting our invitation to participate in this book.

1

Coal, Its Genesis, Its Characteristics

Didier BONIJOLY
Geology, mineral resources and geothermal energy expert,
Retired from BRGM, Orléans, France

1.1. Definition

Coal is a combustible rock, defined as a sedimentary, stratified, brown to black rock, predominantly organogenic, and primarily composed of plant debris (Foucault and Raoult 1980). Coal is a carbonaceous rock with a carbon content exceeding 50% by mass.

This rock forms under very specific conditions, requiring the accumulation of a significant volume of plant material without being decomposed by soil microorganisms (which would oxidize the organic matter, converting it into humus and mineral matter).

Such conditions have occurred during various geological periods, when the right combination of climatic (hot and humid equatorial conditions), tectonic (subsidence of basins where plant debris accumulates), thermal (geothermal gradient enabling organic matter maturation) and biological factors were present.

Coals formed from woody material are known as "humic coals", while those derived from algal material are referred to as "sapropelic coals". The former are abundant and widely distributed across all continents, whereas the latter are much rarer.

For a color version of all of the figures in this chapter, see www.iste.co.uk/vandamme/coal.zip.

Geomechanics of Coal Seams,
coordinated by Matthieu VANDAMME and Jean-Michel PEREIRA.

At any scale of observation, coals are rocks with a complex and heterogeneous composition (Van Krevelen 1993). Their main constituents, known as macerals, are analogous to minerals in inorganic rocks.

The term "maceral" was introduced by M. C. Stopes in 1935 and later standardized in an International Energy Agency (IEA) publication in 1988 (Carpenter 1988).

These macerals possess specific physicochemical properties that determine the behavior of coals. They can be identified under optical microscopy based on their morphology and reflectance (Charcosset and Nickel-Pepin-Donat (1990); see Figure 1.1). They are as follows[1]:

– Vitrinite: this is derived from plant cell walls. It consists of a ligno-cellulosic gel whose polymers undergo physicochemical transformation during biochemical and thermal maturation. Carbon ("C") is the primary component of coal (making up 50%–90% of its mass). Vitrinite is opaque under optical microscopy (in transmitted light) and exhibits increasing reflectance with higher coal ranks (see section 1.5).

– Exinite (or liptinite): this is derived from spores, cuticles and resins. It is mainly found in lignite, cannel coal (oil shales of continental origin) and boghead coal (coals of algal origin rich in volatile matter). This maceral vanishes with increasing temperature during burial and is absent in medium volatile bituminous coals (Pickel et al. (2017); see section 1.5).

– Inertinite: this is derived from oxidized plant material during peat formation. Inertinite is distinguished by its morphology (preserving cellular structures) and very high reflectance.

Although coal is predominantly composed of organic molecules containing carbon, hydrogen and oxygen, it also contains other elements inherited from its original material, such as sulfur, nitrogen and silicate or carbonate minerals, which may contain trace elements like cadmium and mercury.

The structure of coal is complex and has been modeled extensively (Prins et al. 1989; Charcosset and Nickel-Pepin-Donat 1990). Organic molecules form sheet-like structures, whose irregular arrangement creates porosity (see Figure 1.2). As a result, this material has a high porosity and specific surface area, both of which decrease as the coal rank increases.

1 See also: https://www.aquaportail.com/definition-12371-maceral.html and https://coalandcarbonatlas.siu.edu/coal-macerals/coal-macerals-tutorial.php.

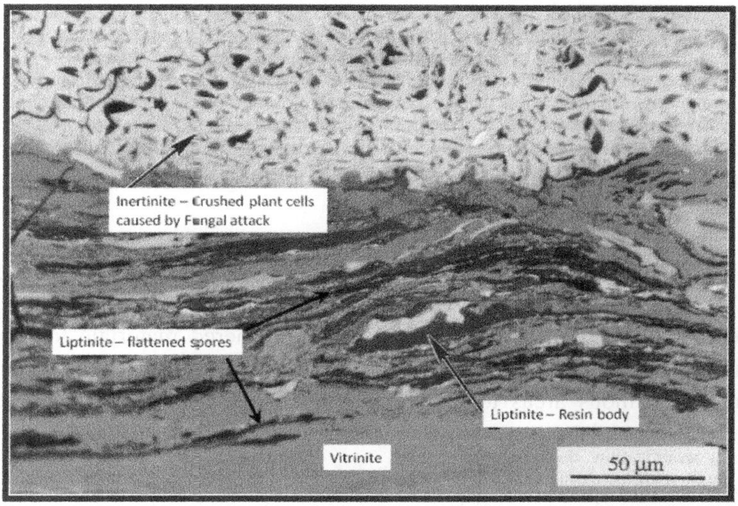

Source Organic Petrology G Taylor, M Teichmuller, A Davis & C Diessel et al. 1998

Figure 1.1. *Organic components of coal (Taylor et al. 1998)*

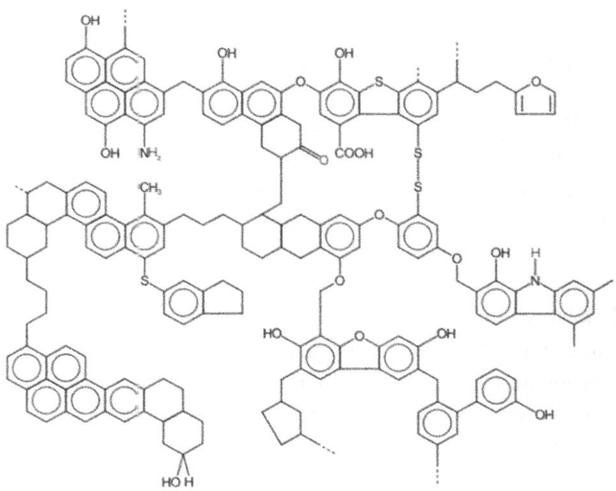

Figure 1.2. *Two-dimensional model of the structure of a coal macromolecule. The illustration shows the molecular network, exchange surfaces and porosity. Adapted from Atkins and Jones (1997). Source: https://commons.wikimedia.org/wiki/File: Struktura_chemiczna_w%C4%99gla_kamiennego.svg (author: Karol Glab. License CC-BY-SA-4.0)*

1.2. Deposition conditions

Coal is formed from the accumulation of plant matter in swamps (peat bogs), lakes or along the margins of ocean basins (lagoons) located in humid tropical climate zones that promote the growth of dense forests. Climatic changes can cause intense fluctuations in rainfall, leading to the flooding and destruction of tropical forests.

The resulting wood then accumulates in depressions, and its preservation depends on how quickly detrital sediments (such as sands and clays) can cover it. These sediments are typically mobilized from weathered materials following the destruction of vegetation cover (erosion). They are transported by rivers from upstream mountainous areas to their deposition in deltas at the mouths of these rivers.

If the available space (the depressions receiving these deposits) is consistently maintained by active tectonics (subsidence), this process can repeat as long as climatic and tectonic conditions are favorable, potentially leading to the formation of coal basins containing numerous layers of coal that can be preserved and fossilized (see Figure 1.3).

In contrast, if the deposits remain exposed to an oxygenated environment, the organic matter will be destroyed by microbial activity.

According to Boulvain (2022), coal deposition environments can be categorized into two main types: paralic environments and limnic environments. Most middle Pennsylvanian (e.g. Westphalian) coals, such as those in the coal basins of Belgium, the Ruhr or Saar-Lorraine, developed in paralic environments, likely of a coastal deltaic nature.

Later coal basins, such as those surrounding the Massif Central (upper Pennsylvanian age, e.g. Stephanian), developed in limnic environments associated with lakes, often located in rift valleys within the Variscan mountain range[2]. The coal basins of the Sillon Houiller (Massif Central, France) are iconic examples of these limnic environments.

2 The Variscan mountain range was formed during the Devonian period (380–340 Ma) as a result of the collision between two supercontinents, Laurasia (or Euramerica) and Gondwana. The erosion of this mountain range generated a large amount of detrital sediments (sands, clays) during the Upper Carboniferous, thereby creating favorable conditions for plant fossilization (see section 1.3).

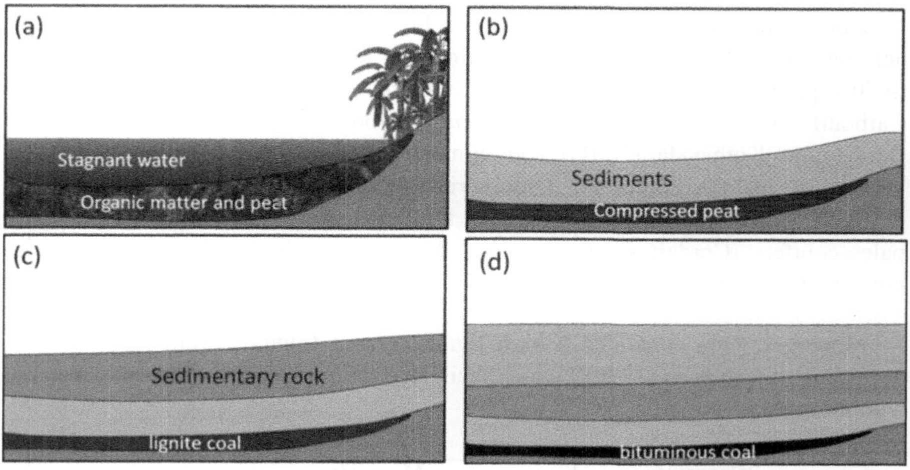

Figure 1.3. *Deposition model of coal measures. Source: https://opentextbc.ca/geology/chapter/20-3-fossil-fuels/ (author: Steven Earle. License CC-BY-SA-4.0)*

1.3. The major periods of coal deposition

Coal deposits are known from all geological eras, spanning from the end of the Primary Era to the end of the Tertiary Era.

However, the majority of these deposits formed in basins dating from the Devonian to the Permian, with a peak occurrence during the Carboniferous, as shown in Figure 1.4.

It was at the end of the Devonian period (−419.2 ± 3.2 to −358.9 ± 0.4 million years ago) that the necessary conditions for the formation of coal layers were first established on Earth (see Figure 1.5). During this time, vegetation began to proliferate on previously barren continents.

The first tree ferns appeared, followed by progymnosperms (precursors to modern conifers), thriving in areas with abundant water. This increased plant growth acted as a significant carbon sink, leading to a paradoxical situation where global temperatures rose, while atmospheric CO_2 levels decreased (Le Hir et al. 2011).

During the Carboniferous period (see Figure 1.6), a combination of tectonic activity (formation of the supercontinent Pangea), a CO_2-rich atmosphere (4,500 ppmv during the Ordovician-Silurian, 500 ppm by the end of the Carboniferous and approximately 280 ppmv before the industrial era according to Bernet and Kothavala (2001)) and climate conditions led to a boom in new vegetation mainly composed (Bournérias and Bock 2014) of pteridophytes (including giant lycopsids like Sigillaria, tree ferns and Calamites) as well as early paleo-conifers (Cordaites). These plants had a much higher bark-to-wood ratio compared to modern vegetation.

Moreover, they contained a high lignin content (ranging from 38% to 58%, according to Robinson (1990)), which facilitated the preservation of woody material – an essential precursor to coal formation.

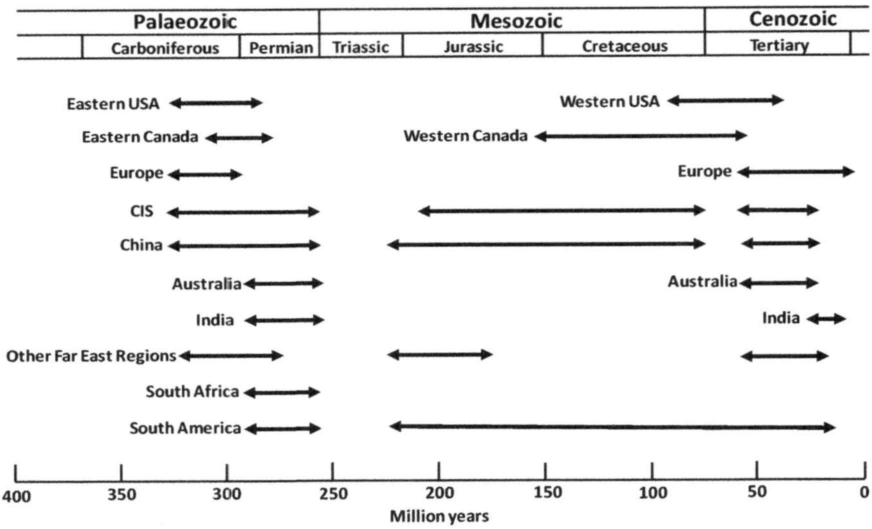

Figure 1.4. *Geological ages and geographical distribution of coal deposits (modified from Walker (2000), International Energy Agency)*

The massive burial of atmospheric carbon during this time is likely responsible for the increase in atmospheric oxygen levels (around 38% according to Berner (1999); 15%–25% according to Lenton (2001), compared to 21% today).

This elevated oxygen content may have contributed to the gigantism seen in both the flora and fauna of this era, with lycophytes reaching heights of 40 m, Calamites growing up to 10 m, Arthropleura (giant millipedes) stretching 2 m long and dragonflies with wingspans of 70 cm.

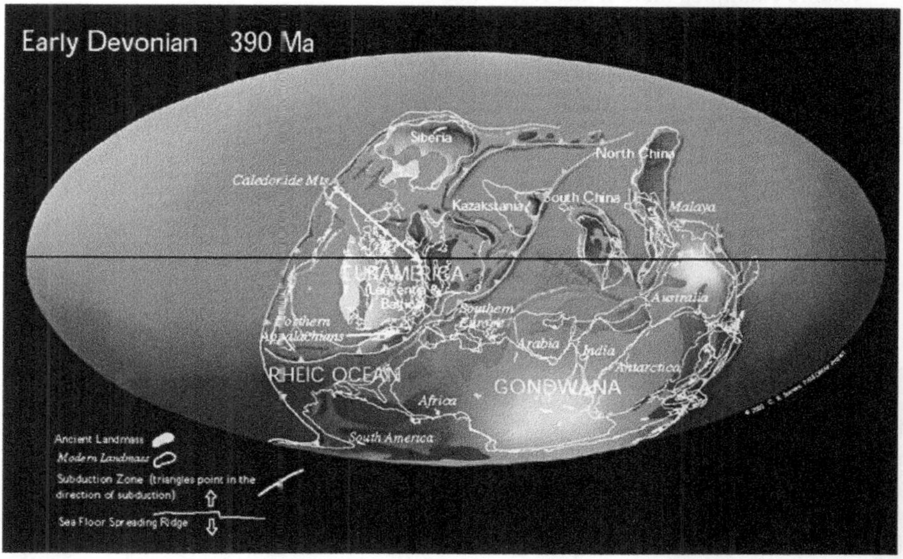

Figure 1.5. *Paleogeography during the Devonian (–390 Ma): Europe and North America collide to form the continent Euramerica, with the resulting suture forming the Caledonian mountain range. Source: http://www.scotese.com/newpage3.htm (credits: C.R. Scotese, Paleomap Project)*

A key question remains unanswered: why was this massive amount of woody material preserved, while today it would be rapidly degraded by microorganisms?

The answer may lie in the absence of lignivorous (wood-decaying) fungi, which did not emerge until the late Carboniferous period (as suggested by molecular clock studies of fungi species capable of lignin degradation; see Floudas et al. (2012) and Garon and Guéguen (2014)).

Even after the Carboniferous, several geological periods provided the conditions necessary for the formation of coal or lignite layers (see Figures 1.4 and 1.7; section 1.4).

Notable examples include the following:

– Lower Permian coals (Autun basin, Lucenay-les-Aix).

– Upper Triassic coals from Vietnam (Tonkin coals, well-known in the early 20th century).

– Cretaceous lignites (Gardanne basin).

– Paleocene coals (65 to 55 million years ago) from Colombia and Venezuela.

– Miocene coals (20 million years ago) from Indonesia.

– Pliocene lignites (Mezos basin).

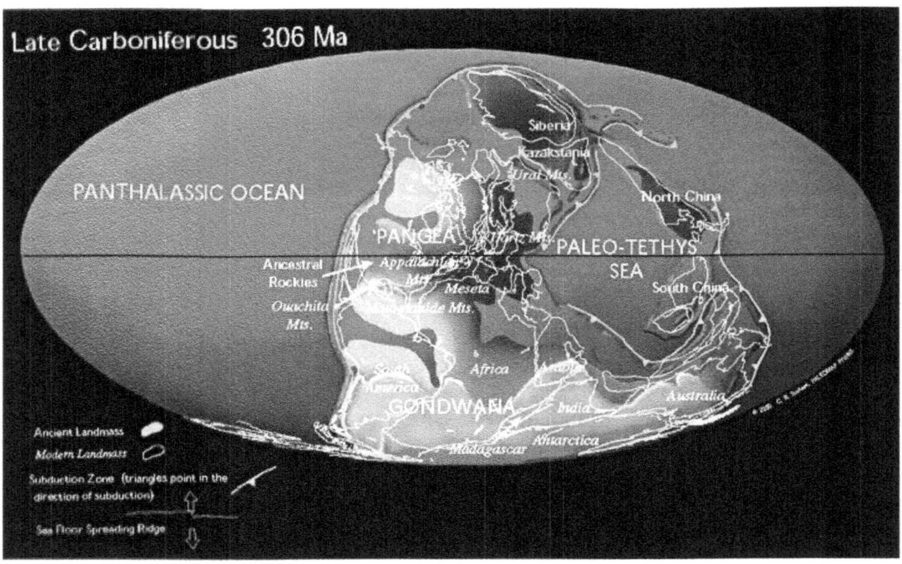

Figure 1.6. *Upper Carboniferous paleogeography (–306 Ma): Gondwana collided with Euramerica, forming the supercontinent Pangea, with the resulting suture forming the Variscan mountain range. Source: http://www.scotese.com/late.htm (credits: C.R. Scotese, Paleomap Project)*

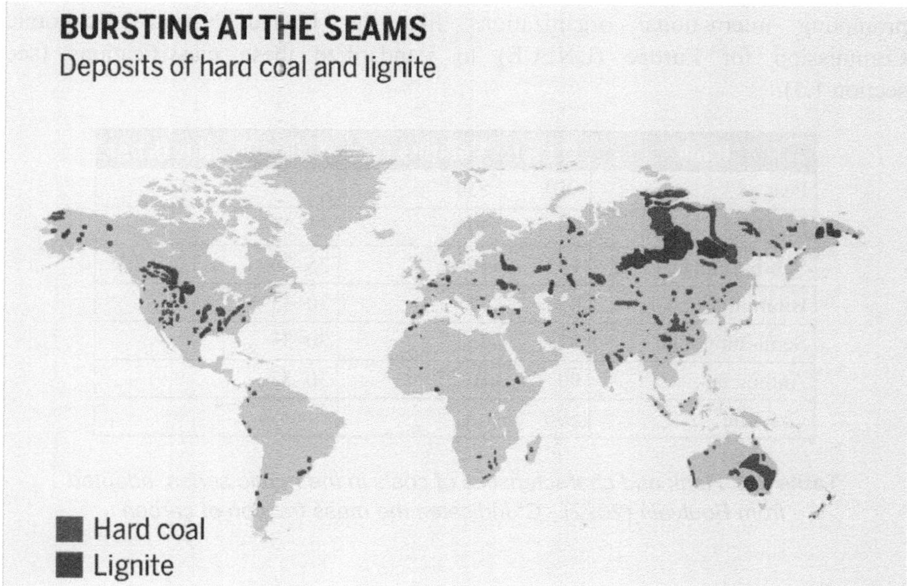

Figure 1.7. *Distribution of coal deposits worldwide. Source: https://eu.boell.org/en/2015/11/18/geology-and-geography-subterranean-forests (author: Heinrich Böll Foundation. CC-BY-SA-3.0 License CC-BY-SA-3.0)*

1.4. Coalification process

The transformation of plant material into coal, known as coalification, involves a series of chemical and physical processes primarily driven by increasing temperature and, to a lesser extent, pressure. These processes are influenced by the geodynamic and tectonic context of the region in question. During the burial of plant material, organic matter gradually loses its water (H_2O) and volatile elements (including N_2, CO_2, H_2S, CH_4, H_2), a process known as the maturation of organic matter (natural carbonization). As this occurs, the material becomes relatively enriched in fixed carbon and produces significant quantities of gas, some of which remain trapped within the rock due to its highly developed microporous structure and very low permeability. The primary gases involved are methane (commonly known as firedamp among miners) and CO_2.

As coalification progresses, coal becomes increasingly enriched in carbon by the elimination of other atoms, known as heteroatoms (Rouzaud and Oberlin 1990). The carbon content is the key determinant for classifying coal rank (see Table 1.1). However, numerous classification systems have been proposed in the literature, prompting international organizations like the United Nations Economic Commission for Europe (UNECE) to standardize these classifications (see section 1.5).

Rank	C (%)	Volatiles (%)	Calorific value (kJ/g)
Peat	<50	>50	
Lignite	60	50	15–25
Sub-bituminous coal	75	45	25–30
Bituminous coal	85	35	30–35
Semi-anthracite	87	25	30–35
Anthracite	90	10	30–35
Graphite	>90	<5	

Table 1.1. *Rank and characteristics of coals in the humic series, adapted from Boulvain (2022). "C" indicates the mass fraction of carbon*

In simplified terms (see Figure 1.8), it can be said that diagenesis[3] transforms peat into lignite (in which plant debris remains visible) at burial depths of up to around 1,000 m. Between depths of 1,000–5,000 m (with temperatures ranging from 100°C to 200°C), increasingly bituminous coals are formed through catagenesis, a process in which plant cells undergo gelification, leading to the development of vitrinite.

Finally, anthracites emerge at the onset of metagenesis[4] at depths of around 5,000 m.

This evolution is summarized in the Van Krevelen diagram (see Figure 1.9) (Van Krevelen 1993), which illustrates the increase in carbon content and the corresponding decrease in hydrogen and oxygen throughout the coalification process.

---

3 Diagenesis is the sedimentary process that transforms sedimentary deposits into rocks (Foucault and Raoult 1980).

4 Metagenesis is part of the process of transforming organic matter during burial. During diagenesis, organic matter loses a significant portion of its water, and plant material is transformed into lignite. During catagenesis (at temperatures above 50°C), crude oil is separated from kerogen, and woody materials are converted into coal. During metagenesis (at temperatures above 150°C and pressures above 1,500 bar), hydrocarbon chains break down, leading to the production of methane and CO_2. For woody materials, coal is transformed into anthracite.

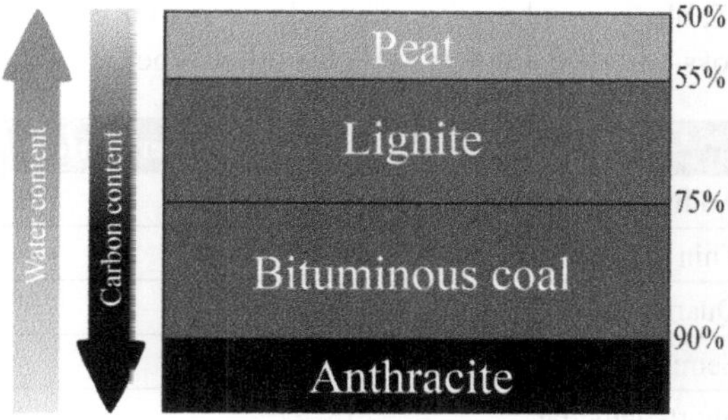

Figure 1.8. *Coal formation. Source: http://commons.wikimedia.org/wiki/ File:CoalTypes.svg (author: Dosto. License CC-BY-SA-3.0)*

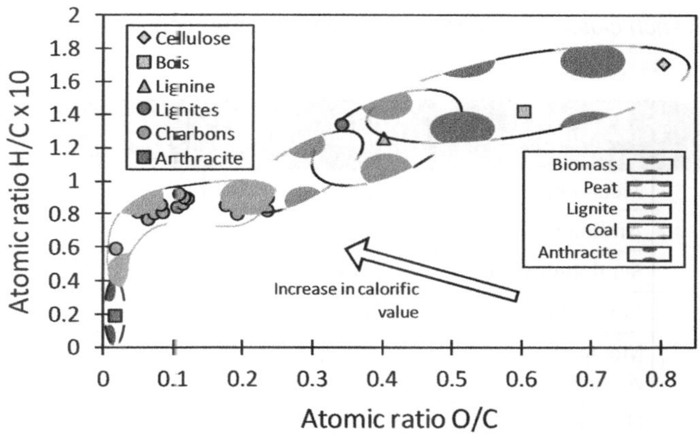

Figure 1.9. *Coalification process (Van Krevelen diagram). Adapted from Van Loo and Koppejan (2007)*

1.5. Coal classification

The French, German, American and international coal classifications are based on different parameters, making direct comparisons challenging.

For instance, the French classification (see Table 1.2) is based on the volatile matter content[5] and the swelling index of the coals, while the American

5 The volatile matter content in coals is measured after pyrolysis of the samples using thermogravimetry and chromatographic analysis of the emitted gases.

classification (see Table 1.3) is based on the volatile matter content for higher grade coals and on the gross calorific value (GCV) for other types.

Coal rank	Volatile matter content
Anthracites	≤8%
Thin and semi-anthracite	8–14%
Quarter-fat	12–16%
Semi-fat	13–22%
Short-flame fat	18–27%
Fatty	27–40%
Fatty flammables	>30%
Dry flammables	>34%

Table 1.2. *French classification based on volatile matter content and swelling index*

Category	Sub-category	Volatile matter	Gross calorific value (MJ/kg)
Anthracite	Meta-anthracite	<2%	
	Anthracite	2–8%	
	Semi-anthracite	8–14%	
Bituminous	Low volatile content	14–22%	
	Medium volatile content	22–31%	
	High volatile content A	>31%	>32.6
	High volatile content B		30.2–32.6
	High volatile content C		26.7–30.2
Sub-bituminous	Sub-bituminous A		24.4–26.7
	Sub-bituminous B		22.1–24.4
	Sub-bituminous C		19.3–22.1
Lignite	Lignite A		14.6–19.3
	Lignite B		<14.6

Table 1.3. *American classification of coals based on volatile matter content and gross calorific value (GCV). Volatile matter contents are expressed as a percentage of the mass of "pure" coal (excluding moisture and mineral matter), while GCV (gross calorific value) values are given on an ash-free basis (including natural moisture)*

The French classification originates from the work of Gruner (1873). The American classification was published by the American Society for Testing and Materials (ASTM) in 1966 and is regularly updated (ASTM 2019).

Given the complexity of comparing these classifications, the United Nations Economic Commission for Europe (UNECE) introduced a standardized classification that has become a key reference. This classification is shown in Figure 1.10.

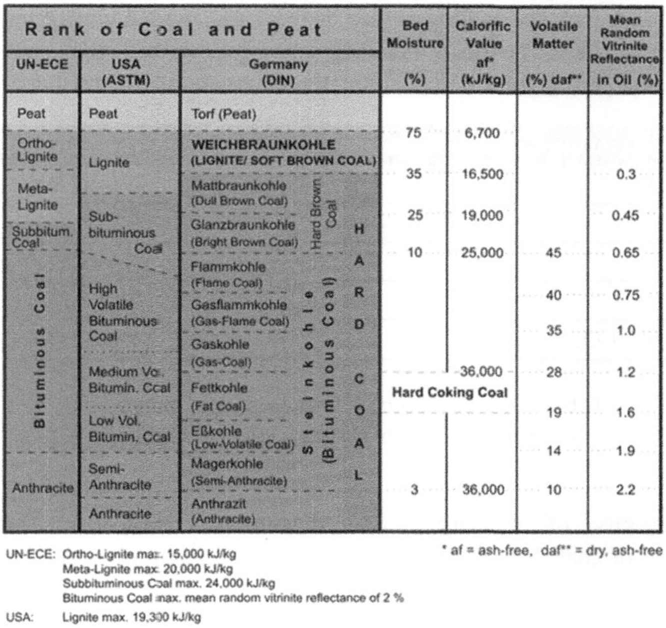

Figure 1.10. *Correlation between UNECE, ASTM and German classifications (source: BGR 2009). a.f. = ash-free; d.a.f. = dry, ash-free. UNECE classification: ortho-Lignite up to 15,000 kJ/kg; meta-lignite up to 20,000 kJ/kg; sub-bituminous coal up to 24,000 kJ/kg; bituminous coal up to 2% average vitrinite reflection*

1.6. Coal gas

1.6.1. *Overview*

The content of this chapter is derived from a report prepared by the Bureau de Recherches Géologiques et Minières (BRGM) and the Institut National de l'Environnement Industriel et des Risques (INERIS) as part of a study commissioned in 2013 by the Directorate General for Risk Prevention and the Directorate General for Energy and Climate of the Ministry of Ecology, Sustainable Development, and Energy (Bonijoly et al. 2013).

As discussed earlier, the decomposition of organic matter by microorganisms, followed by its thermal transformation (coalification), leads to the depletion of volatile components in the material. The gases produced during this process are primarily methane (known as firedamp to miners), potentially accompanied by carbon dioxide, nitrogen, water, hydrogen sulfide, gaseous hydrocarbons (such as ethane and methane) and hydrogen (see Table 1.4). The amount of gas generated depends on various factors, including the genesis of the coal, such as metamorphism, petrographic composition, thermal maturation, and density.

Compounds generated during coalification	Range of gas concentrations in coals
Methane (CH_4)	93.0–99.5%
Ethane (C_2H_6)	0.02–2.8%
Hydrogen	0.00–0.23%
Nitrogen	0.00–3.5%
Carbon dioxide (CO_2)	0.03–3.4%

Table 1.4. *Average composition of coal gases from the British–French–Belgian–Rhine coal basins*

The majority of methane produced during coalification (estimated at 150–200 m^3/t of coal) has, over geological time, migrated out of the coal seam into surrounding areas and/or to the surface.

The residual gas from the coalification process, which can typically range from a few fractions of m^3/t to 10 or even 20 m^3/t, depending on the deposits or parts of deposits, is present in three forms:

– free state: located in fractures and macropores[6] of the rock;

– adsorbed state: within the micropores of the coal;

– dissolved state: in the water contained within the deposit.

6 According to the International Union of Pure and Applied Chemistry (IUPAC) classification, microporosity corresponds to pores with a size not exceeding 2 nm, mesoporosity to pores ranging from 2 to 50 nm, and macroporosity to pores larger than 50 nm.

It is important to note that while coal can exhibit significant porosity (up to 10%), its permeability is typically very low, as most pores are not interconnected (closed porosity, where gases can be trapped). These pores are thus not easily accessible. Permeability values generally range between 0.1 and 50 mDarcys.

In France, most coal basins date back to the Upper Carboniferous or Permian periods (315–275 million years ago). The Westphalian to Stephanian-aged basins are potentially rich in coal gas, particularly in regions such as Nord-Pas-de-Calais, Lorraine, and Alès. Conversely, the Stephanian to Permian coal basins (Decize, Devay, Autun, Lons-le-Saunier) generally contain little to no firedamp.

Coalfields in highly tectonized regions (such as La Mure in the Alps and the Sillon Houiller basins) are exceptions to this general rule, as the intense geological deformation they have undergone (Brown 2002) has facilitated methane leakage.

A particular case worth mentioning is the Gardanne mine in the Arc Basin, which contains low-maturity lignite to sub-bituminous coals dating back to the Late Cretaceous period (around 80 million years ago). These coals have methane contents ranging from 0 to 5 m^3/t.

Figure 1.11 illustrates the relationship between coal rank and gas content.

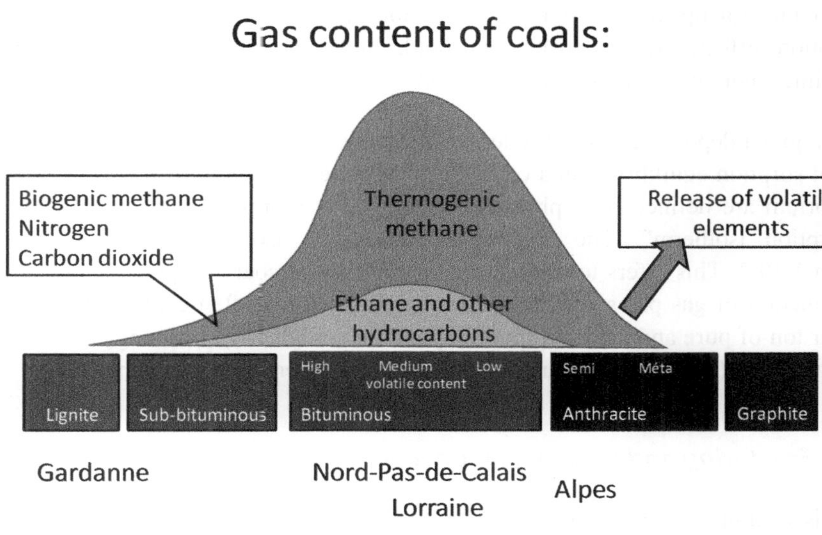

Figure 1.11. *Relationship between coal rank and gas content (adapted from Brown (2002))*

These gases have been known for a long time, but they are difficult to exploit. Two main types are distinguished: CBM (coal bed methane), which refers to the total amount of methane present in the coal, and CMM (coal mine methane), which is the quantity of gas actually extracted from the mine. Firedamp represents a fraction of the CBM.

1.6.2. *Gas–coal interactions*

Coal exhibits a highly complex structure and a diverse range of pore sizes, including nanopores (<2 nm), mesopores (2–50 nm), macropores (50–100 nm) and "cleats" (ranging from 100 nm to several millimeters; Mosher et al. (2013)). The total surface area of these pores is significant: the internal surface area of the "grains" ranges from several tens to several hundreds of square meters per gram of coal. In the adsorbed form, which is predominantly present, gas molecules closely adhere to the surface of the elemental grains that make up the coal structure via physical bonding.

The amount of gas adsorbed is directly proportional to the accessible adsorbent surface area, making coal an excellent gas adsorbent. Consequently, under typical conditions for shallow to moderately deep deposits (depths less than 1,500 m, gas pressure < 50 MPa), the adsorbed form of gas is overwhelmingly predominant. It can account for up to 90% or more of the total gas contained in coal. Due to the adsorption effect, coal deposits can hold, under normal pressure conditions, 6–10 times more methane than a conventional reservoir of equivalent size.

In a given deposit, at a specific temperature, the "coal and gas" system reaches a state of sorption equilibrium at a certain gas pressure. The conditions governing this equilibrium are defined by a physical law, typically represented graphically as an "adsorption isotherm". The concept of adsorption isotherms is discussed in section 1.10.2. This refers to the adsorption capacity of coal at a given temperature as a function of gas pressure. This adsorption capacity is often expressed in m^3 of gas per ton of pure and dry coal (pure organic matter (POM)), meaning coal free of mineral matter (ash) and moisture, under standard temperature conditions.

1.6.3. *Fracturing and mechanical effects*

This section is adapted from Bonijoly et al. (2013).

The natural fracturing of coal (also referred to as the cleat system or diaclases) is not isotropic; it is primarily oriented along the bedding planes of the coal layers and, to a lesser extent, along a perpendicular plane (see Figure 1.12). This fracturing is characterized by the density of the fractures, their connectivity, their aperture sizes and their anisotropic distribution. The mechanical properties of coal are markedly different from those of rocks in conventional gas reservoirs. Coal exhibits significantly higher compressibility (by an average of two orders of magnitude), and its permeability is dependent on the state and orientation of the applied stresses.

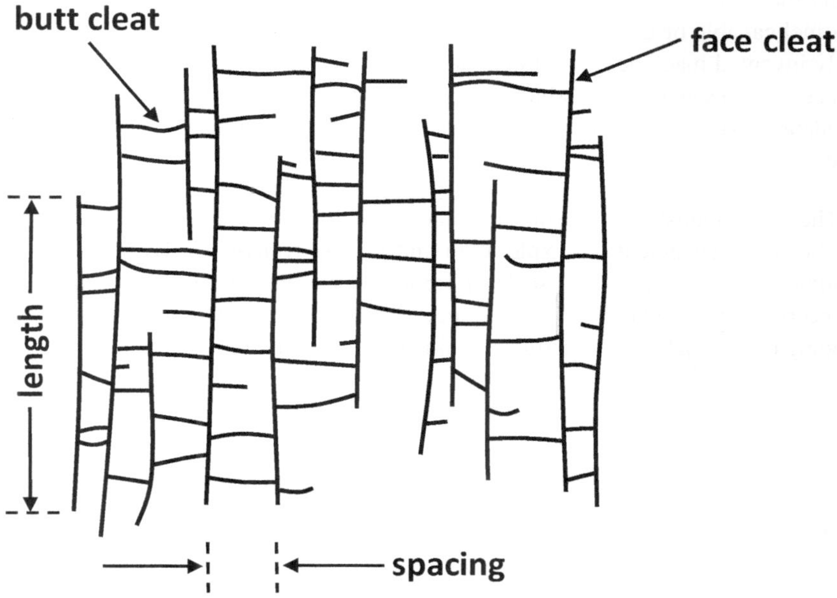

Figure 1.12. *Diagram illustrating the natural fracturing of coal within the bedding plane (Cleat System) (adapted from Moore (2012))*

When conditions such as pressure, temperature and gas composition fluctuate, the coal matrix either adsorbs or desorbs methane. This process induces mechanical stresses (either swelling or contraction), which affect the porosity (either increasing or decreasing it, thereby influencing the opening or closing of the fracture network) which, in turn, alter permeability, fluid behavior and the overall environmental conditions.

In general, fluid flow within the fracture network predominantly governs permeability, significantly facilitating long-distance fluid transport within the reservoir. Meanwhile, diffusion within the coal matrix primarily controls the sorption capacity. The relationship between hydro-mechanical stresses and variations in permeability will be examined in detail in Chapter 2.

1.7. Coal availability, resources and reserves

Coal is a fossil fuel that has been in use since the 11th century. However, it was during the 18th century that its utilization became widespread, as it emerged as the only fuel capable of generating sufficient heat to produce the steam required for the newly invented machines. The Industrial Revolution in the West at the dawn of the 19th century became inseparable from coal, thanks to its exceptional qualities: an abundant energy source that is easy to transport, store and available at a competitive price.

The continuously increasing demand (see Figures 1.13 and 1.14) spurred significant advancements in exploration and mining techniques. By 1841, the first geological map of France at a scale of 1:500,000 was created by two young mining engineers, Armand Dufrenoy and Léonce Élie de Beaumont. The precision of the mapping of the coal-bearing terrains on this map was truly remarkable[7].

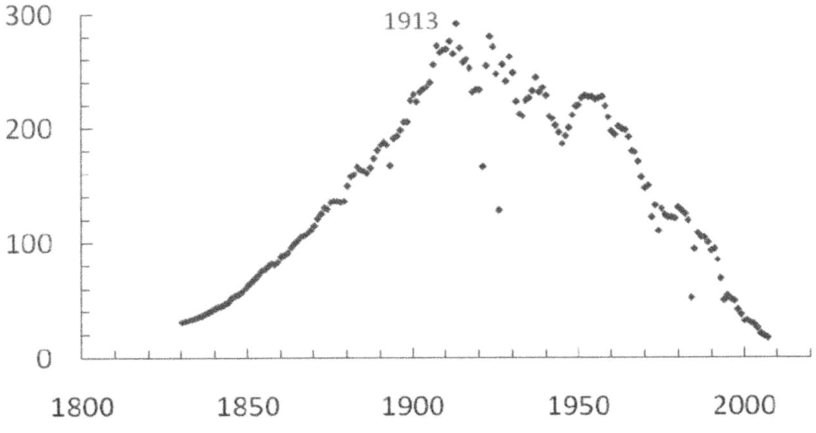

Figure 1.13. *UK coal production, in millions of tonnes. Peak in 1913. Source: Rutledge (2010), cited by Jancovici (2012)*

7 BRGM: https://histoire.brgm.fr/fr/instants/1746-les-premices-de-la-carte-geologique#:~:text=La%20premi%C3%A8re%20carte%20g%C3%A9ologique%20de%20la%20France%20%C3%A0,apparaissent%20dans%20la%20premi%C3%A8re%20moiti%C3%A9%20du%20XIXe%20si%C3%A8cle.

Mining techniques also evolved rapidly. What began as small-scale artisanal operations soon developed into large-scale industrial enterprises to meet the growing demand.

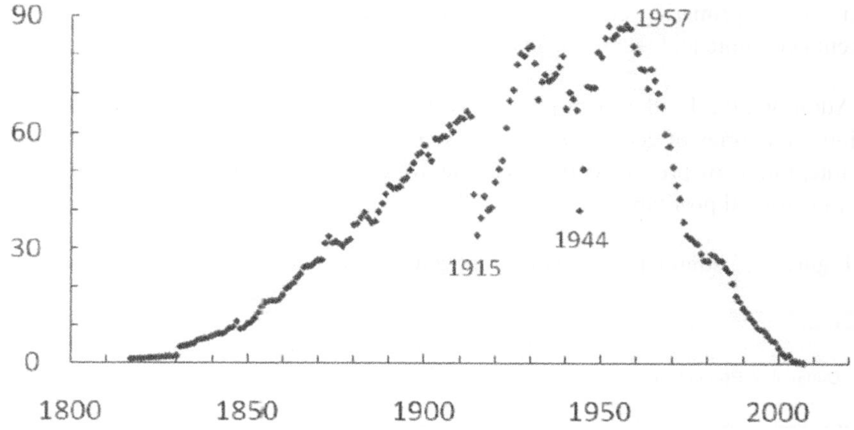

Figure 1.14. *Coal production of France and Belgium combined, in millions of tonnes. Peak in 1957. Source: Rutledge (2010), cited by Jancovici (2012)*

1.7.1. *Coal resources*

Coal resources refer to an estimate of the total amount of coal naturally present in the Earth's crust (coal in place), in a form and quantity such that its extraction for economic purposes is either currently or potentially feasible. These resources are broadly evaluated and not all of them have been discovered. Coal resources encompass both identified resources and undiscovered resources.

Undiscovered resources are further divided into two categories: hypothetical resources and speculative resources, whose estimates are based on very limited geological knowledge.

Hypothetical resources are geologically recognized but not mapped in detail, while speculative resources are merely assumed. The estimation of identified resources incorporates the initial phases of detailed geological exploration.

Reserves are calculated once a resource has been identified and subjected to dedicated and specific exploration work. These calculations aim to determine the economic feasibility of potential extraction. Reserves include inferred reserves, indicated reserves and demonstrated reserves.

For technical and economic feasibility studies, demonstrated and measured reserves are primarily considered. These are calculated by taking into account current economic and technological conditions.

Additionally, development plans may consider presumed reserves depending on various categories of economic feasibility. Fluctuations in market prices can lead to the integration of presumed reserves that were initially excluded from evaluations due to marginal profitability.

Figure 1.15 summarizes the different categories of coal resources and reserves.

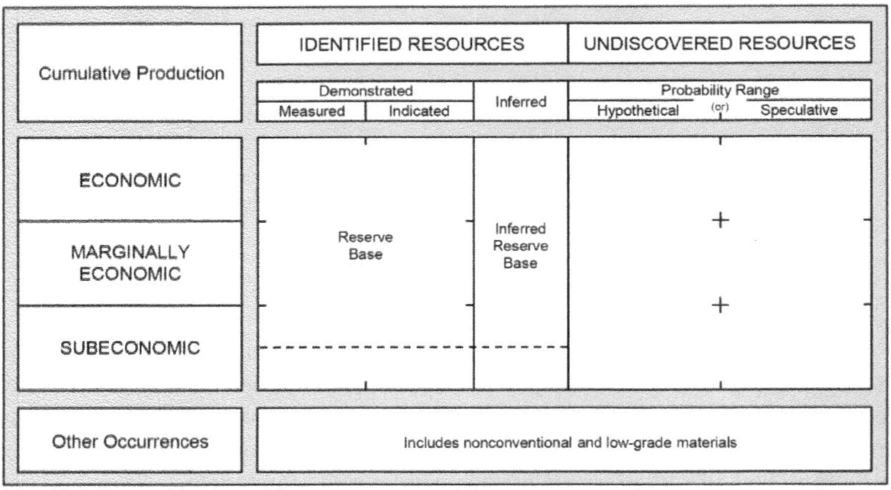

Figure 1.15. *Classification of coal resources and reserves according to geological uncertainty and economic viability (U.S. Geological Survey 2023)*

Coal resources thus encompass the total occurrence of coal, of which the exploitable reserves represent only a small portion, as illustrated in Figure 1.16.

Another equally typical representation (Figure 1.17) visualizes resources and reserves in a pyramid, with the demonstrated reserves at the top.

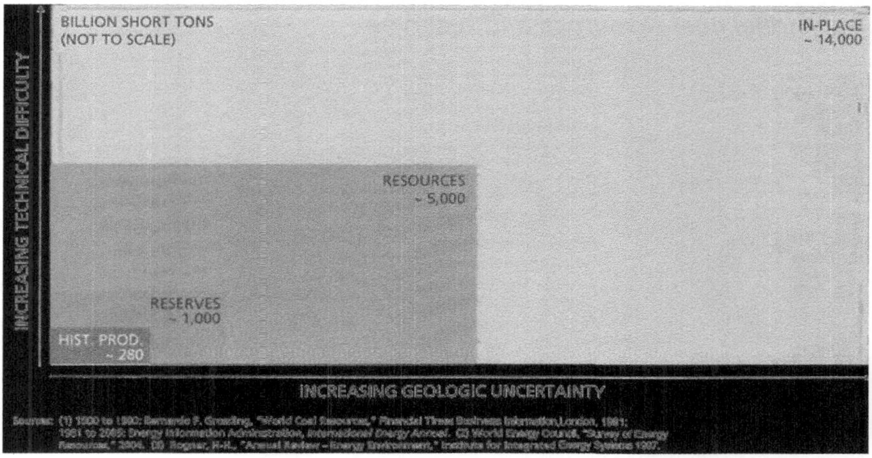

Figure 1.16. *Representation of resources and reserves in a space defined by technological availability versus geological uncertainty. Source: https://jancovici.com/en/energy-transition/coal/what-is-a-coal-reserve/*

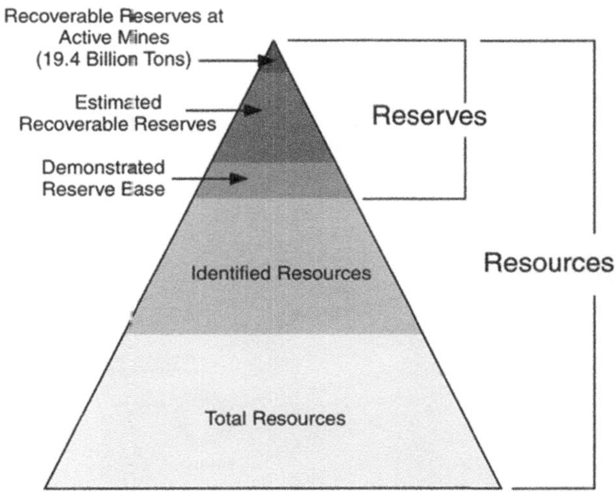

Figure 1.17. *Representation of resources and reserves. The darker the color, the more reliable the data used for the estimation. Source: U.S. Energy Information Administration (1999), cited by National Research Council (2007)*

1.7.2. *Global coal resources and reserves*

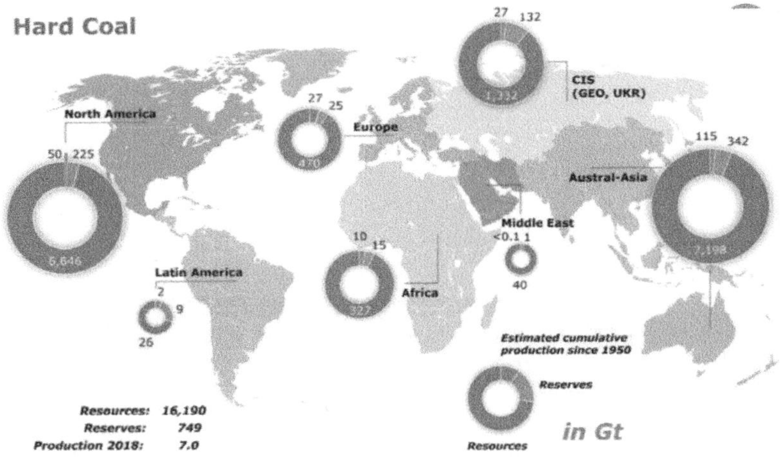

Figure 1.18. *Regional distribution of coal (excluding lignite) worldwide in 2018 (BGR 2020)*

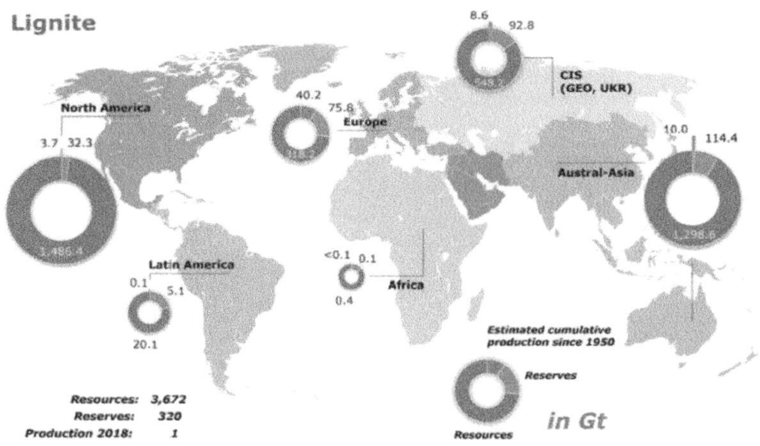

Figure 1.19. *Regional distribution of lignite worldwide in 2018 (BGR 2020)*

Unlike other fossil fuels, coal is found on all continents (see Figures 1.18 and 1.19), and its resources are substantial enough to meet global demand for over a century, based on current consumption levels.

In 2018, global coal resources were estimated at 16,190 Mt for anthracite and bituminous coal and 3,672 Mt for lignite and sub-bituminous coal.

The countries in the Indo-Asian bloc (including Australia) hold by far the largest coal resources globally, with 7,198 Mt: around 44.5% of the world's total, followed by the United States follows (6,646 Mt or 40%), and the Commonwealth of Independent States (CIS) ranks third with 1,332 Mt (BGR 2020).

When comparing the energy content of all available fossil fuel resources, coal accounts for about 80% of the total energy available to humanity.

Coal reserves[8] are distributed as shown in Table 1.5.

Coal and lignite remain crucial fossil energy sources, currently meeting 27.2% of the world's total energy needs. They are the primary source of electricity generation, accounting for 38.3% of global electricity production (International Energy Agency 2019b).

However, coal is also responsible for over 40% of global CO_2 emissions from the energy sector. Its environmental impact is therefore significant, making it one of the major contributors to climate change.

Regions	Hard coal in Gt	Lignite in Gt	Total in Gt	% of global resources	Reserve years
North America	225.0	32.3	257.3	24.1	367
Central and South America	8.7	5.0	13.7	1.3	152
Europe	57.0	78.1	135.1	12.6	244
CIS	100.2	90.4	190.6	17.8	338
Africa and Middle East	15.9	0.06	16.0	1.5	57
Asia-Pacific	342.3	114.4	456.8	42.7	77
World	749.1	320.5	1,069.6	100	132

Table 1.5. *Global coal reserves in 2019 (BP 2020a, 2020b). The reserve years are calculated by dividing the estimated reserves by the production for the given year*

8 By convention, "hard coal" resources and reserves include anthracite and bituminous coals, while "lignite" resources include lignites and sub-bituminous coals.

1.7.3. *Coal resources and reserves in France*

Coal production in France began in earnest at the start of the 20th century with the development of the Nord-Pas de Calais basin and numerous smaller deposits in the southern half of the country.

Production rose rapidly, increasing from 30 Mt in 1900 to 57.7 Mt in 1958 – a record year – with half of this output coming from the Houillères du Nord (colliery) (see Figure 1.14).

After 1958, production steadily declined (9.5 Mt in 1992, 4.9 Mt in 1998) until coal mining in France ceased entirely in 2004, marked by the extraction of the last ton of coal from the La Houve site in the Lorraine basin.

As of 1994, cumulative coal production from French coal basins was estimated at 4,700 Mt. Coal resources were estimated at 1,400 Mt, with reserves amounting to 172 Mt.

Cumulative lignite production in France was estimated at 160 Mt in 1994, with lignite resources totaling 125 Mt and reserves between 45 and 50 Mt.

These resource and reserve estimates significantly increased following a comprehensive coal resource inventory conducted by BRGM on behalf of Charbonnages de France and EDF between 1981 and 1985.

Through geophysical techniques, the following deposits were discovered:

– A Lower Permian coal deposit with substantial resources (250 Mt) and proven reserves of 60 Mt, located beneath a sedimentary cover (thickness: 200–300 m). This deposit was considered for an open-pit mining project in 2005, followed by an underground mining project in 2006, and more recently, in 2013, through an in situ gasification process project (Lucenay-les-Aix deposit, Sud Nivernais).

– Another Lower Permian deposit was found at Devay (northern Massif Central) but with more limited resources (14 Mt).

– A lignite deposit was identified in the southern Landes region at Mézos, where 132 Mt of reserves were confirmed.

Table 1.6 provides detailed information on the production and identified reserves for each of the French coal basins.

Name	Geological Age	Mining Start	Closure	Extraction	Reserves
Nord-Pas-de-Calais	Stephanian	1720	1990	2,500	
Lorraine	Stephanian	1818	2004	850	920
Loire	Stephanian	1267	1983	600	
Cévennes	Stephanian	1230	2001	200	111
Blanzy	Stephanian	16th century	2000	192	
Decazeville	Stephanian	15th century	2001	120	80
Provence	Cretaceous	15th century	2003	80	50
Carmaux		1542	1997	80	116
Saint-Éloy			1978	25	
Decize	Stephanian	1488	1974	20	16.5
Sous-vosgien	Stephanian	1744	1958	17	30–150
Messeix			1989	10	
Laval		1821	1896	7	
Basse Loire	Namurian and Stephanian	13th century		5	
Littry	Saxonian	1741	1949	2.5	3
Boulonnais/Picard	Stephanian	1692	1950	2.2	
Keuperien Saônois	Keuper	1589	1944	1.16	0.5
Vendée		1775	1958	1	
Sincey-lès-Rouvray	Stephanian	1835	1908	0.225	
Le Plessis-Lastelle	Saxonian	1794	1859	0.070	
Vallée de Villé	Stephanian	Mid-18th century	Late 19th century		
Jura	Stephanian	Unexploited			135–250
Épinac et Aubigny-la-Ronce	Stephanian	17th century	1966		
La Chapelle-sous-Dun	Stephanian	18th century	1960		
Dauphiné	Stephanian	17th century	1997		
Brassac	Stephanian	17th century	1979		
Prades	Stephanian	14th century	1963		
Champagnac			1959		
Limousin		17th century	1960		
Mezos	Pliocene	Unexploited			132
Lucenay-les-Aix (Charbonnages de France 1999)	Autunian	Unexploited			67
Devay	Autunian	Unexploited			

Table 1.6. *List of French coal deposits with details on geological age, operating period, closure date, quantity of coal extracted and identified reserves (adapted from Charbonnages de France (1999))*

1.8. The uses of coal

Coal is primarily used in three major industrial sectors: power generation, steelmaking and cement and lime manufacturing.

1.8.1. *Steam coal*

The primary application remains the production of heat through combustion. Thermal power plants use coal as a fuel to produce steam that drives turbines. These plants can accommodate coals of varying quality.

There are two categories of flame-based thermal power plants:

– conventional power plants, which operate with a steam boiler;

– circulating fluidized bed (CFB) power plants, which function similarly to traditional flame-based plants but emit lower levels of nitrogen and sulfur oxides due to their lower combustion temperatures.

It is important to note the existence of a third type of power plant, which does not use coal as a fuel but rather as a raw material to produce synthesis gas (through gasification). These integrated gasification combined cycle (IGCC) plants generate synthesis gas (Syngas[9]) which is used to power a "Combined Cycle Combustion Turbine (TAC)" plant, which integrates both a combustion turbine and a steam turbine.

In 2019, the power generation sector consumed 6.175 Gt of steam coal, making it by far the largest coal consumer across all industrial sectors, accounting for 78% of total consumption (International Energy Agency 2020b).

1.8.2. *Iron and steel industry*

Coal has another significant application: the reduction of iron in steelmaking.

In China, coal was used as early as the 5th-century BCE to produce cast iron and steel. The widespread use of coal began in the 18th century with the development of the blast furnace process, which replaced charcoal with coke for iron reduction.

---

9 The syngas for an IGCC plant mainly contains hydrogen (H_2), carbon monoxide (CO) and carbon dioxide (CO_2). It also contains very small amounts of water (H_2O), methane (CH_4) and nitrogen (N_2). See: https://www.netl.doe.gov/research/coal/energy-systems/gasification/gasifipedia/syngas-composition-igcc.

Coke is a solid material composed of highly pure carbon (over 95%). It offers several advantages over coal, including greater mechanical strength (which prevents crushing and loss of reactivity) and sufficient reactivity to drive the key chemical reaction: $C + CO_2 = 2CO$. Not all coals are suitable for coking; the most effective are bituminous coals with a medium volatile content (see Table 1.4).

The process of making coke involves heating a blend of various coal grades in the absence of air to over 1,000°C in a carbonization process known as coking, which removes all volatile materials.

The temperature in the coke oven is gradually increased, first dehydrating the coal at around 150°C. Then, at 350–500°C, the coal undergoes physical transformation into a coke paste, which is then heated above 1,000°C to eliminate the remaining volatile elements (Rouzaud 1990).

The final product is a solid, porous material – coke – composed almost entirely of pure carbon.

This coke is then mixed with iron ore and fed into a blast furnace[10]. The combustion of coke generates carbon monoxide gas, which reduces the iron oxides. The resulting reduced iron bonds with carbon to form an alloy: liquid cast iron.

To produce 1 Gt of cast iron globally in 2019, the steel industry consumed 627 Mtoe, equivalent to about 1,013 Mt of coal.

The steel industry accounts for 12.7% of global coal production (7.9 Gt in 2019), representing 67% of the total coal consumption across all manufacturing industries (International Energy Agency 2020b). Consequently, the iron and steel industry is the largest industrial consumer of coal among manufacturing sectors worldwide.

1.8.3. *Cement industry*

Coal is traditionally the fuel used to produce clinker (calcium aluminates and silicates) from limestone and clay. This process is highly energy-intensive, requiring the raw material to be heated to 1,450°C. Additionally, coal quality plays a crucial

10 The blast furnace is charged by alternating layers of coke and iron ore. Air is injected through nozzles located at the bottom of the furnace, leading to the following reactions: C (coke) $+ O_2 = CO_2$, followed by C (coke) $+ CO_2 = 2CO$. Then, Fe_2O_3 (iron ore) $+ 3CO = 2Fe + 3CO_2$. The produced CO_2 reacts further as follows: C (coke) $+ CO_2 = 2\ CO$ (Loison et al. 1989; Prins et al. 1989).

role in the calcination process because it provides supplemental silicates through its ash content, with an ideal ash content of 27%. However, this contribution can be substituted by incorporating fly ash from coal-fired power plants, making it possible to replace coal with other fuels, such as petroleum coke, heavy oils, or gas (Brown 2011).

Given that energy costs make up a significant portion of production expenses, cement manufacturers are increasingly turning to alternative fuels derived from supposedly non-hazardous waste (used oils, tires, sewage sludge, animal meal, etc.).

Despite these efforts, the cement industry remains the third-largest consumer of coal globally (3%), following the electricity sector (78%) and the steel industry (12.7%). In 2019, the sector consumed 157 Mtoe, equivalent to approximately 252 Mt of coal (International Energy Agency 2020b), accounting for 17% of global coal consumption in the manufacturing industry.

1.8.4. *Coal gasification*

Coal gasification is a thermochemical process that converts solid fuels (such as coal, wood and straw) into a gaseous fuel through the controlled and limited injection of an oxidizing agent (e.g. O_2, air, CO_2 and steam). This process differs from pyrolysis, which is purely thermal and occurs in the absence of an oxidizing agent, and from combustion, which involves abundant oxidizing agents[11].

The gasification process allows carbonaceous or organic materials to be transformed into a combustible synthesis gas (commonly known as "syngas"), primarily composed of carbon monoxide (CO) and hydrogen (H_2).

Discovered in the late 18th century, this process was industrialized in the late 19th century. Initially used for public lighting (through the distillation of coal gas), its applications expanded to heating and eventually to electricity generation.

During World War II, the gasification process was adapted for hydrocarbon synthesis (using the Fischer–Tropsch method), enabling Germany to circumvent the petroleum embargo. At the same time, small-scale gasifiers were developed to power internal combustion engines in vehicles. Similarly, South Africa began utilizing gasification in 1955 in response to the embargo imposed after the establishment of apartheid.

11 See: https://www.gazeification.info/.

The widespread use of gasification declined with the introduction of low-cost natural gas on the market. However, interest in the process revived for electricity production with the development of IGCC power plants. These coal-fired plants have a much higher efficiency than conventional thermal power plants, reaching up to 42%, with ongoing advancements suggesting potential efficiencies as high as 50%.

Syngas is a versatile raw material that can be chemically processed to produce methanol, methane and synthetic fuels through the Fischer–Tropsch (coal-to-liquid) process.

The types of coal most suitable for gasification are sub-bituminous coals and bituminous coals with medium volatile content (see section 1.5).

As of today, there are 686 gasifiers operating across 274 industrial sites and 238 gasifiers under construction at 74 sites globally. Most of these facilities use coal as the primary feedstock.

Located predominantly in Asia and Australia (see Figure 1.20), these installations supply raw materials for chemical manufacturing, fertilizer production and liquid fuel generation (coal-to-liquid).

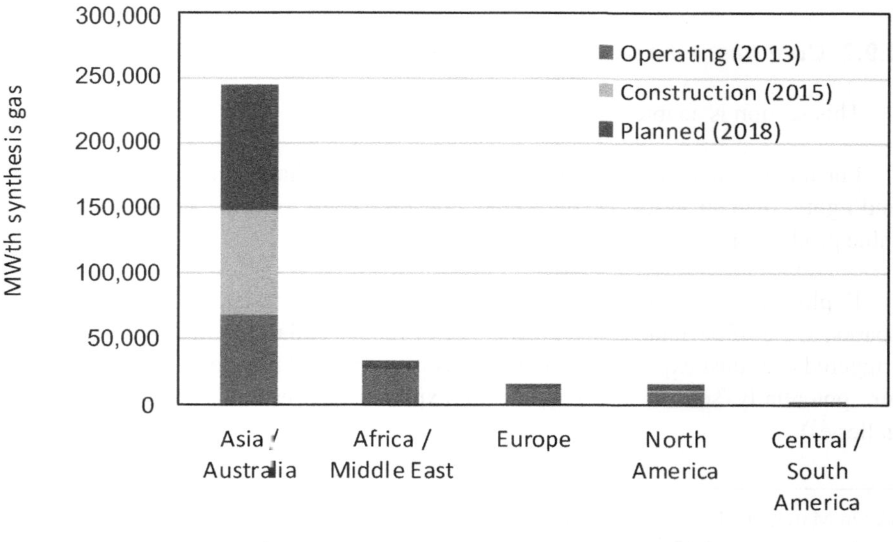

Figure 1.20. *Gasification capacity by region. Adapted from a restricted report by the Global Syngas Technologies Council*

1.9. Coal gas exploitation

This section is adapted from Bonijoly et al. (2013).

As described in section 1.4, coal gas results from the slow process of organic matter maturation (coalification).

1.9.1. *Different types of coal gas*

The literature identifies several types of coal gas (Thielemann et al. 2004):

– *CBM* (Coal Bed Methane) is a coal gas extracted through drilling in areas free from prior mining activities.

– *CMM* (Coal Mine Methane) is a gas that escapes during coal mining operations.

– *CSM* (Coal Seam Methane) is a gas originating from the ventilation of active mines.

NOTE.– "Coal gas" is a manufactured gas produced during the conversion of coal into coke. The term "coal gas" is also sometimes used to refer to desorbed seam gas from coal layers.

1.9.2. *Coal gas: a hazard turned resource*

This section is adapted from Bonijoly et al. (2013).

For many years, coal gas, mainly composed of methane, an odorless, colorless, and lighter than air gas, was considered a health hazard for miners and a risk to coal mine productivity.

Explosions, or "firedamp explosions", caused by the presence of naked flames or sparks, were often fatal and typically led to tunnel collapses. In some cases, they triggered coal dust explosions[12] (explosions of suspended fine coal dust particles and air, potentially set off by a firedamp explosion, spreading through coal mine galleries).

---

12 On March 10, 1906, the pits 2 (Billy-Montigny), 3 (Méricourt), and 4/11 (Sallaumines) of the Courrières Coal Mining Company (Pas-de-Calais) were the site of one of the deadliest mining disasters in history. The tragedy claimed 1,099 lives, excluding missing rescuers, and left 696 injured (see: https://www.universalis.fr/encyclopedie/catastrophe-de-courrieres/#i_51262).

The presence of methane in large quantities was viewed as an obstacle due to the additional costs associated with the necessary ventilation and air circulation measures to prevent firedamp explosions.

Consequently, early research on coal gas focused solely on evaluating the quantity of methane that could hinder coal mining operations. Notably, these investigations were not intended to recover gas as a resource.

However, as global reserves of conventional resources (coal, oil and natural gas) decline and prices rise, coal gas exploitation as an energy resource has gained increasing attention over the past 30 years.

Methane exists in coal seams in four states: adsorbed in micropores of the coal (<2 nm in diameter), trapped in the matrix porosity, dissolved in formation water and, more rarely, in free form in the fractures (cleats).

Gas desorption requires the seam to be desaturated through pumping. When the pressure of formation water drops to (or below) the gas phase pressure, the gas is desorbed from the matrix (Mohanty and Pal 2017) (see Figure 1.21).

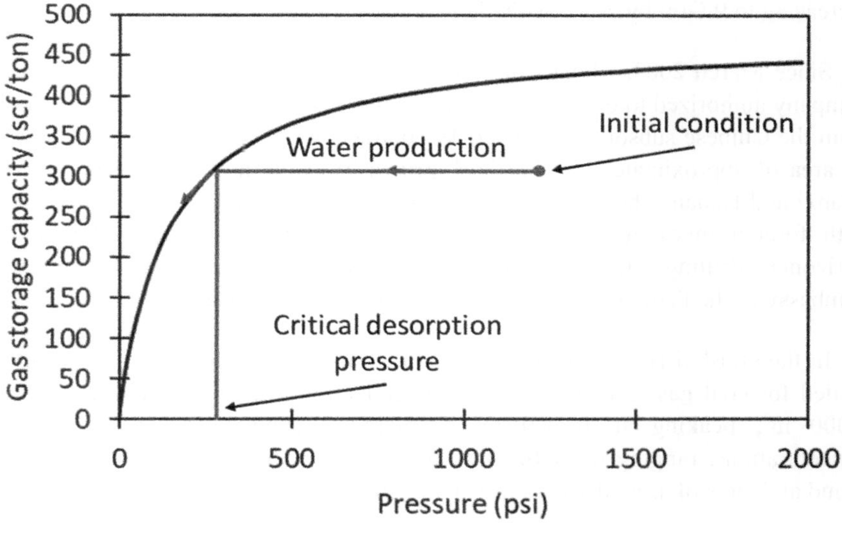

Figure 1.21. *Coal gas production on the Langmuir isotherm: methane desorption occurs as a result of desaturation. When pressure in the fracture system reaches the critical desorption pressure, methane migrates from the matrix to the fractures. Adapted from Mohanty and Pal (2017)*

Globally, methane resources trapped within coal seams are estimated to be between 100 and 260 Tm^3. For reference, global natural gas reserves in 2019 were estimated at 198 Tm^3 (BP 2020a). Canada, Russia and China together hold approximately 80% of the world's coal gas resources. Other significant holders include Australia, India, Ukraine, Germany and Poland (Bessereau and Saniere 2008).

The first experiments to recover coal gas as a resource began in the United States in the late 1970s. In some cases, gas was extracted independently of coal mining through surface drilling, primarily in the Black Warrior (Alabama) and San Juan (Colorado) basins.

In the United States, following the 1993 Climate Plan aimed at reducing methane emissions into the atmosphere and promoting coal gas recovery, coal gas production accounted for about 8% of total natural gas production in 2006, representing 45 Gm^3. The development of new drilling techniques has rendered many projects economically viable; they are mainly found in the Rocky Mountain region.

China possesses one-third of the world's coal gas resources, considering only seams at depths of up to 2,000 m. In 2002, China produced 1.5 Gm^3 of coal gas, increasing to 9 Gm^3 by 2017, with plans to reach 24 Gm^3 by 2020[13].

Since March 2003, China United Coalbed Methane Co., Ltd. (the sole Chinese company authorized to explore, develop, produce, transport, sell and utilize coal gas from the Chinese subsoil) has signed 19 contracts with foreign companies covering an area of approximately 32,000 km^2, primarily in the north-central provinces of Shanxi and Henan. These areas are estimated to contain about 3.4 Tm^3 of coal gas, with foreign investments amounting to \$90 million USD. The technology for horizontal drilling, developed by U.S. companies, is particularly considered (Embassy of the People's Republic of China in the United States 2006[14]).

In the Qinshui Basin (Shanxi Province), around 70 exploration wells have been drilled for coal gas recovery, with daily production averaging between 2,000 and 4,000 m^3, peaking at 16,000 m^3. Exploration campaigns confirm high gas concentrations, ranging from 10 to 37 m^3 per ton, with the highest concentrations found at depths of around 700 m (Su et al. 2005).

---

13 Source: https://www.chinadailyhk.com/articles/218/208/130/1550743774994.html#:~:text=China%20recently%20announced%20plans%20to%20build%20two%20large,annual%20production%20capacity%20of%208.3%20billion%20cu%20m.

14 Source: http://www.china.org.cn/english/international/57254.htm.

In France, mine gas recovery has been ongoing since 1992 in the former Pas-de-Calais mining basin, specifically at Avion, Lourches and Divion. The gas from Avion is purified and injected into the public natural gas distribution network.

The operating company, "La Française de l'Énergie" (formerly Gazonor), is also developing a project in the Lorraine region. The company estimates that coal gas reserves in Lorraine and mine gas in Hauts-de-France could produce the equivalent of 5%[15] of France's annual gas consumption by 2025.

Other countries, such as Germany, have also initiated coal gas recovery. In 2002, 77 stations in the Ruhr basin, with a combined capacity of 70 MW, converted 280 million m^3 of coal gas into 650 GWh of energy, requiring an investment of €85 million.

In Hungary, work in the Mecsek coal basin (in the south of the country, near the border with Croatia) began in 1980 and again in 1993–1994. Despite attempts, the techniques used at depths of around 1,000 m were mostly unsuccessful. Of four exploration wells, only one provided enough gas to be economically viable (850 m depth, 10-m-thick coal seam).

However, technical issues related to well obstruction prevented further progress.

Recent studies conducted in 2003 by the U.S. Geological Survey in collaboration with the Hungarian Geological Survey demonstrated the economic feasibility of coal gas recovery in the Mecsek coal basin, with methane concentrations of around 15 m^3 per ton of in situ coal, despite very low permeabilities (less than 0.00039 mDarcy). Estimated resources amount to 28 Gm^3 (Landis et al. 2003).

In Australia, Arrow Energy has been exploring methane recovery at multiple sites in the Bowen, Surat Clarence Moreton and Nagoorin Graben basins in Queensland. Since 2006, three projects have moved into the production phase using both vertical and horizontal wells.

The most advanced project, North Kogan in the Surat basin, plans to drill 600 wells starting in 2020, delivering approximately 300 TJ per day over 27 years. The entire project includes the completion of 2,500 wells[16].

15 For comparison, biomethane consumption accounted for less than 1% of total gas consumption in 2019. This rapidly growing sector is projected to represent 7% of gas consumption by 2030 (Multiannual Energy Programming, January 2019).

16 See: http://www.arrowenergy.com.au.

1.9.3. *In situ coal gasification (UCG)*

The concept of gasifying coal in situ emerged simultaneously with Sir William Siemens' discovery of the gasification process in 1868. However, it was the Russians who first developed the technology (Couch 2009), driven by Lenin and later by Stalin, who viewed this technological solution as a significant social advancement.

Underground coal gasification (UCG) offered several benefits (Couch 2009), including:

– preserving air quality;

– reducing transportation costs;

– utilizing gas turbines, which were expected to be twice as efficient as steam turbines.

The primary advantage of this technology lies in harnessing the energy contained within coal seams, while avoiding the associated risks and drawbacks. As a result, the targeted seams are typically those located at great depths or those of limited mining interest (e.g. due to low quality, thinness, or challenging depth) or economic interest.

Although numerous coal-producing countries explored this technology, particularly during the 1940s and 1950s, it fell out of favor in the 1960s following the discovery of large natural gas deposits.

The oil crisis of the 1970s rekindled interest in UCG during the 1980s, notably in France and Belgium.

In France, research was conducted from 1979 to 1985 by a consortium led by Charbonnages de France, in partnership with the BRGM, the French Petroleum Institute, and Gaz de France (under the GEGS: Group for the Study of Underground Gasification).

The principle of UCG is as follows: a gasifier is created within a deep coal seam. Two wells are drilled and connected through hydraulic fracturing and/or directional drilling.

Combustion is initiated at the injection well using an electrical system and maintained through the injection of oxygen (or air) and steam. Controlled gasification allows the recovery of a mixture of synthesis gases (CO–H_2–CH_4) at the production well. Figure 1.22 illustrates this process.

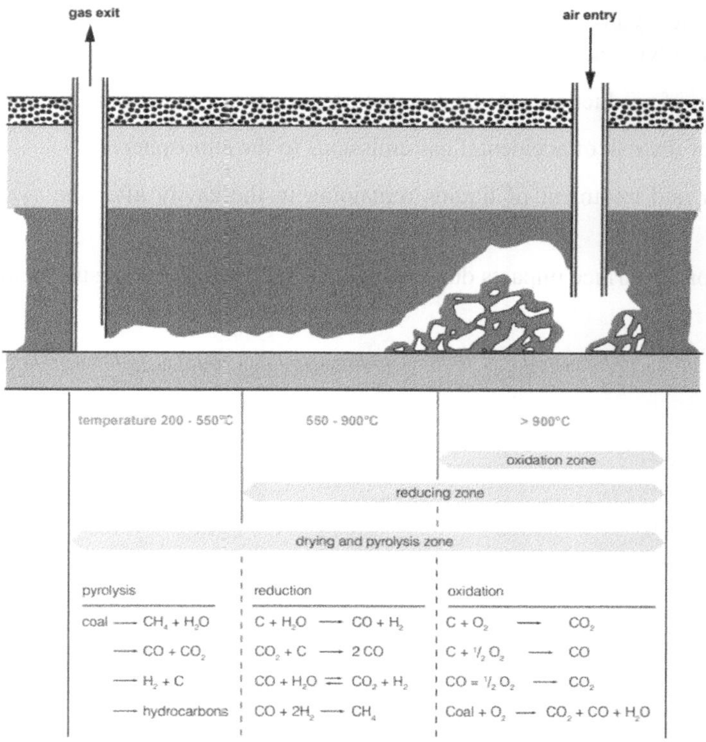

Figure 1.22. *Schematic diagram of the processes involved in in situ gasification (Chaiken and Martin 1992; Beath et al. 2004, cited by Couch 2009). Source: IEA Clean Coal Center*

The efficiency of this technology (see Figure 1.23) is impressive, with estimates suggesting that 80–90% of the coal can be gasified, depending on the size of the pillars preserved within the coal layer to prevent cavity collapse. In terms of energy efficiency, the technology allows for the recovery of 75% of the contained energy when using oxygen combustion and 65% when using air combustion.

Despite its potential, the implementation of this technology remains at the project stage due to challenges in managing its impacts on groundwater quality. Specific measures are necessary to ensure (Couch 2009):

– the presence of an impermeable and sufficiently thick cover to isolate the exploited layer from overlying aquifers;

– a favorable hydrogeological setting that minimizes the risk of contaminant migration to overlying aquifers;

– exclusion of production sites from any current or future water resources;

– control of the risk of accidental gas emissions to the atmosphere;

– recovery and treatment of liquids remaining in the cavity after the pyrolysis process;

– mitigation of surface impacts due to potential roof collapse from subsidence.

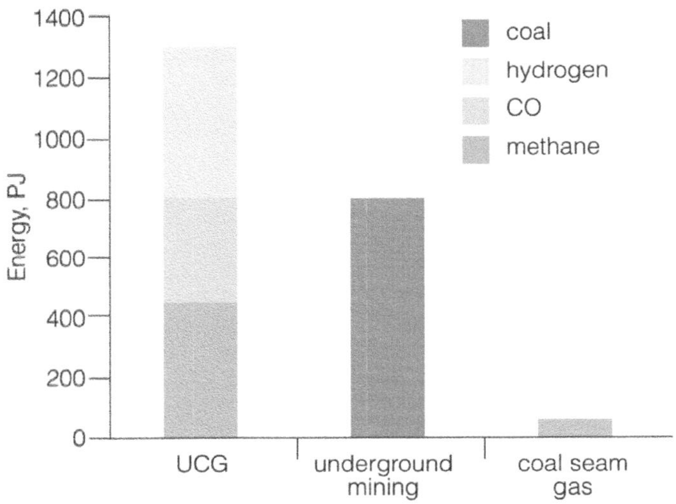

Figure 1.23. *Comparison of energy efficiencies across different exploitation techniques (Mallett 2008, cited by Couch 2009). Source: IEA Clean Coal Center*

An industrial-scale demonstrator has been operational in South Africa (Majuba thermal power plant) since 2002 (Pershad et al. 2018). Since 2007, this plant has supplied 3,000–5,000 m^3/day of gas to a CCGT (combined cycle gas turbine) unit.

Additionally, tests are being conducted in China.

In Uzbekistan, UCG has been in use since 1963. The company Yerostigaz supplies 1 million m^3 of syngas to the Angren power plant.

1.10. Coal and climate change

While coal enabled societies to access abundant and affordable energy, fueling the Industrial Revolution of the 19th century, it has also played a significant role in a major shift in the climate by drastically increasing greenhouse gas concentrations in the atmosphere.

Indeed, the use of coal has been a key contributor to rising CO_2 emissions since the 19th century, as illustrated in Figure 1.24.

Although these emissions have since been joined by those from other fossil fuels (oil, gas), coal remains the primary energy source for electricity production (accounting for approximately 30% of emissions) and the second largest source for primary energy production.

In 2018, a historic peak was reached, with the energy sector emitting around 33 Gt of CO_2 (Figure 1.25).

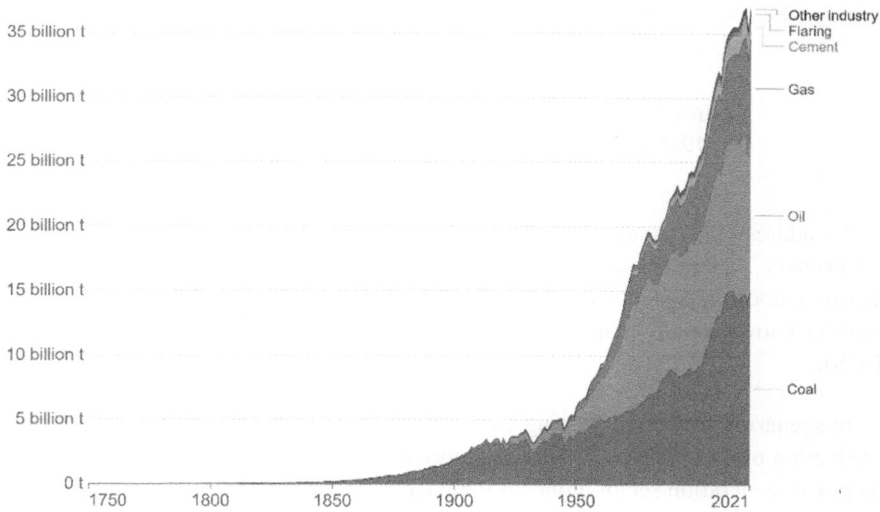

Figure 1.24. *Global carbon emissions by fossil energy source or industry type from 1750 to 2021. Source: Global Carbon Budget (2022). Retrieved from: https://ourworldindata.org/emissions-by-fuel. License CC-BY-4.0. As a reference, one ton of carbon is equivalent to 3.7 tons of carbon dioxide*

This situation can be attributed to the dominance of coal in electricity generation and the high volume of CO_2 emitted per ton of coal burned, which far exceeds emissions from all other fossil fuels (ADEME, https://www.bilans-ges.ademe.fr/):

– coal-fired power plant: 1.058 kg CO_2e/kWh;

– oil-fired power plant: 0.730 kg CO_2e/kWh;

– gas power plant: 0.443 kg CO_2e/kWh.

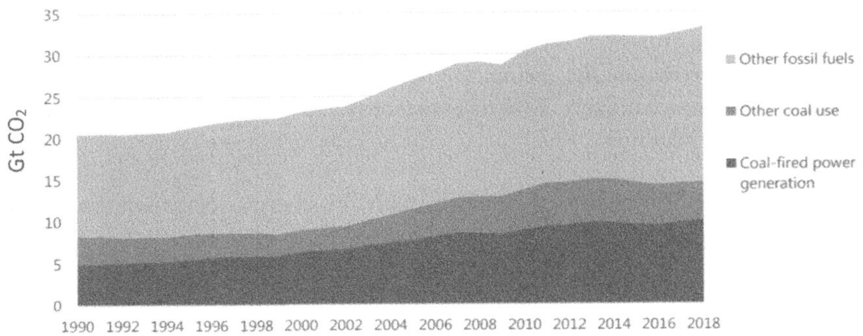

Figure 1.25. *CO_2 emissions by fossil energy sources from 1990 to 2018 (International Energy Agency 2019c)*

To address this situation, various solutions have been proposed. In addition to the primary strategies of promoting energy efficiency and expanding renewable energy, carbon-intensive sectors (electricity, steel, cement, etc.) have put forward a solution known as CO_2 capture, transport and storage (carbon capture and storage (CCS)).

In scenarios aimed at reducing CO_2 emissions, CCS is projected to contribute to a reduction of 6 Gt CO_2 by 2050, with more than half of that reduction coming from the power generation sector alone, as depicted in Figure 1.26.

CCS (Figure 1.27) involves a set of technological processes that extract and capture CO_2 emissions from a gas stream generated by an industrial facility.

The captured CO_2 is then compressed and transported via pipeline, ship or road to a storage site located in deep geological formations (e.g. deep saline aquifers, depleted gas or oil reservoirs and coal beds).

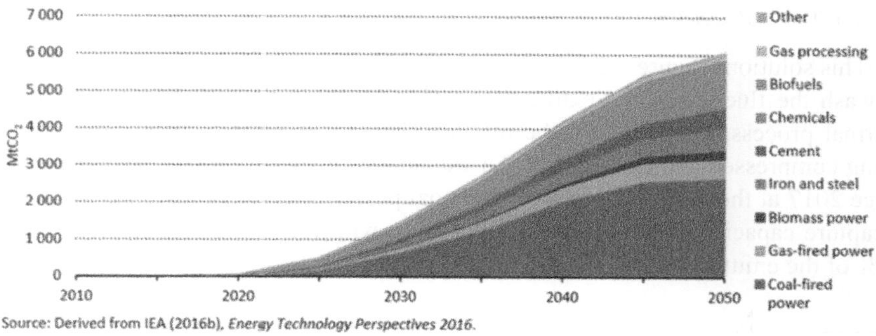

Figure 1.26. *Distribution of CO_2 capture quantities by industrial sector between 2020 and 2050 (International Energy Agency 2016, 2DS scenario)*

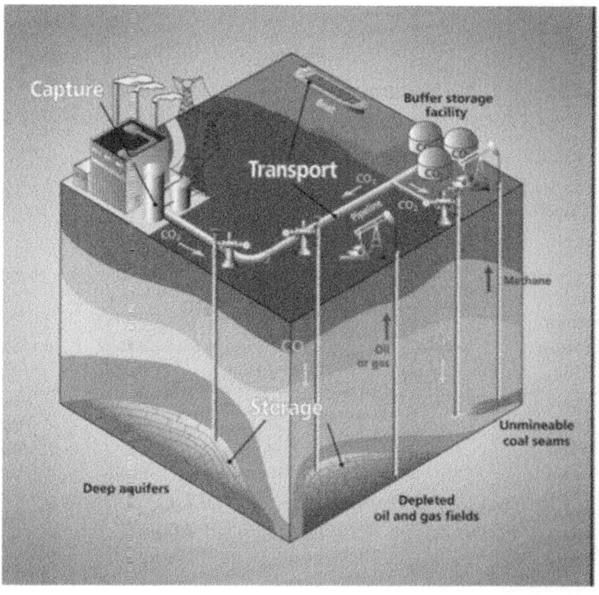

Figure 1.27. *The concept of CO2 capture, transport and storage (BRGM©)*

1.10.1. *Various CO_2 capture solutions for the coal-fired power plant sector*

Three methods are currently available, each with proven feasibility.

1.10.1.1. *Post-combustion CO_2 capture*

This solution (Figure 1.28) is applied to combustion facilities. A solvent[17] is used to wash the flue gases and extract CO_2. The solvent is then regenerated through a thermal process, allowing it to be reused. The CO_2 is purified if necessary before being compressed and transported. An example of this process has been operational since 2017 at the Petra Nova pulverized coal power plant in Texas. This facility has a capture capacity of 1.6 Mt of CO_2 per year, with the system designed to capture 90% of the emitted CO_2.

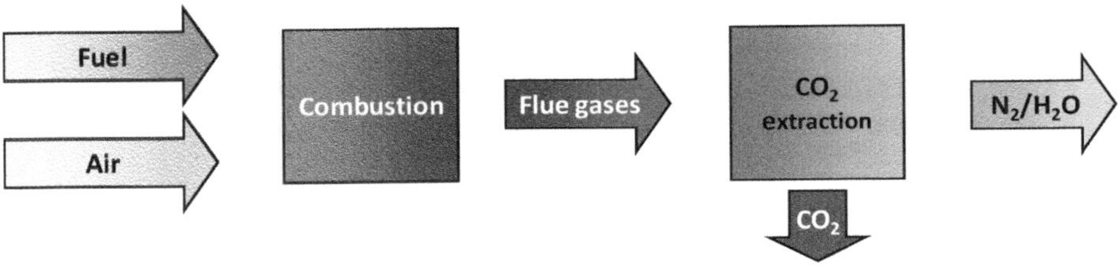

Figure 1.28. *Schematic diagram of post-combustion CO_2 capture. Adapted from an ADEME figure*

1.10.1.2. *CO_2 capture by oxy-combustion*

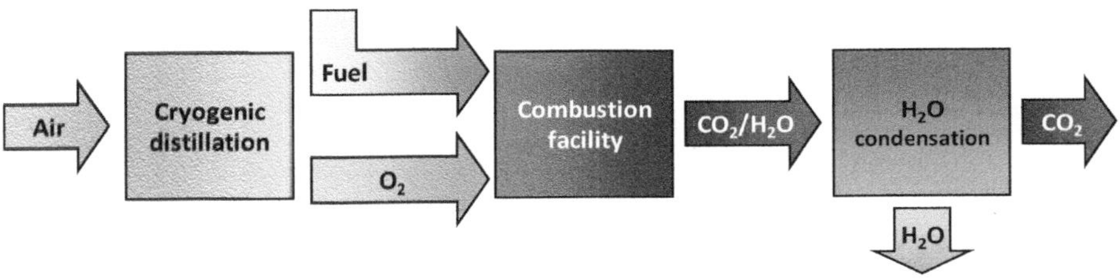

Figure 1.29. *Schematic diagram of CO_2 capture in oxy-combustion. Adapted from an ADEME figure*

In conventional power plants, the primary gases emitted consist of water vapor, nitrogen oxides, sulfur oxides and CO_2 (typically around 7% in conventional plants). The proposed solution to enrich the CO_2 concentration in the boiler's exhaust is to use oxygen as the oxidizer instead of air (Figure 1.29). This approach requires facilities specifically designed for this technology due to the significantly higher

17 The most commonly used solvents are monoethanolamine (MEA) and dimethylethanolamine (DMEA) amines.

temperatures involved. Additionally, an oxygen separation unit using cryogenics is necessary. However, this method results in a much higher CO_2 concentration at the boiler outlet, potentially reaching up to 95%.

1.10.1.3. *Pre-combustion CO_2 capture*

This solution (Figure 1.30) is based on a completely different concept. Instead of capturing CO_2 after fuel combustion, the process involves fuel pyrolysis (coal gasification; see section 1.9) to produce hydrogen and carbon monoxide. The carbon monoxide then reacts with steam (steam reforming) to yield hydrogen and CO_2. The CO_2 is then separated from the gas mixture, and the hydrogen is burned in a gas turbine.

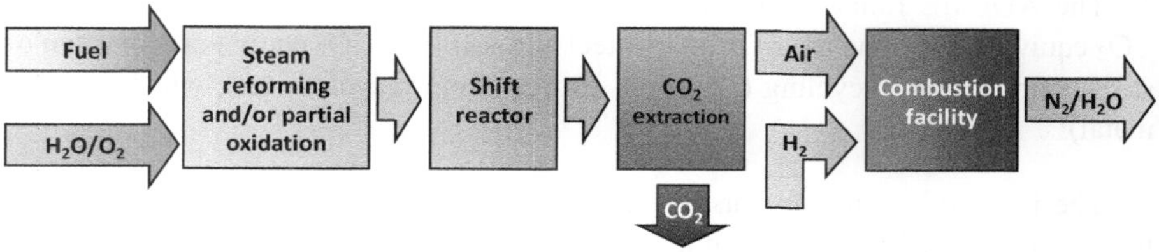

Figure 1.30. *Schematic diagram of pre-combustion CO_2 capture. Adapted from an ADEME figure*

1.10.2. *CCS for carbon-intensive industries*

Like the electricity sector, manufacturing industries that require significant consumption of fossil fuels in their production processes face the same need to reduce CO_2 emissions.

1.10.2.1. *The steel industry*

The steel industry, early on, recognized its environmental impact and continuously sought to mitigate it, initially by improving steel production processes, which significantly reduced energy intensity. However, the necessity of using a reducing agent, such as coal or methane, to reduce iron prevents further progress toward net-zero steel.

According to figures provided by EUROFER, the steel industry is currently responsible for 7% of global anthropogenic greenhouse gas emissions[18] and 31% of industrial emissions.

18 See: https://www.eurofer.eu/publications/position-papers/a-regulatory-framework-for-co2-lean-steel-produced-in-europe.

Nonetheless, the two main steel production routes (blast furnace (cast iron) and electric) have vastly different climate impacts.

While the electric route is low in CO_2 emissions, steel production via blast furnaces is highly CO_2 intensive.

According to data from the World Steel Association, the most efficient blast furnaces emit 1.6 tons of CO_2 per ton of steel produced, while the least efficient emit up to 2.8 tons. In contrast, producing one ton of steel in an electric furnace generates about 600 kg of CO_2.

The ADEME Emission Factor Guide provides an emission factor of 3,190 kg CO_2 equivalent for one ton of primary steel and 1,100 kg CO_2 equivalent for 1 ton of steel entirely from recycling (i.e. made entirely from recycled steel, which is scrap metal)[19].

The industry's commitments to reduce emissions and achieve carbon neutrality by 2050 primarily involve either capturing CO_2 for blast furnace routes or developing breakthrough technologies using either green hydrogen (via water electrolysis) or blue hydrogen (via methane steam reforming coupled with CCS).

Figure 1.31. *Artist's impression of the DMX™ CO_2 capture facilities at the ArcelorMittal steel plant in Dunkirk. Credits: Axens. Source: http://ccu-news.info/cosub2-sub-capture-launch-of-dinamx-an-industrial-pilot-demonstration-project-for-the-dmx-process/*

19 Source: https://www.senat.fr/rap/r18-649-1/r18-649-117.html.

For CO_2 capture, an industrial demonstrator was launched in 2019 by a consortium (including ArcelorMittal, Total and IFPEN).

The goal is to demonstrate that the DMX™ process (a solvent-based CO_2 capture technology patented and developed by IFPEN) is a highly competitive solution that could reduce CO_2 capture costs by 30%.

This industrial pilot, the final step before commercialization, is part of the "3D" project ("DMX Demonstration in Dunkirk", 2019–2023) under the European H2020 program. Launched in May 2019 for four years, the project involves 11 European partners[20].

The project focuses on capturing CO_2 from blast furnace gases at the ArcelorMittal site in Dunkirk (see Figure 1.31) for potential storage in deep geological reservoirs in the North Sea (CCS) or reuse (CCU).

1.10.2.2. *Reducing CO_2 emissions in the cement industry*

The cement industry is responsible for 6% of global CO_2 emissions. Although fuel consumption accounts for only one-third of the CO_2 emissions from the process, the sector is actively researching ways to reduce its carbon footprint through[21]:

– rational resource use, including alternative fuels;

– improving process energy efficiency. Thermal energy consumption has been halved since the 1960s due to facility improvements;

– integrating CO_2 capture, transport and storage to achieve carbon neutrality by 2050.

The sector's flagship project is the European LEILAC (Low Emissions Intensity Lime and Cement)[22] project, which aims to explore the best solutions for reducing CO_2 emissions:

– The LEILAC 1 project demonstrated, at an industrial pilot scale, the feasibility of a process that separates CO_2 from raw materials (calcination via indirect heating). Pure CO_2 is recovered at the reactor's output.

20 Source: https://3d-ccus.com/3d-overview/.

21 See: https://www.construirelawallonie.be/artikel/le-secteur-cimentier-lutte-contre-les emissions-de-co2/.

22 See: https://www.project-leilac.eu/.

– The LEILAC 2 project aims to build an industrial demonstrator that will reduce emissions across the entire cement manufacturing chain by 20%.

In Norway, the NORCEM plant in Brevik is part of the Longship program, which seeks to demonstrate the viability of the CCUS (carbon capture, use and storage) pathway for reducing emissions from European energy-intensive industries[23]. The project, launched in 2013, is expected to test four different technologies: amine capture, membrane capture, solid absorbents and "carbonate looping".

1.10.3. *Enhanced coalbed methane recovery by CO_2 injection (ECBM)*

Coal possesses a remarkably large specific surface area, enabling the adsorption of significant quantities of gas (see section 1.6.1). As a result, not all of the methane produced during coalification is evacuated; some remains adsorbed (see the gas content of coals; section 1.6.1).

This characteristic has been recognized since the 19th century, with Langmuir formalizing this adsorption process using an isotherm in 1918. This curve is derived from laboratory experiments. The sorption phenomenon and its measurement will be detailed in Chapter 3.

The establishment of adsorption isotherms for methane and CO_2 further reveals that coal has a much greater affinity for CO_2 than for CH_4 (see Figure 1.32).

The isotherms also demonstrate that the adsorption capacity of coals varies with coal rank. In high-rank coals, two moles of CO_2 are adsorbed for every mole of CH_4, while in low-rank coals, the ratio is 4:1 (Mohanty and Pal 2017).

Based on these findings, the concept of enhancing methane recovery through CO_2 injection was first implemented in the San Juan Basin, during the 1990s (1995–2001). A total of 335,000 tonnes of CO_2 was injected into the coal seams, resulting in the recovery of 45 million m^3 of gas and the storage of 270,000 tonnes of CO_2.

By the late 1990s, the International Energy Agency Greenhouse Gas R&D Programme (IEAGHG) proposed this solution for CO_2 storage. The sequestration of CO_2 in coal seams became attractive for two reasons: CO_2 injection improved methane recovery (ECBM) and facilitated CO_2 storage. Specifically, this technique allows CO_2 to be stored as "free" gas in the open porosity or fractures of the coal,

23 See: https://ccsnorway.com/capture-norcem/.

adsorbed on the coal surface and dissolved in formation water (Ibrahim and Nasr-El-Din 2015).

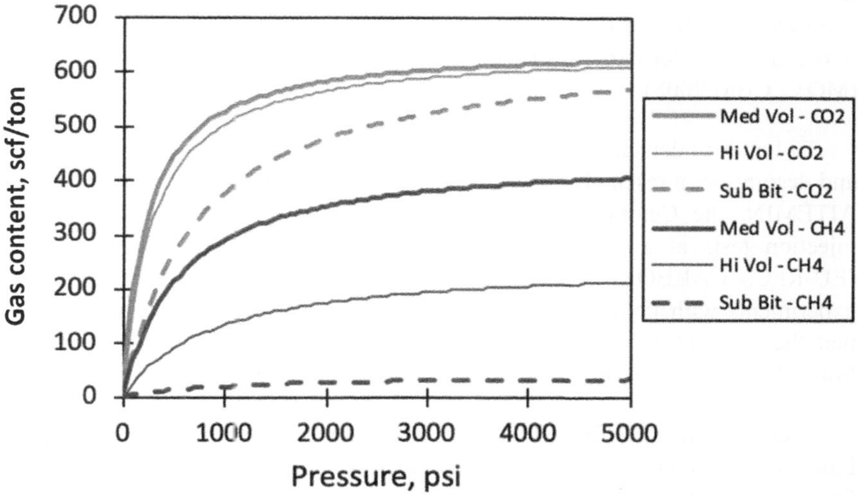

Figure 1.32. *CO_2 and methane adsorption isotherms used in ECBM operation simulations (adapted from Godec et al. (2014))*

Several experiments were subsequently conducted in various basins across the United States and Europe but encountered significant challenges, primarily due to the drastic reduction in permeability (and therefore injectivity) caused by coal creep and swelling induced by CO_2 injection (Mukherjee and Misra 2018). These studies highlighted that multiple factors influence the success of such operations, including (Godec et al. 2014):

– coal rank (with sub-bituminous and bituminous coals high in volatile content being the most favorable);

– coal permeability (in this regard, the San Juan deposit is exceptionally favorable);

– alterations in the mechanical properties of coals under high-pressure CO_2;

– reservoir stimulation, which can enhance permeability but may also be detrimental (e.g. rapid CO_2 breakthrough at the production well);

– the lack of reservoir models capable of predicting methane production and the quantities of CO_2 sequestered.

To recall, the RECOPOL project was a European Commission co-financed project under the FP5 program (2001–2005). Its objective was to assess the feasibility of ECBM under European coal basin conditions and to evaluate the viability and safety of CO_2 storage. A drilling operation injected 760 tonnes of CO_2 into coal seams located at a depth of 1,000 m in the Upper Silesian Basin, Poland (MOVECBM 2008).

To better understand the physicochemical processes involved in gas adsorption and desorption, a research consortium led by BRGM, in collaboration with INERIS, AITEMIN, the Central Mining Institute (GIG) of Poland, and Total, proposed injection tests at a pilot site in a coal mine in Spain operated by HUNOSA (EU-RFCS CARBOLAB project, 2009–2013). The aim was to bridge laboratory experiments with field injection tests (Gal 2014). These in situ tests demonstrated that the total adsorption capacities of CO_2 + CH_4 gas mixtures are significantly lower than the sum of the capacities measured for the pure gas phases.

These findings confirmed that the gases compete for adsorption on coal, and Langmuir's model-based isotherms do not fully represent the experimental results. Consequently, INERIS proposed a new empirical model derived from the Langmuir model (Lafortune et al. 2014).

In conclusion, ECBM shows potential in certain cases because it enables CO_2 storage while enhancing coal gas (methane) recovery. The improvement in recovery is substantial, with conventional CBM production achieving 50% recovery, which can reach up to 90% with CO_2 injection (ECBM). However, the effectiveness of this approach has only been demonstrated in limited cases, primarily under specific reservoir conditions: coals must preferably be of low rank with high permeability. Additionally, several uncertainties remain to be addressed to fully master this technology, such as the mechanical behavior of coals under CO_2 injection, gas mixture desorption phenomena and reservoir models to simulate coal behavior and gas production. The following chapters of this book will review the current state of knowledge on these issues.

1.11. Acknowledgments

My thanks go to Jean Noël Rouzaud, research director, expert in the fields of natural and anthropogenic carbon, RAMAN paleothermometry, and archaeometry, retired from the CNRS - Ecole Normale Supérieure, who carried out a complete, demanding and constructive review of this chapter.

1.12. References

ASTM (1966). Standard definitions of terms relating to coal and coke. D121-62. In *Book of ASTM Standards; Part 19, Gaseous Fuels; Coal and Coke*. ASTM, Philadelphia.

ASTM (1970). *Annual Book: Gaseous Fuels; Coal and Coke; Catalysts; Bioenergy and Industrial Chemicals from Biomass*. ASTM, Philadelphia.

ASTM (2019). D388-19a: Standard Classification of Coals by Rank. ASTM, West Conshohocken.

Atkins, P. and Jones, L. (1997). *Chemistry. Molecules, Matter, and Change*, 3rd edition. W. H. Freeman and Co., New York.

Beath, A., Craig, S., Littleboy, A., Mark, R., Mallett, C. (2004). *Underground Coal Gasification: Evaluating Environmental Barriers*. CSIRO, Kenmore, Queensland.

Berner, R.A. (1999). Atmospheric oxygen over Phanerozoic time. *Proc. Nat. Acad. Sci.*, 96(20), 10955–10957.

Berner, R.A. and Kothavala, Z. (2001). GEOCARB III: A revised model of atmospheric CO_2 over Phanerozoic time. *Am. J. Sci.*, 301, 182–204.

Bessereau, G. and Saniere, A. (2008). *Le point sur : charbon, ressources, réserves et production*. Panorama IFP, Auvergne-Rhône-Alpes.

BGR (2009). *Energierohstoffe 2009. Reserven, Ressourcen, Verfügbarkeit. Erdöl, Erdgas, Kohle, Kernbrennstoffe, Geothermische Energie*. Bundesanstalt für Geowissenschaften und Rohstoffe, Hannover.

BGR (2020). *BGR Energy Study 2019*. Bundesanstalt für Geowissenschaften und Rohstoffe, Hannover.

Bonijoly, D., Didier, C., Fabriol, H., Andres, S., Bazargan, B., Bonnard, R., Defossez, P., Lahaie, F., de Lary, L., Leynet, A. et al. (2013). Synthèse sur les gaz de houille : exploitation, risques et impacts environnementaux. Report, BRGM, Orléans.

Boulvain, F. (2022). *Pétrologie sédimentaire, des roches aux processus*. Éditions Ellipses, Paris.

Bourniéras, M. and Bock, C. (2014). *Le génie des végétaux. Des conquérants fragiles*. Humensis, Paris.

BP (2020a). *BP Energy Outlook*. BP p.l.c., London.

BP (2020b). *Statistical Review of World Energy*, 69th edition. BP p.l.c., London.

Brown, W.T. (2002). Developing CBM in the powder river basin. In *Coalbed Methane Development in the Intermountain West (April 4-5)*. University Press of Colorado, Denver.

Brown, M.A., Cortes-Lobos, R., Cox, M. (2011). Reinventing industrial energy use in a resource-constrained world. In *Energy, Sustainability and the Environment: Technology, Incentives, Behavior*, Sioshansi, F. (ed.). Butterworth-Heinemann, Oxford.

Carpenter, A.M. (1988). Coal Classification, IEACR/12. Report, I.C. Research, London.

Chaiken, R.F. and Martin, J.W. (1992). In-situ gasification and combustion coal. In *SME Mining Engineering Handbook*, 2nd edition, Hartman, H.L. (ed.). Society for Mining, Metallurgy and Exploration, Littleton, CO.

Charbonnages de France (1999). Les combustibles minéraux solides : charbon, lignite. Annals, DGEMP.

Charcosset, H. and Nickel-Pepin-Donat, B. (1990). *Advanced Methodologies in Coal Characterization*. Elsevier, Amsterdam.

Couch, G.R. (2009). Underground coal gasification. Technical Report, CCC-151. IEA Clean Coal Center, Paris.

Encyclopedia Universalis (2010). Major coal deposit of the world [Online]. Available at: https://kids.britannica.com/students/assembly/view/143386.

Floudas, D., Binder, M., Riley, R., Barry, K., Blanchette, R.A., Henrissat, B., Martínez, A.T., Otillar, R., Spatafora, J.W., Yadav, J.S. et al. (2012). The paleozoic origin of enzymatic lignin decomposition reconstructed from 31 fungal genomes. *Science*, 336, 1715–1719.

Foucault, A. and Raoult, J. (1980). *Dictionnaire de géologie*. Masson, Paris.

Gal, F., Proust, E., Leynet, A. (2014). CO_2 injection in a coal seam – Insight from the European CARBOLAB project with focus on water geochemical monitoring. *Energy Procedia*, 63, 5836–5848.

Garon, J. and Guéguen, D. (2014). *Biodiversité et évolution du monde végétal*. EDP Sciences, Les Ulis.

Godec, M., Koperna, G., Gale J. (2014). CO_2-ECBM: A review of its status and global potential. *Energy Procedia*, 63, 5858–5869.

Gruner, L.E. (1873). Pouvoir calorifique et classification des houilles. *Annales des Mines*, 4(7), 169–207.

Ibrahim, A. and Nasr-El-Din, A. (2015). CO_2 sequestration in unmineable coal seams. In *Proc. SPE Asia Pacific Oil & Gas Conference and Exhibition*. Texas A&M University, Bali.

International Energy Agency (2016). Energy Technology Perspectives 2016. Report, IEA, Paris.

International Energy Agency (2019a). Coal 2019 – Analysis and Forecast to 2024. Report, IEA, Paris.

International Energy Agency (2019b). Electricity Information: Overview 2019. Report, IEA, Paris.

International Energy Agency (2019c). Global Energy and C02 Status Report 2018. Report, IEA, Paris.

International Energy Agency (2020a). The Challenge of Reaching Zero Emissions in Heavy Industy. Report, IEA, Paris.

International Energy Agency (2020b). Coal Information 2020. Report, IEA, Paris.

Jancovici, J.-M. (2012). Transition énergétique, le charbon. Blog [Online]. Available at: https://jancovici.com/category/transition-energetique/charbon/.

Kandiyoti, R., Herod, A.A., Bartle, K.D. (2006). *Solid Fuels and Heavy Hydrocarbon Liquids – Thermal Characterization and Analysis*. Imperial College London, London.

Kumar, D. and Kumar, D. (2018). Noncoking coal washing. In *Sustainable Management of Coal Preparation*. Elsevier, Amsterdam.

Lafortune, S.A.-T., Lafortune, S., Adelise, F., Bentivegna, G., Didier, C., Farret, C., Canto-Toimil, N. (2014). An experimental approach to adsorption of CO_2 + CH_4 gas mixtures onto coal (European RFCS CARBOLAB research project). *Energy Procedia*, 63, 5870–5878.

Landis, E.R., Rohrbacher, T.J., Barker, C.E., Fodor, B., Gombar, G. (2003). Coalbed gas in the Mecsek Basin, Hungary. *International Journal of Coal Geology*, 54(1–2), 41–55.

Le Hir, G., Donnadieu, Y., Godderis, Y., Meyer-Berthaud, B., Ramstein, G., Blakey, R. (2011). The climate change caused by the land plant invasion in the Devonian. *Earth and Planetary Science Letters*, 310, 203–212.

Lenton, T.M. (2001). The role of land plants, phosphorus weathering and fire in the rise and regulation of atmospheric oxygen. *Global Change Biology*, 7(6), 613–629.

Loison, R., Foch, P., Boyer, A. (1989). *Coke. Quality and Production*. Elsevier, Amsterdam.

Mallett, C. (2008). Carbon energy's bloodwood creek UCG project. In *CoalTech 2008*. IIR Pty Ltd., Brisbane.

Mohanty, M.M. and Pal, B.K. (2017). Sorption behavior of coal for implication in coal bed methane. An overview. *Int. Journal Min. Sci. Techno.*, 27(2), 307–314.

Moore, T. (2012). Coalbed methane: A review. *Int. J. Coal Geol.*, 101, 36–81.

Mosher, K.J., He, Y., Liu, Y., Rupp, E., Wilcox, J. (2013). The influence of micro- and mesoporosity on methane adsorption: Applications to capacity estimates in coal and gas shale systems. *Int. J. Coal. Geol.*, 109–110, 36–44.

MOVECBM (2008). MOVECBM. Monitoring and Verification of CO_2 Storage and ECBM in Poland. Report, TNO [Online]. Available at: https://cordis.europa.eu/docs/results/38/38967/123991231-6_en.pdf.

Mukherjee, M. and Misra, M. (2018). A review of experimental research on enhanced coal bed methane (ECBM) recovery via CO_2 sequestration. *Earth Science Review*, 179, 392–410.

National Research Council. (2007). *Coal: Research and Development to Support National Energy Policy*. The National Academies Press, Washington, D.C.

Pershad, S., Pistorius, J., van der Riet, M. (2018). Majuba underground coal gasification project. In *Underground Coal Gasification and Combustion*, Klimenko, M.S. (ed.). Woodhead Publishing, Sawston.

Pickel, W., Kus, J., Flores, D., Kalaitzidis, S., Christanis, K., Cardott, B.J., Misz-Kennan, M., Rodrigues, S., Hentschel, A., Hamor-Vido, M. et al. (2017). Classification of liptinite – ICCP System 1994. *International Journal of Coal Geology*, 169, 40–61.

Prins, W., Nater, K.A., Chermin, H.A., Moulijn, J.A. (1989). Coal characterisation for conversion processes. In *International Rolduc Symposium on Coal Science*. Elsevier, Rolduc.

Robinson, J.M. (1990). Lignin, land plants, and fungi: Biological evolution affecting Phanerozoic oxygen balance. *Geology*, 18(7), 607–610.

Rouzaud, J. (1990). Contribution of transmission electron microscopy to the study of the coal carbonization processes. *Fuel Processing Technology*, 24, 55–69.

Rouzaud, J. and Oberlin, A. (1990). The characterization of coals and cokes by transmission electron microscopy. *Coal Sci. Tech.*, 15, 311–355.

Rutledge, D. (2010). Historique de production du charbon. Jancovici [Online]. Available at: https://jancovici.com/transition-energetique/charbon/le-pic-de-production-pour-le-charbon-on-peut-en-voir-quelque-part/.

Stopes, M.C. (1935). On the petrology of banded bituminous coals. *Fuel*, 14, 4–13.

Su, X., Lin, X., Liu, S., Zhao, M., Song, Y. (2005). Geology of coalbed methane reservoirs in the southeast Qinshui basin of China. *Int. J. Coal Geol.*, 62(4), 197–210.

Suárez-Ruiz, I., Diez, M.A., Rubiera, F. (2019). *New Trends in Coal Conversion: Combustion, Gasification, Emissions, and Coking*. Elsevier, Amsterdam.

Taylor, G.H., Teichmüller, M., Davis, A., Diessel, C., Littke, R., Robert, P. (1998). *Organic Petrology. A New Handbook Incorporating Some Revised Parts of Stach's Textbook of Coal Petrology*. Gebrüder Borntraeger, Berlin-Stuttgart.

Thielemann, T., Cramer, B., Schippers, A. (2004). Coalbed methane in the Ruhr Basin, Germany: A renewable energy resource? *Organic Geochemistry*, 35, 1537–1549.

U.S. Energy Information Administration (1999). U.S. Coal Reserves: 1997 Update. Report, AEI [Online]. Available at: http://tonto.eia.doe.gov/FTPROOT/coal/052997.pdf.

U.S. Geological Survey (2023). Mineral Commodity Summaries 2023. Report, U.S. Geological Survey, Nottingham. doi: 10.3133/mcs2023.

Van Krevelen, D.W. (1993). *Coal. Typology, Chemistry, Physics, Constitution*, 3rd edition. Elsevier, Amsterdam.

Van Loo, S. and Koppejan, J. (2007). *The Handbook of Biomass Combustion and Co-firing*, 1st edition. Routledge, Oxford.

Walker, S. (2000). *Major Coalfields of the World*. International Energy Agency Coal Research, Paris.

2

Hydrogeology of Fractured Media – With Special Reference to Coalbeds

Derek ELSWORTH[1,2], Jishan LIU[3], Shimin LIU[1], Brandon SCHWARTZ[4] and Shugang WANG[1]

[1]*Department of Energy and Mineral Engineering, EMS Energy Institute and G3 Center, Pennsylvania State University, University Park, USA*

[2]*Department of Geosciences, Pennsylvania State University, University Park, USA*

[3]*Department of Civil, Environmental and Mining Engineering, School of Engineering, University of Western Australia, Nedlands, Australia*

[4]*Terraforma, State College, Pennsylvania, USA*

2.1. Introduction

The flow of fluids in fractured coal seams exerts control on the accumulation of methane, its storage and then recovery. These fluids comprise both liquids and gases – they are the saturating water, containing small quantities of dissolved native gases, and the sorbed methane stored in the coal seam and any carbon dioxide or nitrogen used for enhanced coalbed methane recovery (ECBM). All of these fluids are stored and transported in the fractured coal seam and conform to similar physical laws: those controlling the storage and transport of fluids in fractured media. In such fractured media, the fractures isolate matrix blocks that provide primary storage of gases within the reservoir while the fractures provide the main pathways for transport. In its native state, coal seams are typically water saturated in the fractures and gas saturated in sorbed form in the organic matrix blocks. Recovery of the sorbed charge of methane in the matrix requires that the fracture network is first

Geomechanics of Coal Seams,
coordinated by Matthieu VANDAMME and Jean-Michel PEREIRA.

desaturated of water to allow gas transport from the depressurized matrix blocks (see Chapter 1). Thus, flows that are driven by a drop in reservoir pressure – with both liquid and gas flows conforming to a similar set of physical principles – are discussed in the following.

2.2. Controls on groundwater flow

The release of fluids from storage within the fractured porous medium of the coal seam results from depressurization and desaturation. This depressurization typically occurs where the seam is pierced by a borehole, dropping the water pressure in the well allows the transport and ultimate desaturation of the fracture system and from that recovery of the gas. The transmission of water or gas within the fracture/cleat system is controlled by the permeability and the release from storage is controlled by the storativity for water drainage and sorption characteristics for gas drainage. These two features – storage (storativity/sorption) and transmission (permeability) – are combined within a governing equation to define the progress of water or gas drainage in terms of rates – are defined as a mass conservation equation. We define the concepts of permeability, storage and related governing equations for flow in the following.

2.2.1. *Transmission of fluids – permeability*

The fluid transmission rate is governed by Darcy's law defining fluid transmission at velocity, q_i, in the ith direction and as a function of the permeability tensor k_{ij} and viscosity of the fluid, μ, as,

$$q_i = \frac{k_{ij}}{\mu}\frac{\partial p}{\partial x_j} \qquad [2.1]$$

where $\partial p/\partial x_j$ is the pressure gradient, ignoring gravity effects. Thus, defining the rate of fluid transmission reduces to an understanding of permeability and how this is influenced by the reservoir structure, inclusive of fractures at scales of centimeters to kilometers (faults). Fractured geologic media are typified as transversely isotropic at all scales, such that the off-diagonal entries of the permeability tensor can usually be neglected. Numerous characterizations are available to describe the form of the permeability tensor for both porous and fractured media.

2.2.1.1. *Porous media flows*

Many descriptions of permeability of porous and fractured rocks are based on pore-scale representation of the porous medium as a bundle of capillary tubes. The oldest and best known of these is the Kozeny–Carman relationship (Kozeny 1927; Carman 1956; Mitchell 1976) where the principal magnitudes of the permeability tensor, given in equation [2.1], are

$$k_{ii} = \frac{1}{\alpha_0 T^2 S_0^2} \frac{n^3}{(1-n)^2} \qquad [2.2]$$

where α_0 is the pore shape factor (~2.5), T is the tortuosity factor (~$\sqrt{2}$), S_0 is the specific surface area per unit volume and n is porosity. Reasonable correspondence is obtained for materials of sand gradation and coarser, although equation [2.2] is more usefully quoted in terms of mean grain size where the first term $(1/\alpha_0 T^2 S_0^2)$ is replaced by $d_m^2/180$, where d_m is the mean grain diameter. In either instance, permeability changes that result from changes in desorption or effective stress may be indexed as a revised permeability, k, relative to an initial permeability k_0 as

$$\frac{k}{k_0} = \frac{n^3}{n_0^3} \qquad [2.3]$$

since all other parameters in equation [2.2] remain functionally constant. There is usually a dominant entry in the permeability tensor corresponding to the bedding-parallel direction (1–2) such that $k_{11} = k_{22} \gg k_{33}$, where k_{33} represents the bedding-normal permeability.

2.2.1.2. *Fracture flow*

Where the permeable matrix is fractured, the equivalent permeability of the mass may be defined using similar capillary concepts as previously described, but where the fundamental building block is that of the transmission characteristics of parallel-plate fractures. The permeability of a single "parallel-plate" fracture in the laminar regime is nominally

$$k = \frac{b^2}{12C} \qquad [2.4]$$

where b is the aperture separating the fracture walls with $C = 1$ for smooth fractures (Lomize 1951; Snow 1965; Louis 1969) and valid to apertures as narrow as $b \sim 0.2$ μm (Romm 1966; Witherspoon et al. 1980).

Such models have been used as the basis for rough (Iwai 1980) and deformable fractures in normal closure (Gangi 1978; Tsang and Witherspoon 1981; Pyrak-Nolte et al. 1987) and shear (Barton et al. 1985; Elsworth and Goodman 1986) in defining evolution of permeability. The basis of these various methods is illustrated in the following.

2.2.1.3. *Fractured media*

For an array of orthogonal fractures (see Figure 2.1), the bulk permeability of the mass may be determined by using the parallel plate model as the fundamental building block. Thus, an array of parallel and continuous fractures (see Figure 2.1(a)) (Snow 1969) with a uniform spacing, s_1, and uniform aperture, b_1, defines permeability as

$$k = \frac{b_1^3}{12s_1} \quad [2.5]$$

where it is assumed that $C = 1$. Since fracture spacing is commonly much greater than the individual aperture of the fractures, the bulk permeability of the matrix may be defined additively as

$$k = \frac{b_1^3}{12s_1} + k_{matrix}(1 - \frac{b_1}{s_1}) \quad [2.6]$$

where the term $1 - b_1/s_1$ approximates unity.

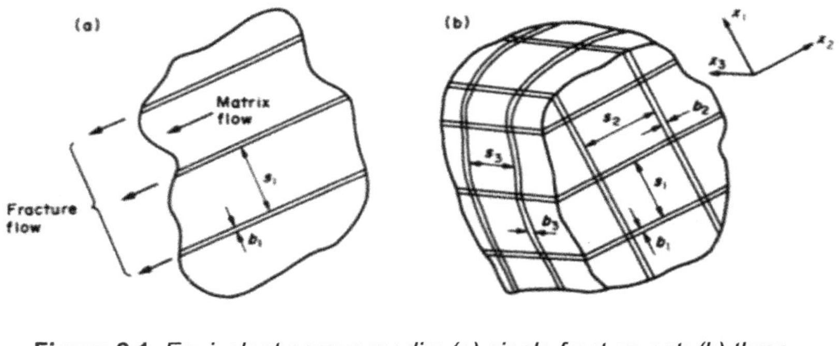

Figure 2.1. *Equivalent porous media: (a) single fracture set; (b) three orthogonal fracture sets (after Elsworth and Mase 1993)*

As further fracture sets are added to produce an orthogonal array, the principal magnitudes of permeability remain coincident with the lines of intersection of the fracture sets. The permeability, k_{11}, within the x_1-direction of Figure 2.1(b) is therefore

$$k = \left(\frac{b_2^3}{12s_2} + \frac{b_3^3}{12s_3}\right) + k_{matrix}(1 - \frac{b_2}{s_2})(1 - \frac{b_3}{s_3}) \quad [2.7]$$

where the terms $1 - b/s$ again approximate unity and the matrix permeability is typically sufficiently small, relative to the contribution of fracture permeability, to be ignored.

Permeabilities in the other two orthogonal directions may be determined from appropriate permutation of the subscripts to yield the principal permeabilities for the block jointed system. Where fracture apertures and spacing between fractures differ for each of the sets, permeability of the fractured porous medium will be anisotropic (specifically, orthotropic) and must be recognized as such if accurate evaluation of the flow regime is to be obtained. Commonly, fracture permeabilities dominate over the matrix permeabilities. Consequently, the second term of equation [2.7] may often be neglected without serious effect.

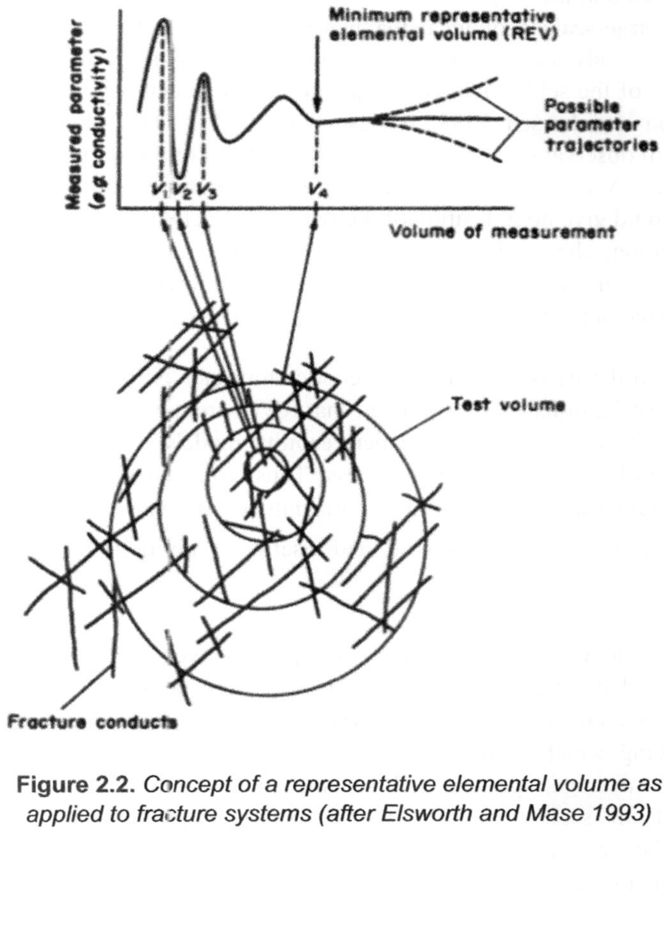

Figure 2.2. *Concept of a representative elemental volume as applied to fracture systems (after Elsworth and Mase 1993)*

Although adequate to determine permeability distributions at large scales, the assumption that fracture sets are continuous and of infinite extent is particularly poor where the scale of interest is small. At small scales, the discontinuous nature of fracturing may be particularly evident (Long and Witherspoon 1985) and therefore continuum models (e.g. equation [2.7]) of permeability are inadequate. This problem is particularly acute, since it suggests a scale effect – where the sample or test volume is small, that volume may not contain a "representative distribution of fractures", and correspondingly the transmission (or storage) parameters evaluated for it will therefore not be representative of the larger scale mass. This concept is illustrated in Figure 2.2, where testing of successively larger volumes results in the eventual establishment of a steady magnitude of permeability beyond the minimum "representative elementary volume" (REV) (Bear 1972) of the system. Although it may be implied that the magnitude of the parameter of interest becomes constant at some REV, large-scale geological features may induce further oscillations in the trace or add a steady inclination or declination to the response. These modifications are indicative of the self-similarity of fracture networks that have been reported by some (Barton and Larsen 1985; Aviles et al. 1987). Thus, changing the scale (or volume) of observation will yield an unaltered form of the network, as illustrated in Figure 2.3. Viewing the fracture network separately in the two frames of Figure 2.3 would give no indication of the scale if the networks were fractal in form. In this situation, the positive correlation of increased aperture (and therefore permeability) with increased fracture trace length would result in a positive gradient to the final trace of permeability versus the volume of Figure 2.2.

Relevant to the transmission characteristics for coals is that the fracturing is at such a small scale, order of a centimeter, that REVs are quite small (order of 10s of centimeters) and as such equivalent permeability models perform well for seams typically an order of a meter in thickness. Thus, equivalent permeability models representing either absolute permeability magnitudes or changes in permeability as, $\frac{k}{k_0} \sim \frac{b^3}{b_0^3}$, are generally representative and useful in defining broad features of response.

The cubic model shown in equation [2.5] has proven useful in characterizing permeability in laboratory measurements. However, the model has two variables that are typically unknown within core samples: the initial fracture aperture and the fracture spacing which both vary between fractures. Permeability measurements represent the average aperture and spacing (Narr 1996; Ortega et al. 2006). However, at all scales, geologic media exhibit a preferential flow network comprising the largest, most conductive pathways. Whether this departure from the model average matters, or not, will depend on the application.

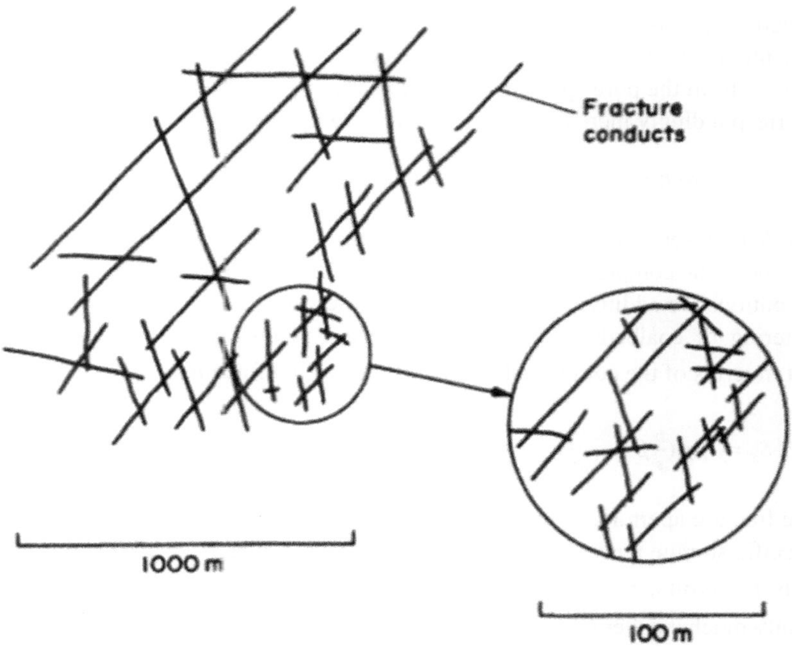

Figure 2.3. *Self-similar or fractal networks of fractures (after Elsworth and Mase 1993)*

2.2.2. *Storage of fluids – specific storage*

Fluid is released from storage as reservoir pressures change and augment the flow driven by the pressure gradient, apparent in Darcy's law. Thus, the conservation of fluid mass requires that the mass released from storage must be both quantified and incorporated into the conservation equation as

$$\frac{\partial q_i}{\partial x_i} = S_s \frac{\partial h}{\partial t} = \frac{S_s}{\rho g} \frac{\partial p}{\partial t} \qquad [2.8]$$

where S_s represents the specific storage defined as volume of fluid per unit volume of aquifer released for a unit change in hydraulic head, h, in a given time, t, and in the unit of 1/m. Since pressure, p, may be defined in terms of fluid density, ρ, and gravitational acceleration, g, as $p = \rho g h$, then the continuity equation may also be defined in terms of pressures.

There are two components to the total fluid volume released from storage, representing (i) fluid expansion under a decrease in fluid pressure and (ii) fluid expulsion from the pore space due to a decrease in pore volume as effective stresses are correspondingly increased. Thus, specific storage may be defined as

$$S_s = \rho g(n\beta + \alpha) \qquad [2.9]$$

where β represents the compressibility (1/Pa) of the fluid present only in the pore space and α the compressibility of the porous medium comprising the aquifer. These contributions are additive. Where the bulk of the storage of accessible water, for the dewatering of coalbeds, is present within the fractures, this relationship may be recast in terms of the compressibility of the individual fractures, k_n, as $\alpha \sim \frac{1}{k_n s}$ as

$$S_s = \rho g(\frac{b}{s}\beta + \frac{1}{k_n s}) \qquad [2.10]$$

where fracture aperture, b, and spacing, s, are as defined previously. The magnitude of specific storage represented in equation [2.10] is typically small in comparison to porous reservoirs, as the liquid-filled porosity ($n \sim \frac{b}{s}$) of the fractured media is typically much smaller ($n << 0.01$) than that of porous media ($n > 0.1$).

2.2.3. *Governing equations*

Substituting Darcy's law of equation [2.1] into equation [2.8] yields the standard pressure diffusion relation as

$$\frac{\partial}{\partial x_i}\left(\frac{k_{ij}}{\mu}\frac{\partial p}{\partial x_j}\right) = \frac{S_s}{\rho g}\frac{\partial p}{\partial t} \qquad [2.11]$$

allowing the evolution of pressures to be determined subject to a description of the permeability field and the application of initial and boundary conditions. For irregular physical geometries and to represent complex geological conditions, the diffusion equation may be solved numerically by finite difference or finite element methods. Such relations may be used to define the transmission of liquids or slightly compressible gases in equivalent single porosity media and the relation provides the basis for various multiporosity models of relevance to gas flow in coal seams. The major challenge lies in identifying a reasonable expression of the in situ flow and storage parameters as input to the model. The fidelity of this representation controls the adequacy of the resulting simulation in matching the real behavior of the system.

2.3. Characterization methods

Understanding rates of fluids recovered from coalbeds requires the permeability and storage characteristics to be accurately and reliably determined at relevant scales. The scale of interest depends on the application but certainly brackets laboratory and field measurements. Laboratory measurements are advantaged in that boundary conditions and measured response can be completed at fine resolution although the representativeness of the sample (sub-REV) and impacts of disturbance may disadvantage the results. Conversely, field measurements may be super-REV and of relevance to field scale behavior but provide less constraint of the true subsurface conditions, typically poorly known and perhaps unknowable.

A variety of laboratory and field techniques are available to quantify the subsurface, all of which have apparent strengths and weaknesses. These are discussed separately in the following.

2.3.1. *Laboratory methods*

Laboratory tests most commonly utilize minimally disturbed core samples that may be free of fractures or contain single or multiple fractures. Their small size exacerbates scale effects, sampling disturbance may transform the hydraulic characteristics from those experienced in situ and permeabilities may be sensitive to recreated in situ stresses. These concerns aside, laboratory testing provides the only method of applying known and carefully controlled boundary conditions in a testing program.

Laboratory tests conducted on core samples conform generally to longitudinal or radial steady tests or transient longitudinal tests. These individual test configurations are illustrated in Figure 2.4. In all configurations, the sample should be pre-saturated with the fluid of interest before beginning the permeability measurements.

Otherwise, the measurement may include phase effects between the flooding fluid and any fluids residing in the sample such as water or air, changes in interfacial tension based on the wettability of the rock, change in dynamic properties due to fluid–rock interaction and manifest as a nonhomogeneous pressure gradient across the sample.

Once a sample has been reintroduced to stress, time-dependent compaction may alter the permeability. This is particularly important in coals and clay-rich shales, where the compaction may not reach its final state for several days (on a laboratory scale). Whether the sample is retracing its deformation and returning to its in situ compactive state is typically unclear, although such displacements or strains may be measured by axial linear variable displacement transducers (LVDTs) or circumferential strain belts.

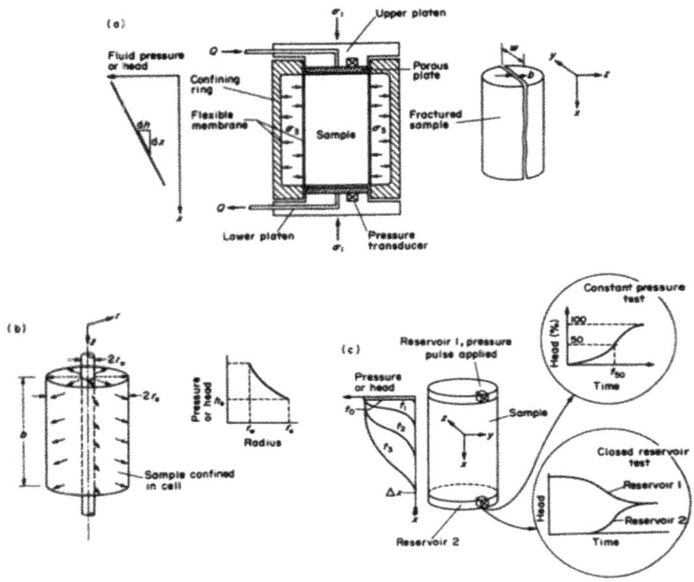

Figure 2.4. *Core and fracture permeameters: (a) longitudinal permeameter; (b) radial permeameter; and (c) longitudinal pulse permeameter (after Elsworth and Mase 2003)*

2.3.1.1. *Linear flow*

In the simplest experimental configuration, a pressure differential $\partial p/\partial x$ is applied along a cylindrical core sample (see Figure 2.5) and the steady behavior measured according to Darcy's law as

$$\frac{k}{\mu} = \frac{Q}{A}\frac{\partial x}{\partial p} \qquad [2.12]$$

to measure a steady flow rate, Q, between the parallel ends of the core of area, A, and separation, ∂x. This recovers the "bulk" permeability of the sample if the circular cross-sectional area of the entire core is used ($A = \pi r_e^2$), or a fracture permeability if the area of the fracture ($A = 2r_e b$) is used. In the event that the upstream and downstream flow rates are not exactly equal, the sample should continue to be flooded to ensure complete saturation. If an average between Q_{up} and Q_{dn} is considered to be acceptable to include in equation [2.12], it should be reported along with the permeability measurement.

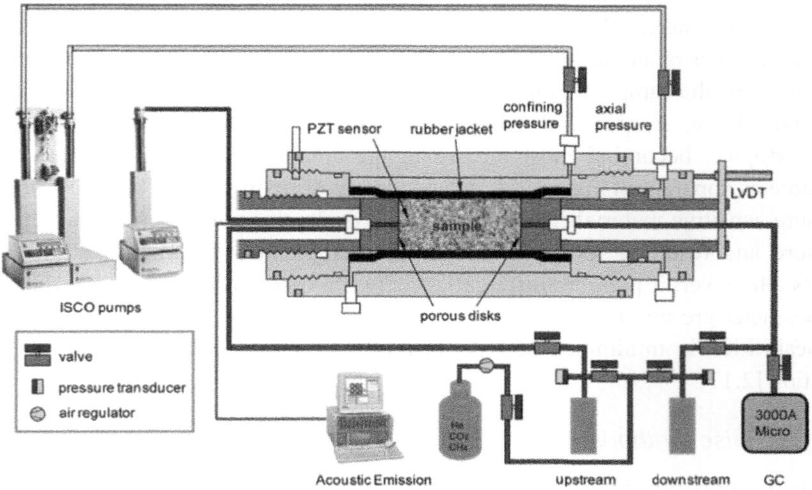

Figure 2.5. *Schematic of a typical linear flow or pulse permeameter apparatus (after Wang et al. (2011) and Wang et al. (2012)). For a color version of this figure, see www.iste.co.uk/vandamme/coal.zip*

Longitudinal permeameters are most easily used in the "confined" testing of samples where a desired isotropic or deviatoric stress field may be created within the sample by sheathing the core in an impermeable membrane. Hydraulic conductivities under a variety of confining stresses and thermal environments may be readily determined as an approximation of the in situ conditions. Examples of such tests range from small (Goodman and Ohnishi 1973) to medium scale (Piggott and Elsworth 1990; Piggott 1990) with the associated testing difficulties increasing with scale.

Although testing becomes increasingly more difficult with size, the profound influence of the scale effect sometimes warrants this further effort. Permeabilities must be sufficiently high that a steady flow regime may be established within a reasonable experimental time and that flow rates are resolvable. This generally limits steady experiments to permeabilities of the order of milliDarcies ($k > 10^{-15}$ m^2).

2.3.1.2. *Radial flow*

The use of radial permeameters is less common, although some contemporary uses have examined hydraulic fracturing in cores from a central blind or fully penetrating wellbore. Analysis requires the use of Darcy's law suitably transformed to examine the steady flow in radial coordinates. The radial flow toward the central well develops a radial seepage force as a result of frictional drag between the fluid and grains of the sample, manifesting as the gradient in pressure. The magnitude of the body force, F, per unit volume is controlled by the pressure gradient as $F = -\partial p/\partial x$, becoming most intense in the area of high gradient close to the wellbore as apparent in Figure 2.4, independently of permeability (Bernaix 1969). In pressure-sensitive materials, such as fractured coals, this nonuniform distribution of pressure and resulting forces in the sample complicates the interpretation of the results. However, if pressure differentials between outer and inner radii in the radial permeameter are minimized, the influence of heterogeneous pressures in modifying permeabilities is minimized and permeability may be determined directly from equation [2.17].

2.3.1.3. *Pulse methods*

The pulse permeameters develop transient longitudinal flows, typically in linear configurations (see Figures 2.4 and 2.5). The transient flow is typically as a pressure pulse between upstream and downstream fluid reservoirs set at an initial pressure differential, as illustrated in Figure 2.4(c) (Wang et al. 2011, 2012). Pressure recovery in the reservoirs is recorded with time following initiation of the test and the results are used to determine the magnitude of the permeability, k, and specific storage, S_s. Two different experimental configurations are possible.

In the first, the upstream reservoir (reservoir 1) is pressurized at time $t = 0$ and this pressure maintained constant with the pressure in the downstream reservoir (reservoir 2) monitored with time. Reservoir 2 must have negligible or no volume and hence "volume compressibility". Plotting the transient behavior enables the time to 50% (t_{50}) or any other convenient degree of pressure equilibration to be determined as illustrated in Figure 2.4(c). The solution of the one-dimensional form of equation [2.11] enables the hydraulic diffusivity, the ratio of hydraulic conductivity K to specific storage, K/S_s to be determined as (Carslaw and Jaeger 1959; Terzaghi 1963)

$$\frac{K}{S_s} = \frac{k}{\mu}\frac{\rho g}{S_s} = \frac{1}{10}\frac{(\Delta x)^2}{t_{50}} \qquad [2.13]$$

where $k/\mu = K/\rho g$. Notably, hydraulic diffusivity alone may be determined from this testing configuration rather than independent evaluation of conductivity and storage. If the test is complemented by a steady longitudinal flow test, then equation [2.12] yields permeability, enabling storage to be evaluated uniquely from equation [2.13]. Conversely, if compressibility, α, is independently measured, then storage is defined from equations [2.9] or [2.10] and permeability may be independently evaluated.

An alternative to this configuration that is especially appropriate for low permeability or low porosity rocks is to use closed reservoirs of known volume at either end of the sample. An instantaneous pressure applied at the upstream end of the core in reservoir 1 is "shut in" at time $t = 0$. As fluid exits reservoir 1, pressure drops and if the disturbance is large enough to reach the downstream reservoir, the pressure in reservoir 2 will climb, as apparent in Figure 2.4. If reservoir volumes are known, hydraulic conductivity and storage may, in most cases, be evaluated for the sample (Hsieh et al. 1980; Neuzil et al. 1981). The sensitivity of the test depends largely on the relative volumes (viz. volume compressibilities) of the reservoirs that may be tuned relative to the sample permeability. The change in pressures in the upstream ($p_{up}(t)$) and downstream ($p_{dn}(t)$) reservoirs with time relative to their respective volumes (V_{up},V_{dn}) are given as (Brace et al 1968; Hsieh et al. 1980)

$$p_{up}(t) - p_{dn}(t) = \left(p_{up}(t_0) - p_{dn}(t_0)\right) e^{-\alpha t} \qquad [2.14]$$

with

$$\alpha = (kA/\mu\beta\Delta x)(\frac{1}{V_{up}} - \frac{1}{V_{dn}}) \qquad [2.15]$$

and permeability, k, sample cross-sectional area, A, dynamic viscosity of the fluid, μ, fluid compressibility, β, and sample length, Δx, are as previously defined. A plot of the decay of the pressure differential $p_{up}(t) - p_{dn}(t)$ with log-time yields the coefficient α from the gradient, enabling permeability, k, to be recovered from the known fluid properties and sample and reservoir dimensions. Such pulse methods are particularly applicable for low permeability samples ($k < 10^{-15}$ m^2), and the permeability of gases and supercritical fluids.

In the case of gas and supercritical fluids, care should also be taken to minimize diurnal temperature fluctuations around the sample and pressure transducers. This can be accomplished by separating the testing apparatus from the rest of the laboratory using a temperature control box, or by measuring during the same period each day with the same temperature. In all cases, the ambient temperature around the testing apparatus should be recorded alongside permeability measurements.

2.3.2. *In situ methods*

By their very nature, field tests sample greater volumes than laboratory tests and have the advantage of incurring minimal disturbance of the in situ structure. The in situ volume sampled in a pumping test will increase with both an increase in the pumping volume or an increase in test duration. The hydraulic parameters evaluated from field tests are aggregated for the volume tested and should correspondingly only be applied over that particular length scale; and in pressure sensitive materials, such as coal beds, be applied only to the particular range of effective stresses experienced.

Pumping tests are usually conducted through straddle packer systems isolating a variable length section of the borehole. The steady or transient hydraulic response may be gauged relative to the solution of the diffusion equation [2.11] to allow for the determination of permeability, k, in steady and transient tests and specific storage, S_s, in transient tests. The axisymmetric geometry of the testing configuration allows equation [2.11] to be represented more conveniently as

$$\frac{k}{\mu}\left(\frac{\partial^2 p}{\partial r^2}+\frac{1}{\mathrm{r}}\frac{\partial p}{\partial r}+\frac{\partial^2 p}{\partial z^2}\right)=\frac{S_S}{\rho g}\frac{\partial p}{\partial t} \qquad [2.16]$$

where r and z are radial and vertical ordinates, respectively. In most applications, the dependence of the solution on vertical gradients, $\partial p/\partial z$, is neglected and the diffusion equation is assumed purely radial. This expression allows solutions to be determined for radial flows to a well for steady ($\frac{\partial p}{\partial t}=0$), quasi-steady ($\frac{\partial p}{\partial t}=0$ but accommodating wellbore storage) and transient ($\frac{\partial p}{\partial t}\neq 0$) pumping tests. For highly compressible fluids (gases at large pressure differentials), the pressure (p) terms on the left-hand side of equation [2.16] may need to be replaced as p^2 or pseudopressure ψ approaches used in the solution.

2.3.2.1. *Steady methods*

When a pumping test is run for a long enough duration such that the drawdown cone either reaches an exterior constant pressure boundary, or the pressure stops changing, the test may be analyzed in the steady state ($\frac{\partial p}{\partial t}=0$). Solving equation [2.16] for pressures p_w and p_t applied at the wellbore radius r_w and externally at r_e enables permeability, k, to be determined from the measured flow rate Q as

$$\frac{k}{\mu}=\frac{K}{\rho g}=\frac{Q}{2\pi L}\frac{\ln(r_w/r_e)}{(p_w-p_e)} \qquad [2.17]$$

where L is the thickness of the open zone in the hole or the coal seam thickness for an open hole (Figure 2.6).

2.3.2.2. *Quasi-steady methods*

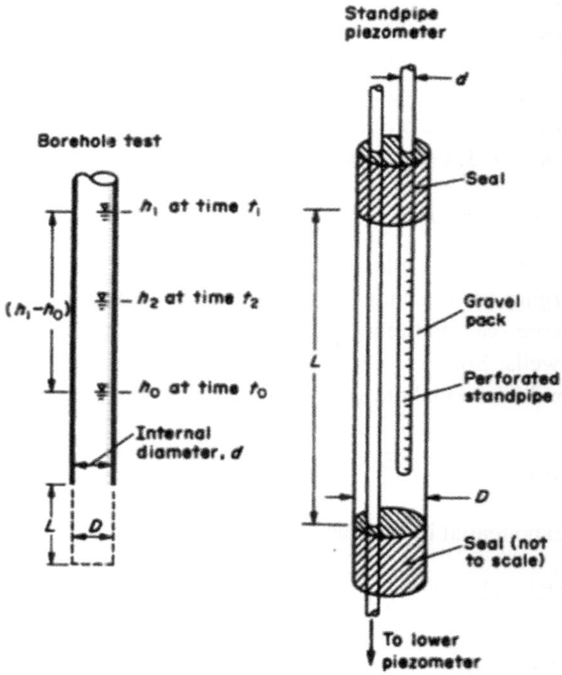

Figure 2.6. *Falling head tests in boreholes and standpipe piezometer nests (after Elsworth and Mase 1993)*

Quasi-steady methods assume steady flow in the reservoir/coal seam but a transient response of the wellbore storage. Typically, such Lugeon of "slug" tests are administered as a "slug" of liquid added to or removed from a packed off section of wellbore, connected via an open standpipe. Where water is removed, the water level will fall and flow will occur from the formation that the wellbore pierces. Solving the steady form of equation [2.16] for flow to a diminishing $\Delta h = \Delta p/\rho g$ at the borehole wall with time, as the wellbore fills, yields the permeability as

$$\frac{k}{\mu} = \frac{A \ln(h_1/h_2)}{\rho g\, F(t_2 - t_1)} \qquad [2.18]$$

where A is the internal cross-sectional area of the casing and h_1 is the water level/hydraulic head at time t_1, as illustrated for open borehole and piezometer tests in Figure 2.6. The shape factor F represents the influence of the geometry of the wellbore base. For a casing flush with the wellbore base, $F = 2.75d$, where d is the internal diameter of the casing. For a long open base section, $F = 2\pi L/ln(2L/D)$, where L is the open length of the borehole and D is the borehole diameter (Hvorslev 1951; Hoek and Bray 1967). Tests of this type may be performed rapidly (tens of minutes) in open jointed rocks and over longer periods (hours) in tighter formations with alternate methods available to recover storage coefficient (Cooper et al. 1967; Papadopoulos et al. 1973; Freeze and Cherry 1979).

2.3.2.3. *Transient methods*

Long-term pumping tests allow the pressure within a reservoir to propagate far from the discharging (or injecting) borehole. Thus, such tests sample the reservoir broadly and at large scale but require significant infrastructure to sample drawdown at intermediate wells. Solving equation [2.16] for a constant flow rate, Q, yields a solution for a nondimensional drawdown, h_D, with nondimensional time, t_D as

$$h_D = \int_{1/4t_D}^{\infty} \frac{e^{-y}}{y} dy = -E_i(-\frac{1}{4t_D}) \qquad [2.19]$$

where E_i is the exponential integral and

$$h_D = \frac{k}{\mu} \frac{4\pi L[h(t_0)-h(t)]\rho g}{Q} \; ; \; t_D = \frac{Kt}{S_S r^2} = \frac{k}{\mu} \frac{\rho g}{S_S} \frac{t}{r^2} \qquad [2.20]$$

where $h(t_0)$ is the water-level in the sampling well at $t = 0$. This well is separated by a dimensionless radius $r_D = \frac{r}{r_w}$, where r_w is the radius of the pumping well. If the monitoring well is more than 10 radii distant, then the solutions collapse onto a single curve (Theis 1935; Kruseman and DeRidder 1983).

Pumping test results, plotted as log drawdown in head versus log time may be superimposed on Figure 2.7 to determine a single matchpoint (say $t_D = 1$ and $h_D = 1$) and recover permeability and storage from equation [2.20]. If a single pumped well is used with no external observation well, then the type curve for $r_D = 1$ allows for the estimation of reservoir and fracture flow parameters.

However, ambiguous matches pose a problem for $r_D = 1$ since the type curve is relatively featureless with only a shallow inflection point as a discriminating feature. Hence, multiple ambiguous matches are possible with corresponding uncertainty in the recovered fluid transport and storage parameters. Solutions may be extended to accommodate more complex aquifer systems including confined leaky (Neuman and Witherspoon 1969a, 1969b), unconfined (Neuman 1972, 1973, 1975) and partially penetrated (Hantush 1964) aquifers together with constant pressure tests with open and closed boundaries (van Everdingen and Hurst 1949; Uraiet 1979; Doe and

Osnes 1985; Elsworth and Doe 1986) and pulsed methods (Crosnier et al. 1985) and directional permeability measurements (Javandel 1983).

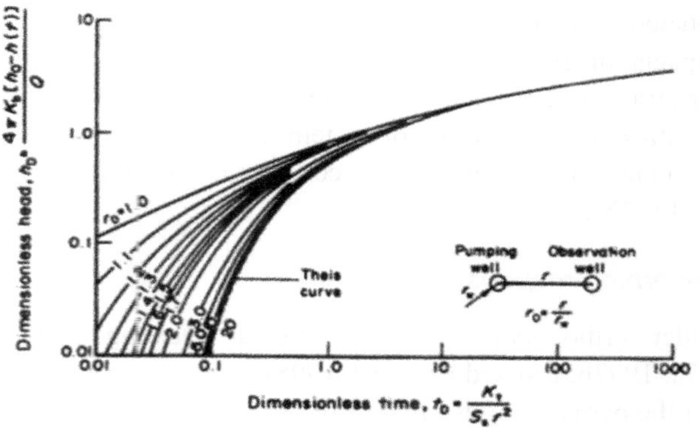

Figure 2.7. *Temporal head response for a constant rate pumping test in a confined aquifer (Theis 1935). Also shown are the results for finite radius wells ($r_d \ll \infty$) (after Elsworth and Mase 1993)*

2.4. Permeability evolution models

A key feature in understanding the hydrogeology of coal seams, particularly the recovery of gases from such units is the extreme stress sensitivity of permeability. Thus, during water desaturation of primarily the fracture network at the beginning of the recovery process, increases in effective stress typically result in closure of individual fractures comprising the fracture network.

As the fracture system desaturates and methane is recovered, desorption of gas from the coal matrix then proceeds, likewise resulting in compaction, but this time sourced in the matrix and potentially in competition with the compaction of the surrounding fractures. It is this competition between the impacts of gas desorption and matrix shrinkage versus changes in effective stresses in the fractures and fracture compaction that are the essence of controls on permeability evolution of coal seams, hence the effective recovery of natural gas.

We consider the utility of both strain-based and stress-based models in defining the innate sensitivity of the permeability of fractured coal seams to deformations resulting from changes in stress.

2.4.1. *Strain-controlled models*

We consider these impacts by first exploring the behavior of non-sorbing fracture-dominated reservoirs (low gas charge seams) to clearly demonstrate the important impacts of sorption and swelling on permeability response. This is completed for strain-based models. We then consider a broad range of behaviors under stress control, including those of systems still evolving under non-equilibrated stress/pressure conditions before finally considering the influence of supercritical fluids used for ECBM.

2.4.1.1. *Non-sorbing response*

We consider orthogonally fractured media (Ouyang and Elsworth 1993; Liu et al. 1997a, 1997b; Liu and Elsworth 1998) where normal strains applied within a plane impact the permeabilities of fracture sets transecting that plane, as illustrated in Figure 2.8. Assuming that the rock matrix is functionally impermeable, then the change in permeability may be determined if the overall strains applied to the rock mass are appropriately partitioned between fracture and solid matrix. For recoverable strains, linear mechanical behavior in matrix and fracture and with a closure cutoff, then permeability changes may readily be related to a modulus reduction factor. This represents two end-member conditions. The first is where the modulus of fractures and matrix are uniform, and then strains will be uniform and the strain-based change in permeability will be minimum (uniform in both matrix and fracture). The second is where the fracture is infinitely soft matrix infinitely stiff, then the entire strain will be concentrated on the fracture and the change in fracture strain/aperture will be a maximum. In this second case, the change in permeability will be a maximum.

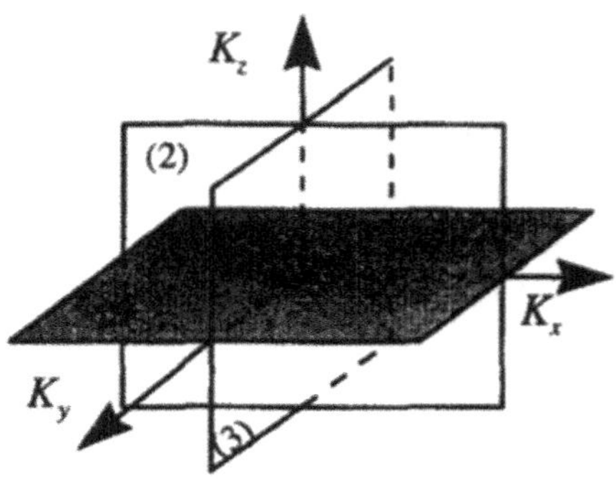

Figure 2.8. *Conceptual model for the effect of fractures on the principal permeabilities. The bracketed numbers represent three sets of orthogonal fracture planes (after Liu and Elsworth 1998)*

Thus, a rock mass modulus reduction ratio (R_m) is defined as the ratio of modulus of the softer rock mass (E_{mass}) containing fractures to the stiffer matrix (E_{matrix}) without fractures as $R_m = E_{mass}/E_{matrix}$. This results in the partition of strains and changes in permeability as illustrated in Figure 2.9.

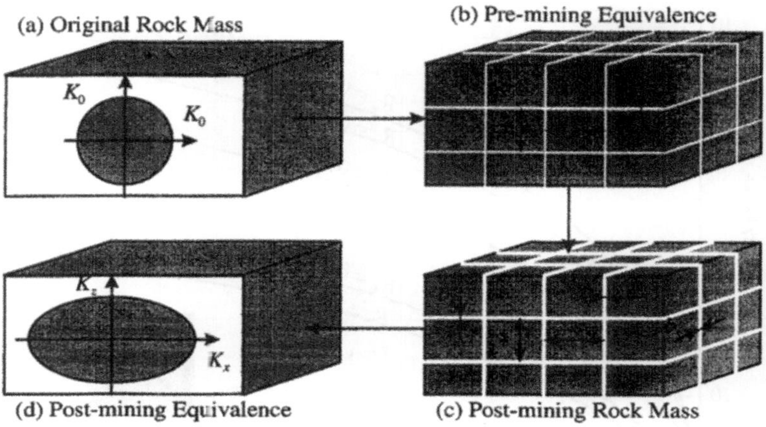

Figure 2.9. *Evolution of permeability as a result of strains applied by mining (changes in total stress) or by pressure depletion (changes in effective stress). Initial isotropic permeability evolves to anisotropic due to the directionally heterogeneous strain field (after Liu and Elsworth 1998)*

This allows changes in permeability, post-deformation (k_x) to pre-deformation (k_{x0}), to be defined as

$$\frac{k_x}{k_{x0}} = \frac{1}{2}[(1 + \beta_y \Delta\epsilon_y)^3 + (1 + \beta_z \Delta\epsilon_z)^3] \qquad [2.21]$$

and

$$\beta_y = 1 + \frac{1 - R_m}{b_y/s_y} \qquad [2.22]$$

where changes in normal strains in the y-direction are $\Delta\epsilon_y$ and the rock mass reduction factor acts through the term β_y. Note that the changes in permeability in the x-direction are due to strains in the orthogonal y- and z-directions, as apparent in Figure 2.9. Similar expressions for k_y/k_{y0} and k_z/k_{z0} and for β_x and β_z are available through permutation of subscripts. The resulting sensitivity of permeability to deformation is apparent in Figure 2.10 where the permeability ratio k_x/k_{x0}

changes over five orders of magnitude for a strain of 0.05. Where the fractures are soft and the matrix stiff ($R_m \rightarrow 0$), the change in permeability is maximum, and where the fractures have similar stiffness to the matrix ($R_m \rightarrow 1$), then the change in permeability is small.

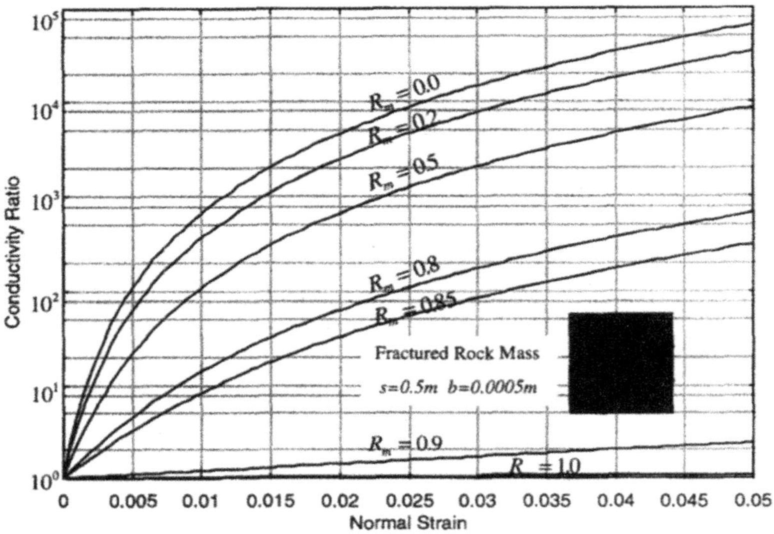

Figure 2.10. *Relationship among conductivity or permeability ratio (k/ko), modulus reduction ratio, R_m, and normal strain, $\Delta\epsilon$, for fracture spacing $s = 0.5\,m$ and aperture $b = 0.005\,m$ (after Liu and Elsworth 1998)*

2.4.1.2. *Sorbing response*

The permeability evolution of coal and other sorptive, geologic media results from an ensemble response to competing strains. Increasing fluid pressure causes fracture and pore expansion proportional to the fracture or pore stiffness and dependent on external boundary conditions. This mechanical dilation typically increases permeability. If the fluid is sorptive, there is a competing strain caused by the swelling of the sorptive solid, which constricts flow paths and reduces permeability. This leads to a typical U-shaped curve in coal permeability with increasing pore pressure. At low pressures, sorptive swelling is the dominant mechanism and flow paths experience net closure. The rate of sorptive swelling decreases as fewer sites remain available for adsorption, allowing dilation from poromechanical expansion to outpace sorptive swelling close to the Langmuir pressure. This outpacing leads to net permeability enhancement after the inflection point and the permeability curve exhibits a standard U-shaped form.

The same is true of desorptive processes, the most common being pressure depletion. As fluid pressure decreases, sorptive fluids desorb from the coal. While pores and fractures close at lower fluid pressures, the constriction of flow paths from sorptive swelling is also relaxed. This can lead to permeability enhancement, depending on the relationship between the virgin in situ pressure of the coal seam and its Langmuir pressure: methane desorption may result in two- to five-fold increases in permeability, or larger, in laboratory experiments (Harpalani and Schraufnagel 1990) and field observations (Mavor and Vaughn 1998).

Changes in strain within coalbed reservoirs result both from the manifestation of changes in effective stress, principally resulting in large changes in deformation across fractures, but also as a result of sorption/desorption-related changes in matrix volume, also impacting the response of the fractures (Izadi et al. 2011; Wang et al. 2011, 2012).

Swelling strains are typically represented as a Langmuir-type relation where the resulting strain, ϵ_S, is defined at a given matrix pressure, p, as

$$\epsilon_S = \epsilon_L \frac{p}{p+p_L} \qquad [2.23]$$

where ϵ_L is the Langmuir strain and p_L is the Langmuir pressure. The Langmuir strain is a material constant defining the maximum strain resulting when sorption/desorption is complete with the Langmuir pressure defining the matrix pressure when strain is 50% complete.

A simple analysis involving a cracked solid is useful to illustrate the dependency on various parameters. Consider an array of fractures shown in Figure 2.11. The geometry is such that overall strains are null due to the mutual interaction of the array of flaws. The closure of the fracture aperture, Δb, that results from the swelling at pressure p can then be defined as it competes with the dilation due to effective stresses as shown in Figure 2.12. In this, the initial reduction in permeability occurs due to swelling before this is countered by the high fracture pressures dilating the fracture. It is clear that lower permeability media are more sensitive to changes in permeability resulting from swelling deformations. This is to be expected as similar magnitudes of swelling strain impart greater relative deformations/closure on small aperture fractures in low permeability aggregates. These amplify the changes in permeability.

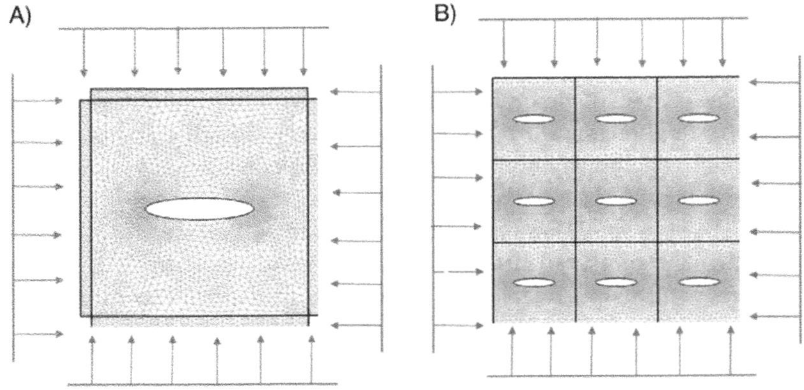

Figure 2.11. *Single and multiple flaws present within a sorbing/swelling matrix (after Izadi et al. 2011)*

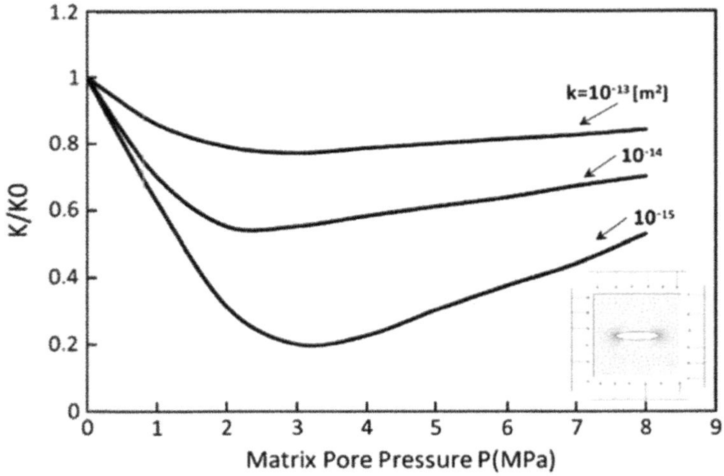

Figure 2.12. *Relation between fracture permeability ratio and matrix permeability for differing beginning permeabilities (after Izadi et al. 2011)*

Indeed, this response may be demonstrated with a simple scaling model for multiple fractures that are fully displacement constrained (null strain as in Figure 2.11). For fractures of initial length, a, spacing, s, and initial aperture, b_0, and assuming simple volumetric swelling for a rectangular crack ($b_0 \times a$), then the resulting change in permeability (k/k_0) may be defined as

$$\frac{k}{k_0} = \left(1 + \frac{\epsilon_S s^2}{ab_0}\right)^3 = \left(1 + \left(\frac{\epsilon_L s^2}{ab_0}\right)\frac{p}{p+p_L}\right)^3 \qquad [2.24]$$

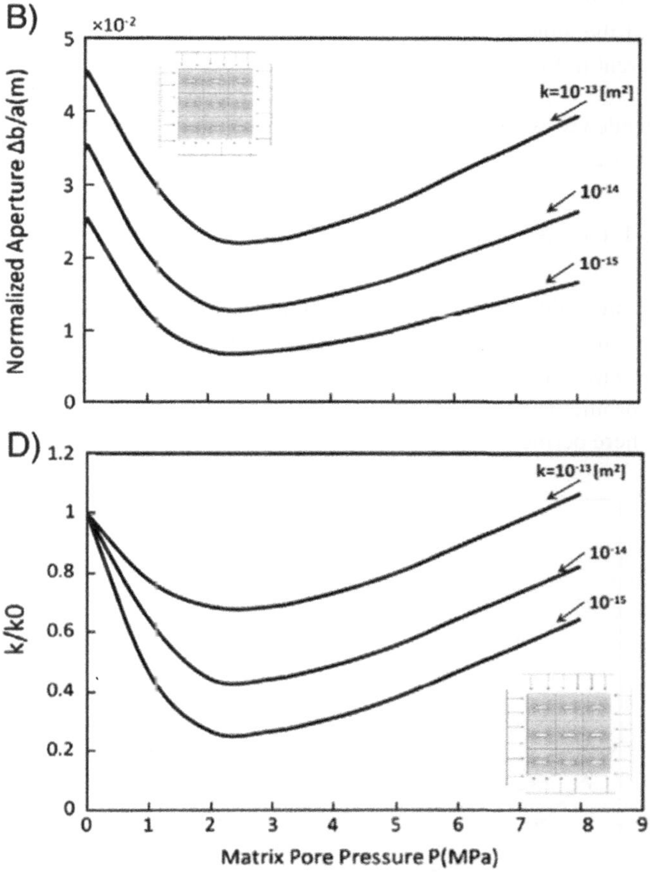

Figure 2.13. *Changes in aperture as a function of swelling (low pressures) and countered by changes in effective stress (high pressures). Note that (B) changes in apertures are similar with changes in pressure (parallel but offset curves) but result in (D) amplified changes in permeability for initially low permeability materials (after Izadi et al. 2011)*

Importantly, this demonstrates that within the amplifying effects of the cubic law relationship for fracture permeability, then permeability scales linearly with Langmuir strain and matrix pressure but with the square of spacing and inversely with fracture length and aperture.

Since initial fracture aperture is related to permeability as $k_0 = \frac{b_0^3}{12s}$, then the dependency of the amplification in change in permeability on initial permeability is clear, as apparent in Figure 2.13.

This dependency with permeability is realized since the prefactor in equation [2.24] defines the relation

$$\frac{k}{k_0} = \left(1 + \left(\frac{\epsilon_L s^2}{a\sqrt[3]{12k_0 s}}\right)\frac{p}{p+p_L}\right)^3 \qquad [2.25]$$

and the apparent scaling of increased sensitivity of the change in permeability with increasing magnitude of the nondimensional parameter $s' = \epsilon_L s^2 / a\sqrt[3]{12k_0 s}$ as apparent in Figure 2.14. Such controls are consistent with observations for both the injection of sorbing gases, where permeabilities are reduced and for the recovery of natural gas where permeability generally increases.

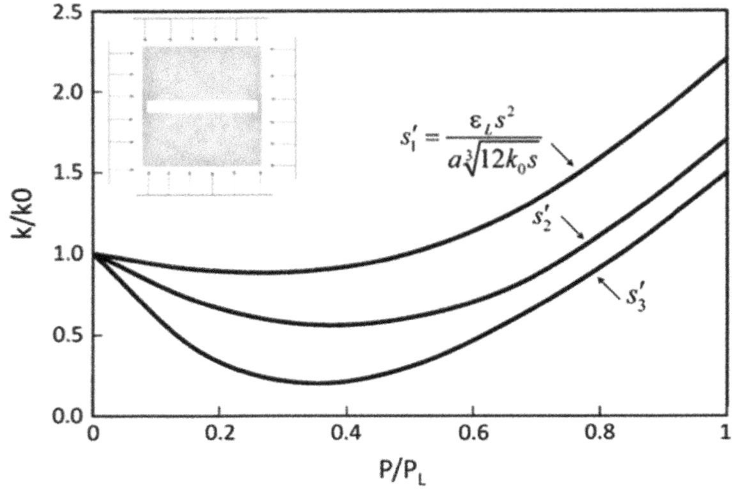

Figure 2.14. *Control of the nondimensional parameter s′ on the change in permeability resulting from swelling (after Izadi et al. 2011)*

2.4.2. *Stress-controlled models*

The evolution of permeability as a result of injection or depletion may also be represented by models with stress control. These examine the impacts of transient effects in contributing to non-monotonic evolution of permeability as gases invade the matrix from fractures and the state of stress within the REV is nonuniform and transient. In addition, the form of the boundary conditions also potentially impacts the form of the permeability evolution, exerting different impacts for stress-controlled experiments conducted under either constant confining pressures (CCPs) or with constant effective stresses (CESs) (Shi et al. 2018, 2020).

2.4.2.1. *Transient effects*

The contrasting transport characteristics of fractured coals, with low permeability matrix and high permeability fractures, inevitably results in a transient evolution to a new equilibrium once injection or depletion is begun. This time lag will scale with a diffusive time typically of the order of $t_D = \frac{ct}{l^2} \sim 1$. In this, l is a characteristic length, t is time since injection and c is an effective diffusivity. For equilibration at the cleat scale, the appropriate characteristic length is the cleat spacing (0 cm) and the hydraulic diffusivity is $c = \frac{K}{S_S} = k/\mu n\beta$. For standard gases at 1 MPa and centimeter-scale cleats, the equilibration time for nanoDarcy coal is of the order of 1 day (~10^5 s) but only minutes (~10^2 s) for microDarcy coal.

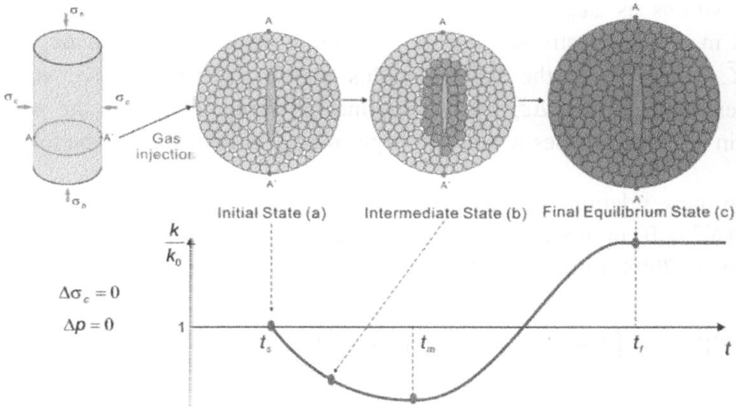

Figure 2.15. *Transient evolution of permeability with diffusion into matrix from a single fracture under stress-controlled conditions (after Shi et al. 2018). For a color version of this figure, see www.iste.co.uk/vandamme/coal.zip*

Transient impacts are important as experimental measurements taken before full equilibration will be incorrect, as illustrated in Figure 2.15. Permeability evolves from the initial inflation of the embedded fracture. As gas diffuses from the fracture into the surrounding matrix, the fracture will close as the swelling zone is mechanically restrained by the outer material and the permeability reduces (t_m, see Figure 2.15). As the gas further permeates and equilibrated gas pressures result, the swelling is uniform and the fracture will dilate and return to an elevated permeability, potentially including an initial offset due to the decrease in effective stress (t_f, see Figure 2.15). Thus, readings taken prior to t_f will not reflect the true equilibrium change in permeability and be in error.

2.4.2.2. *Constant confining pressure or effective stresses*

The magnitude of the potential error in permeability measurement, engendered by transient effects (see above), may be explored by placing bounds on the permeability evolution resulting from experiments conducted under various stress-controlled conditions and reconciled with available laboratory observation data.

These bounds on behavior are recovered from the definition of change in permeability under stress control. Where the fracture is mechanically soft in comparison to the matrix, then $R_m \to 0$ and equation [2.21] for a single fracture set reverts to $\frac{k}{k_0} = (1 + s/b\,\Delta\epsilon_e)^3$, where the effective strain $\Delta\epsilon_e$ is the difference between total strain ($\Delta\epsilon_v$) and gas sorption induced ($\Delta\epsilon_s$) and gas diffusion ($\Delta\epsilon_d$) induced strains as $\Delta\epsilon_e = \Delta\epsilon_v - \Delta\epsilon_s - \Delta\epsilon_d$. Change in strains may be related to changes in confining stresses, $\Delta\overline{\sigma}$, and injection pressures, Δp, as $\Delta\epsilon_e = -(\Delta\overline{\sigma} - \overline{\alpha}\Delta p)/\overline{K}$, where $\overline{K}$ is the bulk modulus and $\overline{\alpha}$ is the effective stress (Biot) coefficient. Substituting $\Delta\epsilon_e$ into the permeability relation yields the permeability change in terms of stresses where two experimental conditions may be followed.

From these relations, changes in permeability for experiments completed either at CCP ($\Delta\overline{\sigma} = 0$) or at variable confining pressure but CES ($(\Delta\overline{\sigma} - \overline{\alpha}\Delta p) = 0$) yield (Zhang et al. 2008; Liu et al. 2011a, 2011b; Pan and Connell 2012):

$$\text{CCP: } \frac{k}{k_0} = \left(1 + \frac{s}{b}\frac{\overline{\alpha}\Delta p}{\overline{K}}\right)^3 = \left(1 + \frac{\overline{\alpha}}{\phi_{f0}}\frac{\Delta p}{\overline{K}}\right)^3 \quad \text{CES: } \frac{k}{k_0} = 1 \qquad [2.26]$$

where $\phi_{f0} = b/s$ represents the fracture porosity. These expressions define the upper bounds for permeability change resulting at CCP and CESs as illustrated in Figures 2.16 and 2.17.

Conversely, when experimental data are taken prior to true equilibration within an experiment (t_m, see Figure 2.15), then a separate but spurious lower bound permeability is feasible. This intermediate result may also be evaluated by defining this lower bound in terms of mechanical constraints. If the permeability lower bound results when swelling immediately adjacent to the fracture constrains outward expansion of the near-fracture sheath (t_m, see Figure 2.15), then this can be approximated by limiting the total strain to zero ($\Delta\epsilon_v = 0$). Thus, the change in effective strain for $\Delta\epsilon_v = 0$ reduces to $\Delta\epsilon_e = -\Delta\epsilon_s - \Delta\epsilon_d$ and substitution into $\frac{k}{k_0} = (1 + s/b\ \Delta\epsilon_e)^3$ yields

$$\text{CCP and CES: } \frac{k}{k_0} = \left(1 + \frac{s}{b}\Delta\epsilon_e\right)^3 = \left(1 - \frac{1}{\phi_{fo}}(\Delta\epsilon_s + \Delta\epsilon_d)\right)^3 \qquad [2.27]$$

to define the lower bound permeability ratio.

Thus, the changes in permeability are bound by upper and lower bounds for each of CCP (Figure 2.16) and CES (Figure 2.17) experiments. The upper bound represents the permeability for the true pressure equilibrated and the lower bound for intermediate measurements before the final state is reached. Convincingly, the laboratory data are restricted to and fully populate the regions between these limits, as illustrated in Figures 2.16 and 2.17, emphasizing the importance of bringing such experiments to full completion if representative results are desired.

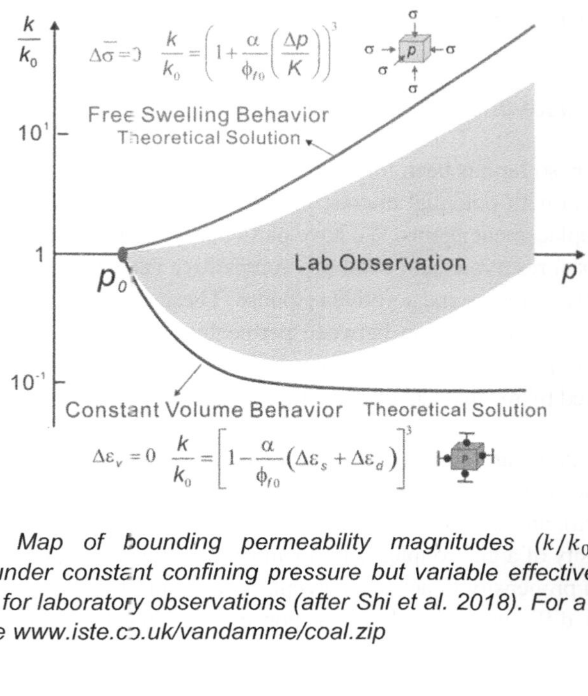

Figure 2.16. *Map of bounding permeability magnitudes (k/k_0) for injection experiments under constant confining pressure but variable effective stresses. See original paper for laboratory observations (after Shi et al. 2018). For a color version of this figure, see www.iste.co.uk/vandamme/coal.zip*

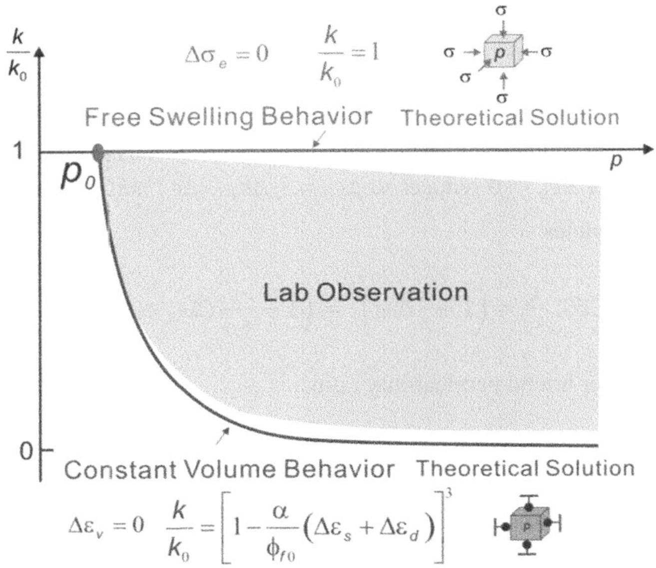

Figure 2.17. *Map of bounding permeability magnitudes (k/k_0) for injection experiments under constant effective stresses. See original paper for laboratory observations (after Shi et al. 2018). For a color version of this figure, see www.iste.co.uk/vandamme/coal.zip*

2.4.3. *Supercritical fluids*

The discussion so far has been for subcritical gases and liquids: water and methane primarily, but also with potential interest in carbon dioxide and nitrogen as reservoir stimulants and replacement agents. We have particularly focused on the sorptive gases. These fluids impact reservoirs and elicit this competitive response between mechanical impacts of effective stresses and sorptive response. These are demonstrated to impact a non-monotonic U-shaped relation between permeability and gas pressure (Zhi et al. 2019). Sorptive impacts are initially strong before reaching a maximum impact and then being counted by mechanical effects of fracture dilation at high pressures.

Supercritical gases are considered to be potential injectates for carbon storage or for ECBM. Under expected injection conditions, carbon dioxide will be supercritical (30°C and 7 MPa) and has the characteristics of a supercritical phase: the density of a liquid, the viscosity of a gas coupled with its miscibility with other supercritical fluids, its reactivity and propensity to act as a solvent. As a result of this last attribute, carbon dioxide exerts a different permeability response when injected into fractured coals.

Experiments across the supercritical phase transition show that the permeability curve is W-shaped with an additional minimum permeability to the normal U-shaped curve, as shown in Figure 2.18. This permeability minimum occurs close to the critical pressure and is two orders of magnitude lower than the initial permeability.

Observed irreversible changes in mechanical properties driven by the supercritical CO_2 ($SCCO_2$) phase transition are linked to the observed permeability reduction. First, observed reductions in coal strength and stiffness after $SCCO_2$ saturation indicate a significant weakening effect. $SCCO_2$ acts as a strong plasticizer enabling rearrangement of the physical structure and adding molecular mobility. Second, the $SCCO_2$ permeability is further reduced by a large increase in sorption-induced swelling observed after the transition. Measured swelling strains indicate a > 90% increase in $SCCO_2$ sorption-induced swelling relative to the Langmuir response at 13 MPa. The enlarged $SCCO_2$ adsorption capacity presumably results from the slightly polar nature of the $SCCO_2$ dimer and the increased SCCO2 density. Permeability reduction due to plasticization and increased swelling is shown to dominate over permeability increases driven by brittle damage.

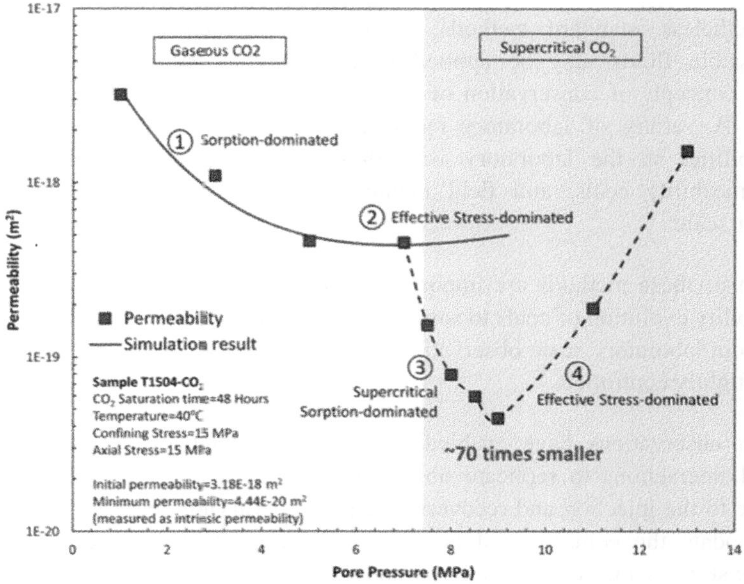

Figure 2.18. *Permeability evolution with pore fluid pressure across the supercritical CO_2 transition. Red squares represent experimental measurement and the red solid line the results of numerical modeling to fit the data (after Zhi et al. 2019). For a color version of this figure, see www.iste.co.uk/vandamme/coal.zip*

2.5. Summary

Contemporary uses of coalbeds relate to their degasification for mining safety, as a resource, and for their potential use for the sequestration of CO_2 and storage of H_2. All of these uses require an understanding of their response to the transmission of fluids: either liquids during dewatering or gases during production and injection.

Thus, the transmission of fluids in coal seams may be viewed through the prism of "hydrogeology", largely without regard to the specifics of the fluid – aqueous or nonaqueous – liquid or gas. The governing principles are largely the same.

The particular behavior of coals, with the discrete structure of the matrix and cleats conditions its response as a dual porosity material. Principal storage within the matrix and the high permeability of the fractured network define its particular response, which is complicated by the presence of sorbing gases that drive large volume changes and the sensitivity of the permeability to changes in effective stresses. These twin features of sorption/swelling and sensitivity to changes in effective stresses make coal a challenging material to characterize.

Nonetheless, standard methods of defining flow and transport of slightly compressible fluids may be applied to describe the fractured-porous medium through concepts of conservation of mass and the definition of Darcian flow and storage. A variety of laboratory methods are also available to define Darcian permeabilities in the laboratory with these differentiating between high- and low-permeability coals, and field methods to recover appropriate behavior at reservoir scale.

Each of these methods are important, although much of our constraint on the permeability evolution of coals to sorption/swelling and changes in effective stresses come from laboratory scale observations where boundary conditions and responses may be tightly controlled.

Such observations have enabled simple physical models reflecting the key physical interactions to replicate observed controls on permeability evolution in response to the injection and recovery of gases. To varying degrees, these properly accommodate the complex and heterogeneous interaction in space and time of effective stresses and sorptive swelling and deliver reliable estimates of response for use in the recovery of resources from coal seams.

2.6. References

Aviles, C.A., Scholz, C.H., Boatwright J. (1987). Fractal analysis applied to characteristic segments of the San Andreas fault. *J. Geophys. Res.*, 92, Bl, 331–344.

Barton, C.C. and Larsen, E.L. (1985). Fractal geometry of two-dimensional fracture networks at Yucca Mountain, south western Nevada. In *Proc. Int. Symp. Fundamentals Rock Joints, Bjorkliden, Sweden*. Centek, Lulea.

Barton, N., Bandis, S., Bakhtar, K. (1985). Strength, deformation and conductivity coupling of rock joints. *International Journal of Rock Mechanics and Mining Sciences & Geomechanics Abstracts*, 22(3), 121–140.

Bear, J. (1972). *Dynamics of Fluids in Porous Media*. Elsevier, New York.

Bernaix, J. (1969). New laboratory methods of studying the mechanical properties of rock. *International Journal of Rock Mechanics and Mining Sciences & Geomechanics Abstracts*, 6(1), 43–90.

Brace, W.F., Walsh, J.B., Frangos, W.T. (1968). Permeability of granite under high pressure. *J. Geophys. Res.*, 73(6), 2225–2236.

Carlslaw, H.S. and Jaeger, J.C. (1959). *Conduction of Heat in Solids*, 2nd edition. Oxford Science Publications, Oxford.

Carman, P.C. (1956). *Flow of Gases through Porous Media*. Academic Press, New York.

Cooper Jr., H.H., Bredehoeft, J.D., Papadopoulos, I.S. (1967). Response of a finite-diameter well to an instantaneous charge of water. *Water Resour. Res.*, 3(1), 263–269.

Crosnier, B., Fras, G., Jouanna, P. (1985). Reconnaissance of fractured media with several systems of fractures by means of harmonic techniques. *Rock Mech. Rock Eng.*, 18, 77–105.

Doe, T.W. and Osnes, J.D. (1985). Interpretation of fracture geometry from well test. In *Proc. Int. Symp. Fundamentals Rock Joints, Bjorkliden, Sweden*. Centek, Lulea.

Elsworth, D. and Doe, T.W. (1986). Application of nonlinear flow laws in determining rock fissure geometry from single borehole pumping tests. *International Journal of Rock Mechanics and Mining Sciences & Geomechanics Abstracts*, 23(3), 245–254.

Elsworth, D. and Goodman, R.E. (1986). Characterization of rock fissure hydraulic conductivity using idealized wall roughness profiles. *International Journal of Rock Mechanics and Mining Sciences & Geomechanics Abstracts*, 23(3), 233–243.

Elsworth, D. and Mase, C.R. (1993). An overview of groundwater in rock engineering. In *Comprehensive Rock Engineering: Principles, Practice and Projects*, Hudson, J.A. (ed.). Pergamon Press, London.

Freeze, R.A. and Cherry, J.A (1979). *Groundwater*. Prentice Hall, New Jersey.

Gangi, A.F. (1978). Variation of whole and fractured porous rock permeability with confining pressure. *International Journal of Rock Mechanics and Mining Sciences & Geomechanics Abstracts*, 15(5), 249–257.

Goodman, R.E. and Ohnishi, Y. (1973). Undrained shear tests on jointed rock. *Rock Mechanics*, 5, 129–149.

Hantush, M.S. (1964). Hydraulics of wells. *Advances in Hydroscience*, 1, 281–432.

Harpalani, S. and Schraufnagel, R.A. (1990). Influence of matrix shrinkage and compressibility on gas production from coalbed methane reservoirs. In *SPE Annual Technical Conference and Exhibition*, September 23–26, New Orleans. doi: 10.2118/20729-MS.

Hoek, E. and Bray, J.W. (1977). *Rock Slope Engineering*. Institution of Mining and Metallurgy, London.

Hsieh, P.A., Tracy, J.V., Neuzil, C.E., Bredehoeft, J.D., Silliman S.E. (1980). A transient laboratory method for determining the hydraulic properties of 'tight' rocks – I. Theory. *International Journal of Rock Mechanics and Mining Sciences & Geomechanics Abstracts*, 18(3), 245–252.

Hvorslev, M.S. (1951). Time lag and soil permeability in groundwater measurements. *U.S. Corps of Engineers Waterways Experiment Station Bulletin*, 36, 50.

Iwai, K. (1976). Fundamentals of fluid flow through a single fracture. PhD Thesis, University of California, Berkeley.

Izadi, G., Wang, S., Elsworth, D., Liu, J., Wu, Y., Pone, D. (2011). Permeability evolution of fluid-infiltrated coal containing discrete fractures. *International Journal of Coal Geology*, 85(2), 201–211.

Javandel, I. (1983). Techniques for measuring the vertical hydraulic conductivity of flood basalts at the basalt waste isolation project site. Technical Report, Lawrence Berkeley National Laboratory (LBNL), Berkeley.

Kozeny, J. (1927). Veber kapillare Leitung des Wassers im Boden. *Wien. Akad. Wiss.*, 132(2a), 271–306.

Kruseman, G.P. and Ridder, N.A. (1983). *Analysis and Evaluation of Pumping Test Data*, 3rd edition. International Institute for Land Reclamation and Improvement (ILRI), Wageningen.

Liu, J. and Elsworth, D. (1998). Three-dimensional effects of hydraulic conductivity enhancement and desaturation around mined panels. *International Journal of Rock Mechanics and Mining Sciences*, 34(8), 1139–1152.

Liu, J., Elsworth, D., Matetic, R.J. (1997a). Evaluation of the post-mining groundwater regime following longwall mining. *Hydrological Processes*, 11(15), 1945–1961.

Liu, J., Elsworth, D., Brady, B.H. (1997b). Analytical evaluation of post-excavation hydraulic conductivity field around a tunnel. *Int. J. Rock Mech. Min. Sci.*, 34(3–4), 181.e1–181.e7.

Liu, J., Wang, J., Chen, Z., Wang, S., Elsworth, D., Jiang, Y. (2011a). Impact of transition from local swelling to macro swelling on the evolution of coal permeability. *International Journal of Coal Geology*, 88(1), 31–40.

Liu J., Chen Z., Elsworth, D., Miao, X., Mao, X. (2011b). Evolution of coal permeability from stress-controlled to displacement-controlled swelling conditions. *Fuel*, 90(10), 2987–2997.

Lomize, G.M. (1951). *Flow in Fractured Rocks* (in Russian). Gosenergoizdat, Moscow.

Long, J.C.S. and Witherspoon, P.A. (1985). The relationship or the degree of interconnection to permeability in fracture networks. *J. Geophys. Res.*, 90(B4), 3087–3098.

Louis, C. (1969). A study of groundwater flow in jointed rock and its influence on the stability of rock masses. Rock Mechanics Research Report, Imperial College, London.

Mavor, M.J. and Vaughn, J.E. (1998). Increasing coal absolute permeability in the San Juan Basin fruitland formation. *SPE Reservoir Evaluation and Engineering*, 1(03), 201–206.

Mitchell, J.K. (1976). *Fundamentals of Soil Behavior*. John Wiley & Sons, New York.

Narr, W. (1996). Estimating average fracture spacing in subsurface rock. *AAPG Bulletin*, 80(10), 1565–1585.

Neuman, S.P. (1972). Theory of flow in unconfined aquifers considering delayed response of the water table. *Water Resour. Res.*, 8, 1031–1045.

Neuman, S.P. (1973). Supplementary comments on theory of flow in unconfined aquifers considering delayed response of the water table. *Water Resour. Res.*, 9(4), 1102–1103.

Neuman, S.P. (1975). Analysis of pumping test data from anisotropic unconfined aquifers considering delayed gravity response. *Water Resour. Res.*, 11(2), 329–342.

Neuman, S.P. and Witherspoon, P.A. (1969a). Theory of flow in a confined two-aquifer system. *Water Resour. Res.*, 5(4), 803–816.

Neuman, S.P. and Witherspoon, P.A. (1969b). Applicability of current theories of flow in leaky aquifers. *Water Resour. Res.*, 5(4), 817–829.

Neuzil, C.E., Cooley, C., Silliman, S.E., Bredehoeft, J.D., Hsieh P.A. (1981). A transient laboratory method for determining the hydraulic properties of 'tight' rocks – II. Application. *International Journal of Rock Mechanics and Mining Sciences & Geomechanics Abstracts*, 18(3), 253–258.

Ortega, O.J., Marrett, R.A., Laubach, S.E. (2006). A scale-independent approach to fracture intensity and average spacing measurement. *AAPG Bulletin*, 90(2), 193–208.

Ouyang, Z. and Elsworth, D. (1993). Evaluation of groundwater flow into mined panels. *International Journal of Rock Mechanics and Mining Sciences & Geomechanics Abstracts*, 30(2), 71–80.

Pan, Z. and Connell, L.D. (2012). Modelling permeability for coal reservoirs: A review of analytical models and testing data. *International Journal of Coal Geology*, 92, 1–44.

Papadopoulos, I.S., Bredehoeft, J.D., Cooper, H.H. (1973). On the analysis of slug test data. *Water Resour. Res.*, 9(4), 1087–1089.

Piggott, A.R. (1990). Analytical and experimental studies of rock fracture hydraulics. PhD Thesis, Pennsylvania State University, University Park.

Piggott, A.R. and Elsworth, D. (1990). Laboratory studies of transport within a single fracture. In *Proc. Int. Symp. on Rock Joints*. Balkema, Loen.

Pyrak-Nolte, L.J., Myer, L.R., Cook, N.G.W., Witherspoon, P.A. (1987). Hydraulic and mechanical properties of natural fractures in low permeability rock. In *Proc. 6th Int. Congr. Rock Mech., Montreal*. Balkema, Rotterdam.

Romm, E.S. (1966). *Flow Characteristics of Fractured Rocks* (in Russian). Nedra, Moscow.

Shi, R., Liu, J., Wei, M., Elsworth, D., Wang, X. (2018). Mechanistic analysis of coal permeability evolution data under stress-controlled conditions. *International Journal of Rock Mechanics and Mining Sciences*, 110, 36–47.

Shi, R., Liu, J., Wang, X., Elsworth, D., Liu, Z., Wei, M., Liu, X., Wang, Z. (2020). Experimental observations of heterogeneous strains inside a dual porosity sample under the influence of gas-sorption: A case study of fractured coal. *International Journal of Coal Geology*, 223, 103450.

Snow, D.T. (1965). A parallel plate model of fractured permeable media. PhD Thesis, University of California, Berkeley.

Snow, D.T. (1969). Anisotropic permeability of fractured media. *Water Resour. Res.*, 5(6), 1273–1289.

Terzaghi, K. (1963). *Theoretical Soil Mechanics*. John Wiley & Sons, New York.

Theis, C.V. (1935). The relation between the lowering of the piezometric surface and the rate and duration of discharge of a well using groundwater storage. *Eos, Transactions American Geophysical Union*, 16(2), 519–524.

Tsang, Y.W. and Witherspoon, P.A. (1981). Hydromechanical behavior of a deformable rock fracture subject to normal stress. *J. Geophys. Res.*, 86(B10), 9287–9298.

Uraiet, A. (1979). Transient pressure behavior in a cylindrical reservoir produced by a well at constant bottom hole pressure. PhD Thesis, University of Tulsa, Tulsa.

Van Everdingen, A.F. and Hurst, W. (1949). The application of the Laplace transformation to flow problems in reservoirs. *J. Pet. Technol*, 1(12), 305–324.

Wang, S., Elsworth, D., Liu, J. (2011). Permeability evolution in fractured coal: The roles of fracture geometry and water-content. *International Journal of Coal Geology*, 87(1), 13–25.

Wang, S., Elsworth, D., Liu, J. (2012). A mechanistic model for permeability evolution in fractured sorbing media. *J. Geophys. Res.*, 117(B6), B06205.

Witherspoon, P.A., Wang, J.S.Y., Iwai, K., Gale, J.E. (1980). Validity of cubic law for fluid flow in a deformable rock fracture. *Water Resour. Res.*, 16(6), 1016–1024.

Zhang, H., Liu, J., Elsworth, D. (2008). How sorption-induced matrix deformation affects gas flow in coal seams: A new FE model. *International Journal of Rock Mechanics and Mining Sciences*, 45(8), 1226–1236.

Zhi, S., Elsworth, D., Liu, L. (2019). W-shaped permeability evolution of coal with supercritical CO_2 phase transition. *International Journal of Coal Geology*, 211, 103221.

3

Measurement of Coal Physical Properties: Poromechanical and Adsorption Testing

D. Nicolas ESPINOZA
Hildebrand Department of Petroleum and Geosystems Engineering, University of Texas at Austin, USA

3.1. Introduction

Unmineable coal seam reservoirs present unique features that make them different to most hydrocarbon-bearing reservoir rocks. These unique features make original gas in-place estimations and gas flow rates in coal bed methane (CBM) different from conventional gas reservoirs. Furthermore, CO_2 storage in coal seams is also different when compared to conventional reservoirs.

First, the original gas-in place is strongly dependent on gas adsorption (Diamond and Schatzel 1998; Moore 2012). Organic matter has a preferential affinity for N_2, CH_4 and CO_2 molecules opposed to other non-sorbing gases. Hence, these gas molecules can adsorb on organic matter surfaces and significantly increase the amount of gas in-place when compared to bulk storage in pores at given pressure and temperature (Busch and Gensterblum 2011). Unlike shales, coal seams are relatively shallow (maximum depth ~1.5 km, pore pressure < 15 MPa and temperature < 80°C) (Moore 2012). Hence, most of the gas-in-place is in an adsorbed state and a minor amount is in bulk conditions.

For a color version of all of the figures in this chapter, see www.iste.co.uk/vandamme/coal.zip.

Geomechanics of Coal Seams,
coordinated by Matthieu VANDAMME and Jean-Michel PEREIRA. © ISTE Ltd 2025.

For example, gas canisters (in which a recently drilled coal core is placed in a sealed sample container where gas desorbs into the bulk space connected to a graduated cylinder to measure produced gas; Diamond et al. 1986) were first developed to estimate gas in-place (mostly adsorbed) in coal mines and much later were adapted to coalbed methane and gas shales.

Second, gas adsorption results in changes of surface energy, which manifest as solid strains and stresses (Ravikovitch and Neimark 2006; Vandamme et al. 2010; Gor et al. 2017). This coupled process is well known in some adsorption processes but it was largely ignored in subsurface energy engineering until coal seam reservoirs started to show atypical responses during gas production. Gas wells producing from coal seams show production decline curves that depart from conventional expectations (Palmer and Mansoori 1996; Moore et al. 2011). This atypical response can be interpreted as an increase of reservoir permeability (up to one-hundred times) with decreasing bottomhole and reservoir pressure, that is, with increasing time (Pan and Connell 2012). Such behavior is contradictory to the expectations of higher effective stress and smaller fracture aperture with decreasing reservoir pressure (Segall 1989). The reason behind this is the coupled mechanical–adsorptive response of coal (see Chapters 2 and 4).

This chapter reviews laboratory experimental methods used to characterize coal and permit estimating original gas in place and the coupled response of coal seams to gas production and CO_2/N_2 injection. The methods are also applicable to other adsorbents (including activated carbon, zeolites and shale rock). These measurements are critical to run coupled numerical simulations and predict the coupled chemo–hydro–mechanical response of coal seams, which is otherwise impossible to capture through conventional reservoir simulation.

The chapter also highlights the typical applications for these measurements and laboratory procedures in CBM and enhanced CBM applications.

3.2. Pressure, stress and strains in geological formations

Coal is a natural porous material similar in many aspects to other rocks. Let us review the in situ pressure and stress conditions and deformational properties of non-sorbing rocks that will help us better understand laboratory tests and measured parameters to predict coal seam response in situ. Additional details are available elsewhere (Espinoza 2021).

Consider an onshore scenario in which the surface coincides with the water table. The fluid pressure in rock pores P at depth z under hydrostatic conditions is:

$$P = \rho_w g z \qquad [3.1]$$

where ρ_w is the water mass density (assumed constant) and g is the gravity acceleration. The "total" vertical stress σ_v acting on a rock element due to the weight of overlying rock is:

$$\sigma_v = -\rho_{bulk} g z \qquad [3.2]$$

where ρ_{bulk} is the bulk rock mass density including weight of minerals and fluids within pores (assumed constant for simplicity). Compression is assumed to have a negative sign in this chapter. The difference between total vertical stress and pore pressure is called Terzaghi's effective vertical stress:

$$\sigma'_v = \sigma_v + P \qquad [3.3]$$

Effective stresses in the subsurface are mostly compressive, hence, $\sigma'_v < 0$. Stress is a second order tensor with six independent components: σ_{ij} (with i and j ranging from 1 to 3). A complete definition of stress requires knowledge of the three principal stresses and the three principal directions. Let us assume that vertical stress is a principal stress; hence, the other two principal stresses are horizontal. These are usually referred to as the maximum total horizontal stress σ_{Hmax} and the minimum total horizontal stress σ_{hmin}.

Horizontal stresses can be smaller or greater than the vertical stress. Their determination requires field testing and/or field data. Total horizontal stresses are usually larger in magnitude than the pore pressure, such that $|\sigma_{Hmax}| > |\sigma_{hmin}| > P$. The principal effective stresses control the likelihood of the rock to incur into inelastic deformation and fail. Within the elastic regime, the relationship between rock stress σ_{ij} and strain ε_{ij} can be better predicted by Biot's effective stress:

$$\sigma''_{ij} = \sigma_{ij} + \delta_{ij} \alpha P \qquad [3.4]$$

where α is the Biot coefficient and δ_{ij} is the Kronecker delta. For example, for a linear isotropic poro-elastic material:

$$\sigma''_{ij} = \lambda \varepsilon_{vol} \delta_{ij} + 2\mu \varepsilon_{ij} \qquad [3.5]$$

where ε_{vol} is the volumetric strain and the two independent constitutive parameters are the Lamé coefficients λ and μ that can also be expressed in terms of Young's modulus E and Poisson's ratio v:

$$\begin{cases} \lambda = \frac{\nu E}{(1+\nu)(1-2\nu)} \\ \mu = \frac{E}{2(1+\nu)} \end{cases} \qquad [3.6]$$

The determination of constitutive parameters E, v and α can be achieved with a "triaxial" cell with independent control of stresses σ_{ij}, pore pressure P and measurement of strains ε_{ij}. Figures 3.1(a) and (b) show a typical axisymmetric triaxial cell and a cylindrical coal sample for testing. The horizontal stress is controlled through pressure of the confining fluid in the triaxial (pressure) vessel by means of an external pump. This pressure, usually known as confining pressure, converts to total stress as it transfers to the porous rock through an impermeable membrane surrounding the rock sample.

The vertical stress is controlled through a piston that goes into the triaxial cell. Vertical stress can be applied with a load frame or through fluid pressure with a third pump. Other triaxial cell models may apply vertical stress as a combination of confining pressure and a deviatoric stress. Fluid (pore) pressure is controlled with another external pump that connects to the rock sample through conduits in the endcaps that apply vertical stress. Strain measurement can be achieved through various modes (see section 3.4).

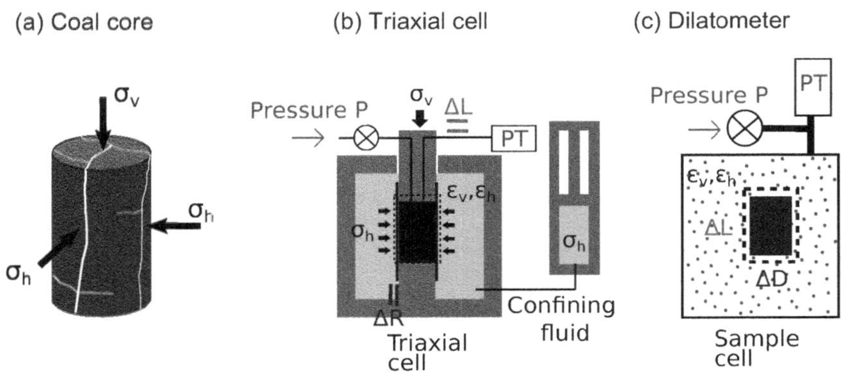

Figure 3.1. *(a) Schematic of coal sample cored perpendicular to bedding for triaxial testing. (b) Schematic of triaxial cell with control of stresses and pore pressure and measurement of strains. (c) Dilatometer cell with control of (pore) pressure only and measurement of strains, also known as unjacketed test in poromechanics. PT: pressure transducer*

Let us assume a coal core (of initial length L_0, radius R_0 and diameter $D_0 = 2R_0$) drilled perpendicular to bedding, such that the vertical direction coincides with the axial direction of the triaxial axisymmetric vessel. The application of vertical stress while keeping horizontal stress and pore pressure constant permits measuring the rock Young's modulus and Poisson's ratio by definition:

$$\begin{cases} E = \left.\dfrac{\partial \sigma_v}{\partial \varepsilon_v}\right|_{\sigma_h,P} \\ \nu = -\left.\dfrac{\partial \varepsilon_h}{\partial \varepsilon_v}\right|_{\sigma_h,P} \end{cases} \qquad [3.7]$$

where $\varepsilon_v = \Delta L/L_0$ is the vertical strain and $\varepsilon_h = \Delta R/R_0 = \Delta D/D_0$ is the horizontal strain. A simultaneous and equal increase in vertical and horizontal stresses at constant pressure permits a direct determination of the (volumetric) bulk modulus:

$$K = \left.\frac{\partial \sigma_m}{\partial \varepsilon_{vol}}\right|_P \qquad [3.8]$$

where $\sigma_m = (\sigma_v + 2\sigma_h)/3$ is the mean total stress and $\varepsilon_{vol} = \varepsilon_v + 2\varepsilon_h$ is the volumetric strain. E, ν and K are bulk properties of the porous rock, including compressibility of solids, grain contacts, fracture surfaces and pores. The combination of equations [3.5], [3.6] and [3.8] yields $K = E/[3(1-2\nu)]$. These properties usually depend on the chosen sample size in geological materials.

Large variations of fluid pressure require incorporating the Biot effective stress correction for calculating bulk strains (equation [3.4]). The measurement of the Biot coefficient can be achieved through alternating variations of the total mean stress σ_m and the pore pressure P in a triaxial cell (Gueguen and Bouteca 1999). Equations [3.4] and [3.5] reduce to the following in terms of mean stress and volumetric strain:

$$\sigma_m = \varepsilon_{vol} K - \alpha P \qquad [3.9]$$

Hence, variation of σ_m and P with simultaneous measurement of ε_{vol} permits obtaining a dataset (σ_m, P, ε_{vol}) that can be adjusted to best-fit equation [3.9] and solve for K and α at the same time. The physical meaning of α is a correction of bulk strain to account for solid-phase compressibility with changes of pore pressure. The possible range of the Biot coefficient is $\alpha = [\varphi_o, 1)$, where φ_o is the rock initial porosity.

Triaxial cells are popular in the rock mechanics and geoscience communities but their use is limited in other disciplines. A more popular experimental setup in the adsorption science community is the dilatometer test (Figure 3.1(c)). In such a setup, also known as an unjacketed test in poromechanics, a fluid pressure is applied directly on a porous sample without a separating membrane or jacket. Hence, the total (external) stress is equal to the fluid pressure in pores $\sigma_m = -P$ and equation [3.9] results in:

$$\varepsilon_{vol} = -\frac{1-\alpha}{K}P \qquad [3.10]$$

The direct application of (non-sorbing) fluid pressure on a solid results in compression of the solid phase (negative strain). For a porous solid with all connected porosity, this results in a reduction of solid volume proportional to the applied pressure P and inversely proportional to the solid "skeleton" bulk modulus K_s (without accounting for pores):

$$\varepsilon_{vol} = -\frac{1}{K_S}P \qquad [3.11]$$

A direct comparison of equations [3.10] and [3.11] yields $\alpha = 1 - K/K_s$, where K_s is the volumetric modulus of the mineral for a porous solid with all connected porosity and made of a single solid phase/mineral. Hence, dilatometer tests with non-sorbing fluids can be used to determine K_s and calculate α. Section 3.4 shows the use of dilatometer tests to measure adsorption-induced strains. The main limitations of "unjacketed" dilatometer tests is that they correspond to zero Terzaghi effective stress condition (equation [3.3]), a situation that does not exist in real coal seams at depth. Results from jacketed tests in triaxial tests (at non-zero effective stress) are closer to real coal seam behavior than unjacketed tests. The discrepancies between results from unjacketed and jacketed tests increase as the sample size increases and cleats and microfracture number increases in tested samples. Dilatometer testing remains as a great tool for fundamental research but yields results applicable to the coal matrix only, with limited application to prediction of stress variation and permeability changes in naturally fractured coal cores or coal seams subjected to pressure changes. Sections 3.4 and 3.5 show the use of triaxial testing for measuring coal core properties relevant to coal seam conditions.

3.3. Total gas-in-place and adsorption measurement

The total amount of gas available in a coal seam n^T is a combination of gas in bulk conditions and gas in adsorbed conditions (Figure 3.2). The excess adsorption amount n^{exc} is defined as the amount of gas in-excess of the gas that should be present in bulk conditions (gas bulk density ρ_F) in the same pore volume, $V_p = \varphi V_{bulk}$, where V_p is the pore volume and V_{bulk} is the sample volume. Hence:

$$n^T = V_p\rho_F + n^{exc} \qquad [3.12]$$

The excess adsorption amount is usually a positive quantity since the density of the adsorbed phase is mostly larger than the density of the bulk phase (Keller and Staudt 2005). A recurrent problem with equation [3.12] is that the porosity of microporous media is not well defined and varies according to the fluid used for testing. A common practice is to report helium porosity, but that does exclude the possibility of other types of fluid molecules going "outside" the pore space available for helium. Coal seams usually contain water in addition to gas. The presence of water decreases the fraction of pore volume available for gas.

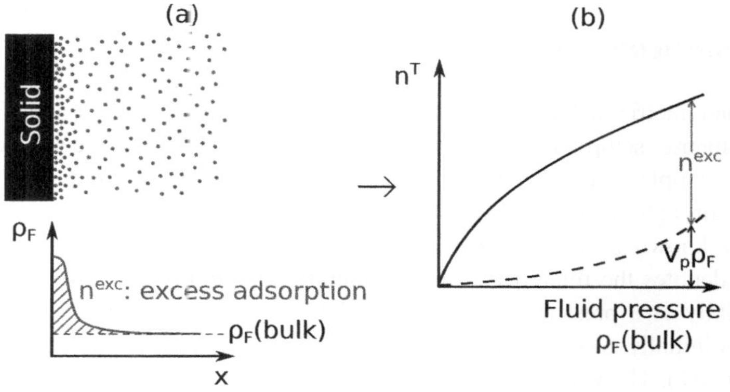

Figure 3.2. *Total fluid amount accounting for adsorption. (a) Pore-scale interpretation. (b) Bulk-scale implications*

The adsorption science community often reports excess adsorption as moles per unit of dry mass of sorbent (without water or other fluids). These values do not need any definition of bulk volume or porosity of the adsorbent material. The geoscience community often reports adsorption amounts per unit of bulk volume. This is convenient to calculate original-gas-in-place (OGIP) with a known reservoir volume or to perform mechanical calculations using the undeformed bulk volume as the reference state to define strains. The dry mass of a coal mass is equal to the bulk volume, times the dry bulk mass density:

$$m_{coal} = V_{bulk}\rho_{bulk}^{dry} \quad [3.13]$$

Adsorption amount affects directly induced strains and stresses. Hence, we will place emphasis on excess adsorption in the following subsection, which revisits fundamentals and principles of adsorption measurement. A conversion from excess adsorption to total gas in place is possible through equation [3.12]. The estimation of "total adsorption amount" requires knowledge of the adsorbed phase volume.

The following section will also place emphasis on adsorption measurements rather than desorption assuming no hysteresis in the process.

Section 3.5 includes a desorption case – closer to field conditions during gas production. The reader will find additional corrections and special cases, including the presence of water and hysteresis, in a thorough review available elsewhere (Gensterblum et al. 2010; Busch and Gensterblum 2011). A detailed explanation of adsorption measurement methods is available elsewhere (Keller and Staudt 2005).

3.3.1. *Manometric method*

The manometric method is based on Boyle's gas expansion law. Figure 3.3 shows the manometric setup, including the reference cell, the sample cell and a solid adsorbent sample with skeletal volume V_{sk}. The skeletal volume V_{sk} is usually measured through gas expansion with a non-sorbing fluid, such as helium, a process commonly known as helium pycnometry. The reference cell hosts the testing gas where V_1 denotes the reference volume including tubing above the connection valve VL1, initially at a pressure P_i (subscript 'i' for initial conditions). The testing gas pressure is initially contained within V_1 by two valves and monitored with a pressure transducer (PT). The volume V_2 accounts for the sample cell including tubing after the valve VL1, where gas is at lower pressure P_i^s ("s" superscript denotes the sample cell).

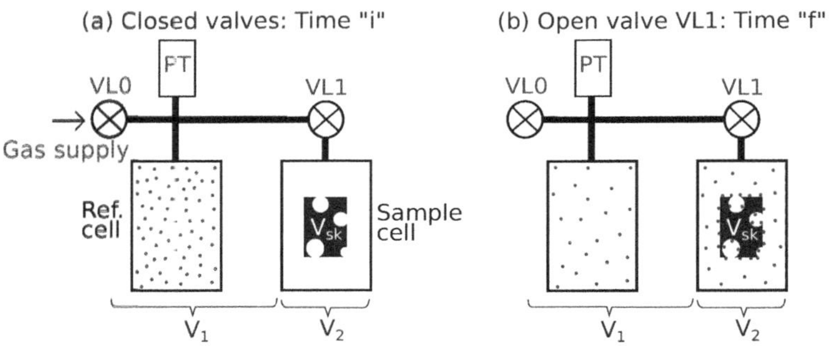

Figure 3.3. *Manometric adsorption setup. Pressure monitoring in the reference and sample cells permit measuring gas moles according to Boyle's law. PT: pressure transducer; VL0: gas supply valve; VL1: sample cell valve*

Hence, for a single gas, the initial number of moles of gas in the entire system is:

$$n_i^T = \frac{P_i V_1}{z_i R T_i} + \frac{P_i^S (V_2 - V_{sk})}{z_i^S R T_i} + n_i^{exc} \quad [3.14]$$

where the first term on the right-hand-side is the gas in bulk conditions before VL1, the second term is the gas expected in bulk conditions in the cell plus pore volume after VL1, and the third term n_i^{exc} is the excess adsorption amount at initial conditions. For example, the second and third terms on the right-hand-side are zero for initial vacuum conditions. The gas equation is corrected by the compressibility factor $z_i(P,T)$ to account for non-ideal gases. The same principle can be extended to other equations of state. Note that this excess amount is the result of ignoring the volume of the adsorbed phase by considering $V_2 - V_{sk}$ (usually measured with helium) as the total "empty" volume available for the gas/adsorbate. Equation [3.14] assumes ideal conditions with equal temperature in the reference and sample cells T_i, even though the equation can be easily extended to different temperatures. Equal temperatures in both cells is usually achieved with a thermal bath or through an oven with high-precision temperature control.

After opening the sample cell valve VL1, gas will flow from the reference cell (at higher pressure) to the sample cell and find equilibrium at a pressure P_f (subscript "f" for final conditions):

$$n_f^T = \frac{P_f V_1}{z_f R T_f} + \frac{P_f (V_2 - V_{sk})}{z_f R T_f} + n_f^{exc} \quad [3.15]$$

Letting enough time for equilibration such that $T_i = T_f = T$, and considering mass balance $n_i^T = n_f^T$, the excess adsorption amount at final conditions is:

$$n_f^{exc} = n_i^{exc} + \frac{1}{RT}\left[\left(\frac{P_i}{z_i} - \frac{P_f}{z_f}\right)V_1 - \left(\frac{P_f}{z_f} - \frac{P_i^S}{z_i^S}\right)(V_2 - V_{sk})\right] \quad [3.16]$$

This procedure can be repeated with different combinations of P_i and P_i^s to obtain a range of pressure P_f that spans the pressure interval of interest. The test can also be performed in incremental steps of P_f to take advantage of testing the same sample and setup. The method can be extended to gas mixtures with an appropriate gas mixture law and measurement of bulk gas molar fractions at equilibrium. Gas mixture composition can be measured with a gas chromatograph (Ottiger et al. 2008). All adsorption tests should start with a vacuumed sample to avoid contamination with other gases. N_2 and CO_2 in air at atmospheric pressure can significantly alter adsorption isotherm results if the sample is not properly vacuumed.

The adsorption amount as a function of pressure at a given temperature is called an "adsorption isotherm". Typical coal seam applications require testing for temperature range from 20°C to 80°C and pressure up to ~15 MPa, opposite to conventional low-pressure and low-temperature measurements (e.g. N_2 adsorption isotherms for specific surface measurements). Hence, medium pressure cells, tubing and fittings are required to accommodate these gas pressures. Figure 3.4 shows an example of excess adsorption isotherms per unit of mass measured through the manometric method for activated carbon with N_2, CH_4 and CO_2.

An extension of this procedure to determine total-gas-in-place, usually referred to as total uptake in laboratory conditions, is straightforward upon knowledge of the initial bulk volume of the tested sample. Bulk volume measurement is simple for coal cores cut in a prescribed geometry (cylinder, cube, etc.), but it is not easy to determine in powders or granular media, which are preferred in adsorption testing because they achieve thermodynamic equilibrium much quicker than large cores. Hence, adsorption isotherms are usually reported per unit of mass for samples with unknown or difficult-to-determine bulk volume.

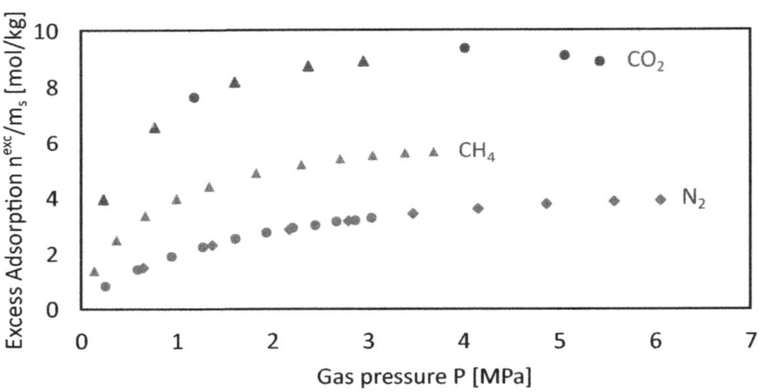

Figure 3.4. *Excess adsorption in granular activated carbon samples with mass m_s. Example results for two experiments involving CO_2, one for CH_4 and two for N_2 conducted in the author's laboratory*

3.3.2. *Volumetric method*

The volumetric method consists of injecting a gas at constant pressure into a sample cell with an initially lower and known pressure (Figure 3.5). Injection is usually performed with a high-pressure pump (pressure–displacement controller) with PID (proportional–integral–derivative) control for achieving constant fluid pressure. The injected volume required to pressurize the sample cell at the prescribed target pump pressure allows calculating the amount of injected fluid and adsorption amount.

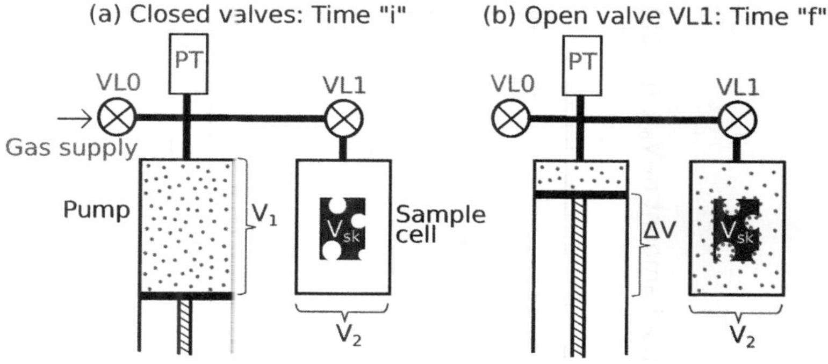

Figure 3.5. *Volumetric adsorption setup. Volume and pressure measurement at pump permit measuring excess adsorbed amount based on mass balance. PT: pressure transducer; VL0: gas supply valve; VL1: sample cell valve*

Let us consider initial conditions with a pump pressurized at pressure P^p, isolated from the sample cell at pressure P_i^s through valve VL1. The total number of moles at initial conditions is:

$$n_i^T = \frac{P^p V_1}{z_i R T_i} + \frac{P_i^S (V_2 - V_{sk})}{z_i^S R T_i} + n_i^{exc} \quad [3.17]$$

The first term on the right-hand side is the number of moles in the pump, originally at a volume V_1. The second and third terms make up the moles of gas in the sample cell at initial conditions. After opening the connection valve VL1 and keeping the pump pressure constant and equal to P^p, the total number of moles in the system is:

$$n_f^T = \frac{P^p (V_1 - \Delta V)}{z_f R T_f} + \frac{P^p (V_2 - V_{sk})}{z_f R T_f} + n_f^{exc} \quad [3.18]$$

where ΔV is the change of volume at the pump operated at a constant pressure. Hence, the final excess adsorption amount of gas assuming equal initial and final temperatures $T_i = T_f = T$ is:

$$n_f^{exc} = n_i^{exc} + \frac{1}{RT}\left[\left(\frac{P^p}{z_f}\right)\Delta V - \left(\frac{P^p}{z_f} - \frac{P_i^S}{z_i^S}\right)(V_2 - V_{sk})\right] \quad [3.19]$$

The simplest case is for vacuum in the sample cell ($P_i^s \sim 0$ Pa), such that the injected volume is equal to volume gone into the empty space of the sample cell, plus the excess adsorption on adsorbent surfaces.

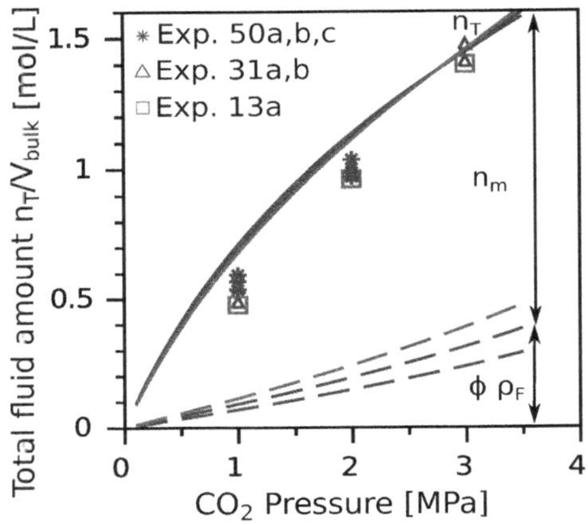

Figure 3.6. *Example of total fluid amount measurement through the volumetric method for granular coal packed at various stresses (13, 31 and 50 MPa). Total fluid amount accounts for excess sorption and gas amount in bulk conditions. The calculation of excess sorption requires knowledge of porosity (equation [3.12]). The model results (lines) in this example allow discriminating of the total amount in gas in the intergranular space and gas within coal grains n_m, which is mostly excess adsorption but also includes gas in bulk conditions in grain mesopores. Data adapted from Espinoza et al. (2016)*

This method can also be adapted to measure total uptake in cases where either the sample cell determines the bulk sample volume V_{bulk} (e.g. packed granular adsorbents in a stiff cell/container) or the sample cell adapts to the sample bulk volume (e.g. in a triaxial cell where endcaps and a membrane enclose the sample; Figure 3.1(b)).

In these cases, $V_2 = V_{bulk} + V_{piping}$ where V_{piping} is known and usually much smaller than V_{bulk}. Assuming $V_{piping} << (V_{bulk}-V_{sk})$ in equation [3.19], the term $P^p \Delta V/(RTz_f)$ represents the total amount of fluid gone into the sample to achieve the target pressure P^p.

Figure 3.6 shows an example of total fluid amount, or uptake, measured for granular coal packed in a cylindrical vessel with constant volume. Note the total uptake measurement does not need a measurement of V_{sk} with helium.

However, the interpretation of where fluid goes, either to adsorption places or remains in bulk conditions in large pores/fractures, requires knowledge of V_{sk}, where $\varphi = (V_{bulk} - V_{sk})/V_{bulk}$. Since V_{bulk} is known, these measurements can be reported as moles per unit of (bulk) volume.

3.3.3. *Gravimetric method*

The gravimetric method is based on the buoyant weight of the solid submerged in a gas (fluid) with known bulk properties at given values of pressure and temperature. These measurements are usually performed with a high precision magnetic balance that supports high pressure and high temperature in the sample cell. The following summary presents the underlying physical mechanisms. Additional details are available elsewhere (Di Giovanni et al. 2001; Ottiger et al. 2008).

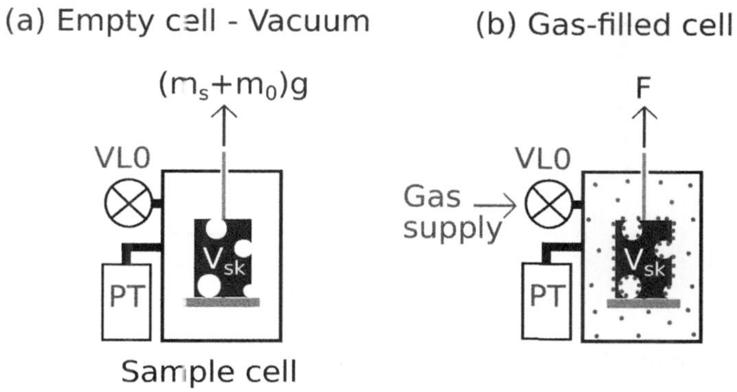

Figure 3.7. *Gravimetric adsorption principle. Apparent weight (F) measurement permits quantifying excess adsorbed amount based on buoyant weight of the coal mass and adsorbed phase. PT: pressure transducer; VL0: gas supply valve*

In Figure 3.7, a high-pressure and high-temperature sample cell is presented, which is able to measure weight of a calibrated holder with mass m_0 (and volume V_0) and a coal sample with mass m_s. At vacuum conditions, the total weight is:

$$F_0 = (m_s + m_0)g \quad [3.20]$$

where g is the gravity acceleration.

Injecting a fluid results in a change of the force measured externally, such that the apparent weight F is the weight of the sample and the core holder plus the adsorbed phase minus the weight of displaced fluid (gas) at the given pressure and temperature conditions. Hence, the following holds:

$$F = [m_s + m_0 + m_{ads} - \rho_F(V_{sk} + V_{ads} + V_0)]g \quad [3.21]$$

where m_{ads} is the mass of the adsorbed phase and V_{ads} is the volume of the adsorbed phase. By definition, the excess adsorbed mass is the difference between the total adsorbed mass and the mass occupied by the volume of adsorbed phase in bulk conditions, that is, $m_{ads} = m_{exc} + \rho_F V_{ads}$. Using this relationship in equation [3.21] and the molar mass M_m of the adsorbate, results in:

$$n^{exc} = \frac{m_{exc}}{M_m} = \frac{1}{M_m}\left[\frac{F}{g} - m_s - m_0 + \rho_F(V_{sk} + V_0)\right] \quad [3.22]$$

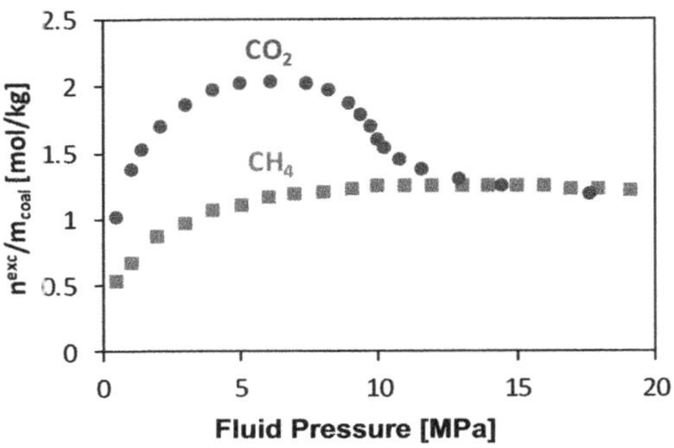

Figure 3.8. *Example of excess sorption measurement with the gravimetric method. Data from Ottiger et al. (2008)*

Note that the only unknown is the apparent weight. All other masses and volumes (including m_0, m_s, V_{sk} and V_0) are previously measured. The bulk density of the fluid phase ρ_F can be calculated from an equation of state or measured directly with additional instrumentation in the magnetic balance.

Additional corrections can be applied to improve the calculations of excess adsorption (Ottiger et al. 2008). Figure 3.8 shows examples of excess sorption measurements with this method. The drop in CO_2 excess adsorption near the critical pressure (around 7.4 MPa) is due to the high mass density of CO_2 near the critical point and the finite density of the adsorbed phase (Goodman et al. 2004; Gensterblum et al. 2009; Khosrokhavar et al. 2014).

Similar to the manometric method, the gravimetric method is better suited for powders or granular samples since the high precision magnetic balance cannot be easily adapted to triaxial mechanical frames. Hence, the results are usually shown as adsorption amount per unit of dry coal mass.

3.3.4. *Sorption models and extension to total adsorption amount*

Sorption isotherms are usually fitted to theoretical adsorption models. Among all models, the Langmuir model is the most used theoretical model for N_2, CH_4 and CO_2 adsorption on coals at medium pressure (<5 MPa). The basic Langmuir model consists of a two-parameter equation where total adsorption amount n is:

$$n = n_{max} \frac{P}{P+P_L} \quad [3.23]$$

and the two parameters are the asymptotic maximum adsorption capacity n_{max} and P_L is the Langmuir pressure. Rigorously, the Langmuir isotherm predicts the fraction of total occupied adsorption sites. The approximation to predict excess amount $n^{exc} \sim n$ is reasonable when the mass density of the adsorbate phase is much larger than the mass density of the adsorbate fluid in bulk conditions. The three-parameter Langmuir isotherm can be used when the fluid bulk mass density approaches the mass density of the adsorbate:

$$n^{exc} = n_{max}^{exc} \frac{P}{P+P_L} - V_{ads}\rho_F \quad [3.24]$$

where the additional parameter V_{ads} is the volume of the adsorbed phase (Gensterblum et al. 2010). The accuracy in fitting this third parameter increases as the excess adsorption amount deviates from a monotonic increasing trend of excess adsorption with pressure, for example, CO_2 isotherm in Figure 3.8. Since adsorption amounts are usually reported by unit of dry mass, the third parameter can be converted to V_{ads}/m_{coal}.

3.4. Adsorption-induced strains

A process of adsorption can develop either contraction or dilation strains on solid adsorbents, although only dilation is observed in practice for the specific case of coal geomechanics. The magnitude of adsorption-induced strains in coal by CH_4 and CO_2 for a dilatometer test at typical unmineable coal seam pressures up to 1–10 MPa is usually much larger than the strain expected solely by fluid pressurization of the solid skeleton (equation [3.11]). Before exploring the procedures and devices to measure adsorption-induced strains, let us explore the theory that explains this response in order to better understand the experimental results.

Equation [3.9] can be generalized for a porous solid that develops adsorption-induced strains and stresses as:

$$\sigma_m = K\varepsilon_{vol} - s(P) \quad [3.25]$$

where $s(P)$ is a function of fluid/adsorbate pressure. For example, by analogy with equation [3.9], $s(P_{He}) = \alpha P_{He}$ for a coal subjected to helium pressure without adsorption, where α is the Biot coefficient. However, the stress function $s(P)$ has to be determined for porous adsorbent solids exposed to adsorbates (Ravikovitch and Neimark 2006; Espinoza et al. 2013). In the case of coal, exposed to CH_4 and CO_2, $s(P)$ is highly nonlinear, and is linked to excess adsorption and $s(P) > P$. Volumetric strains in a porous adsorbent can be caused by changes of gas pressure $s(P)$ and external total stresses σ_m (equation [3.25]). In fact, coal seams are multiporosity systems and distinct pressures may exist depending on pore size and connectivity, which make the porous solid deformational process more complex. The following discussion puts emphasis on testing of small coal samples (a few centimeters or less) where we can assume that the coal matrix is being tested with small influence of fractures, unless explicitly stated.

3.4.1. *Dilatometer tests – swelling isotherms*

An unjacketed test with $\sigma_m = -P$ in equation [3.25] yields the adsorption-induced strain for a dilatometer test (Figure 3.1(c)):

$$\varepsilon_{vol} = \frac{1}{K}[s(P) - P] \quad [3.26]$$

Since $s(P) > P$ for coals in the presence of an adsorbate, equation [3.26] predicts solid expansion with increasing adsorption amounts. A swelling isotherm is the adsorption-induced strain in a dilatometer cell as a function of pressure for a given temperature. Swelling isotherms usually report volumetric strains, although some adsorbents develop anisotropic strains (Hol and Spiers 2012).

The adsorption-induced strain in a dilatometer test can also be interpreted as the combination of adsorption-induced strain and the response expected for a non-sorbing fluid such as helium. Let us define the stress function $s(P)$ as the summation of the "bulk adsorption stress" $s^a(P)$ and the response expected for a non-sorbing fluid, such that $s(P) = s^a(P) + \alpha P$, where α is the Biot coefficient measured for helium (Guerrero and Espinoza 2022).

A rigorous extension of Gibbs excess adsorption thermodynamics interprets $s^a(P)$ as the "derivative of the excess grand thermodynamic potential of the adsorbed phase with respect to the variation of the sample porosity" (Neimark 2017). Hence, equation [3.26] can be written as:

$$\varepsilon_{vol} = \frac{1}{K}[s^a(P) + \alpha P - P] = \frac{1}{K}s^a(P) - \frac{1-\alpha}{K}P \quad [3.27]$$

Using the expression of the Biot coefficient $\alpha = 1 - K/K_s$ permits rewriting equation [3.27] as:

$$\varepsilon_{vol} = \frac{1}{K_s}\left[\frac{s^a(P)}{1-\alpha}\right] - \frac{1}{K_s}P \qquad [3.28]$$

where the first term is the strain solely caused by (excess) adsorption and the second term is the shrinkage expected solely from fluid pressure acting on the solid phase. The term $[s^a(P)/(1-\alpha)]$ is the "skeletal" or "matrix" adsorption stress, which is in the order of ~10^{+2} MPa for coals and activated carbons at adsorbate fluid pressure of several MPa (Espinoza et al. 2013, 2014, 2016). Similar to K_s, the skeletal adsorption stress is a property independent of the degree of compaction or porosity. A corollary of this observation is that a dilatometer (unjacketed) test should yield the same adsorption-induced strain independently of the degree of porous medium compaction, quantified by the Biot coefficient α. The excess adsorption amount approaches a maximum when the adsorbed phase density approaches the bulk phase density. Equation [3.28] captures a decreasing influence of adsorption on strains at asymptotic values of $s^a(P)$ observed experimentally (Hol and Spiers 2012). A formal link between adsorption amount and adsorption induced strain/stress is outside the scope of this chapter but available elsewhere (Brochard et al. 2012; Espinoza et al. 2014; Neimark 2017). An adaptation of the theory of elasticity has been used to model adsorption-induced strains and reservoir implications (Palmer and Mansoori 1996). Such adaptation has several limitations, including no link between adsorption amount and adsorption stress, but has been proven to be useful in many practical applications.

Strains can be measured with a variety of transducers in dilatometer cells (Figures 3.9(a)–(c)). In all cases, we use the definition of small linear strains:

$$\varepsilon = \frac{L-L_0}{L_0} \qquad [3.29]$$

where L is the current sample length in a given direction and L_0 is the original undeformed length in such direction. This definition can be extended to the volumetric strain as:

$$\varepsilon_{vol} = \frac{V-V_0}{V_0} \qquad [3.30]$$

where V is the current bulk sample volume and V_0 is the original undeformed bulk sample volume. Strains can be measured directly through calibrated transducers such as strain gauges and fiber optic, which yield a strain value, or indirectly by measuring lengths and volumes and using equations [3.29] and [3.30]. Displacement transducers placed around the sample can measure change of length in different directions when subjected to mechanical or pore fluid loadings. These measurements are usually done by tracking points either on the tested sample or affixed to the sample (Figures 3.9(a)–(c)). Ellipsometry and XRD methods are also used but are outside the scope of this chapter.

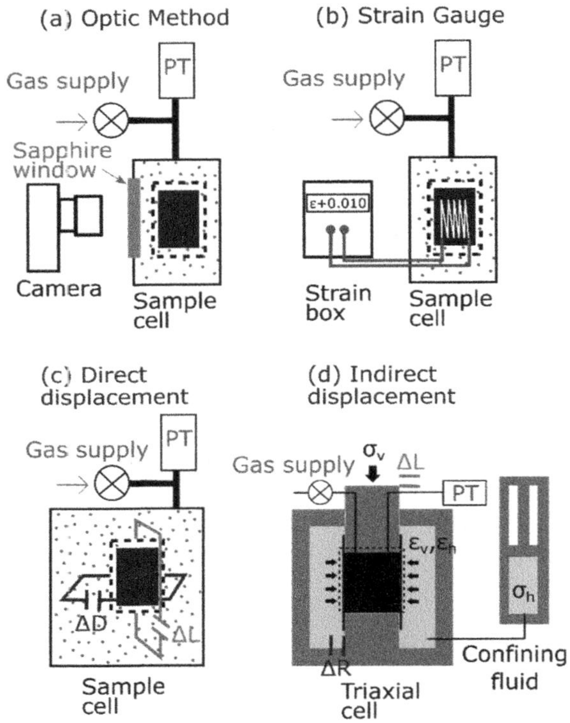

Figure 3.9. *Methods to measure adsorption-induced strain. (a–c) Dilatometer methods use unjacketed samples. (d) Triaxial cells use jacketed samples and permit applying total stress different and independent from fluid pressure*

3.4.1.1. *Optical or visualization methods*

Direct observation of adsorption-induced strains in the optical range at medium and high fluid pressure is possible with a see-through dilatometer cell (Figure 3.9(a)). These cells are usually equipped with a sapphire window that withstands high pressure and permits observing inside high-pressure environments. The change of size can be tracked externally as a 2D projection through a digital camera ideally assisted by digital image correlation. The resolution of this method depends on the image resolution and sample size. For example, a coal sample sized 10 mm can develop linear swelling strains in the order of 0.01. Hence, the change of length would be ($L - L_0$ = 10 mm × 0.01) = 100 μm. Capturing a trend with ~10 datapoints would require a 10 μm precision, and therefore, at least 10 μm pixel size. The image size should have at least (10 mm/10 μm) = 1,000 pixels, or 1,000 by 1,000 pixels, that is, 1 MP (MegaPixel). Current digital cameras have over 10 MP capacity and are suitable for such measurements. However, deformations <0.001 are challenging to capture through optical methods.

Figure 3.10(a) shows an example of strain measurement of coal subjected to He, CH_4 and CO_2 pressure within a see-through dilatometer cell. The strain measurement precision is ±0.0025. The experimental error depends on image resolution, image quality, focus and image processing techniques. Note that the precision shown in Figure 3.10(a) is not enough to accurately calculate the skeletal volumetric stiffness with helium (linear fit to helium data yields K_s = 2.51 GPa). CH_4 and CO_2 adsorption induces a maximum volumetric strain ranging from ε_{vol} = +0.020 to +0.038 at the tested pressures in this coal sample. The limitations of this methodology are strain resolution, testing only at unjacketed conditions (direct exposure of sample to fluid), and two-dimensional measurement. Three-dimensional full-field strain measurements can be achieved through time-lapse X-ray computed tomography.

3.4.1.2. *Strain gauges*

High-precision strains measurements are possible through strain gauges. Strain-gauges attached directly to a coal matrix can monitor adsorption induced strains when subjected to a pressurized gas environment within a dilatometer cell. Similar to the optical methods, a high-pressure cell holds the gas/adsorbate at high pressure. Instead of a see-through window, high-pressure electrical feedthroughs allow cables from the strain gauges to exit and connect to the "strain box" (Figure 3.9(b)).

Strain gauges can be attached in different directions and locations to measure strains in different directions and capture the three principal strains to measure volumetric strain and also quantify anisotropy. The strain gauge precision depends on the quality of the bonding with the solid sample and electronic equipment associated with the Wheatstone bridge box. A high-quality setup can measure strains under 10^{-6}. Figure 3.10(b) shows an example of this type of measurement. A linear fit to helium data yields K_s = 8.17 GPa. Methane and carbon dioxide adsorption induces a maximum volumetric strain ranging from ε_{vol} = +0.007 to +0.012 at the tested pressures. Systematic errors can be introduced with this method if strain gauges are glued incorrectly to samples or if the glue softens and creeps in the presence of pressurized gases. Rigorous quality control should be performed with other mechanical loadings to verify the strain gauge reading is accurate.

3.4.1.3. *Displacement transducers*

The change of length measurement can be done with various transducers within a pressure vessel including (1) linear variable differential transducers, (2) strain gauge-calibrated cantilever arms and (3) inductive eddy current sensors among others (Hol and Spiers 2012; Espinoza et al. 2014). These transducers can be re-used for testing multiple samples and are not subjected to possible artifacts, such as loss of adhesion, as long as correctly mounted. The transducer's precision varies widely, ranging from 10 to 0.050 μm, with inductive eddy current sensors being the transducers with the highest precision.

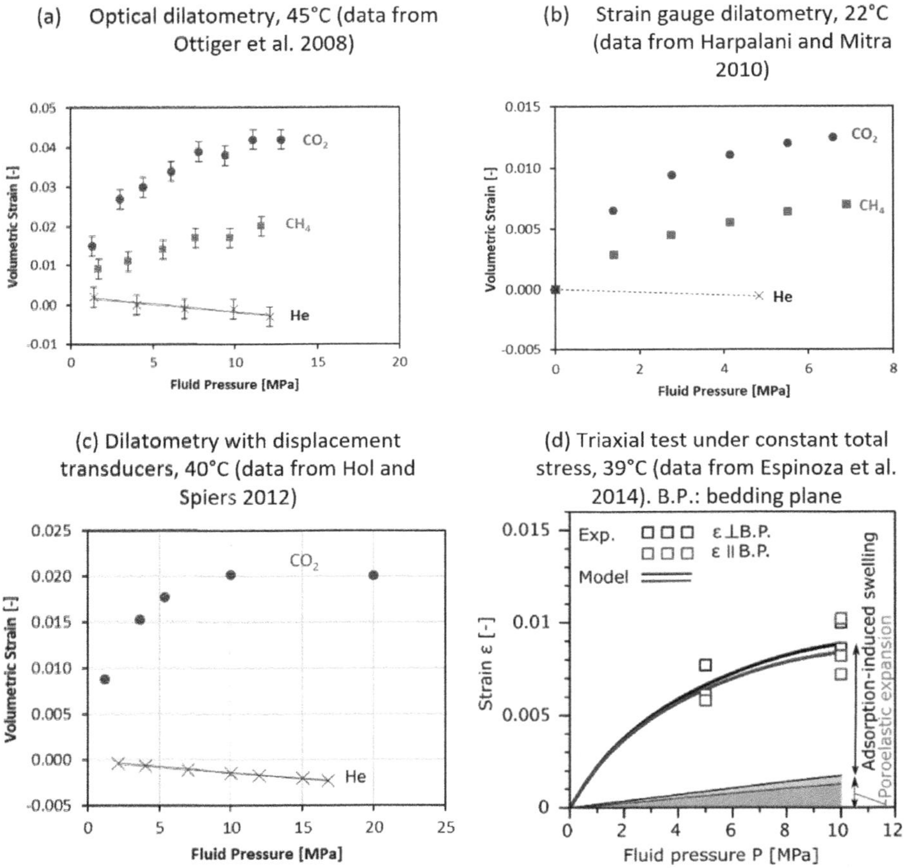

Figure 3.10. *Examples of adsorption-induced strain measurement in coals. (a–c) Dilatometer unjacketed tests. (d) Triaxial test*

Figure 3.10(c) shows an example of this type of measurement for coal samples subjected to pressurized He and CO_2. A linear fit to helium data yields K_s = 7.63 GPa. Carbon dioxide adsorption induces a maximum volumetric strain ε_{vol} = +0.020. Continued pressure increases above 20 MPa results in sample volume reduction (data up to 100 MPa available in Hol and Spiers (2012)).

3.4.2. *Adsorption-induced strains in triaxial testing*

A jacketed triaxial test makes possible the application of total stress different than fluid pressure. In fact, $|\sigma_m| > P$ in triaxial tests to reproduce the typical case in geological formations at equilibrium. Furthermore, the total vertical stress is usually larger than the horizontal stress $|\sigma_v| > |\sigma_h|$ in most tectonically passive subsurface environments, a state of stress that can also be reproduced in triaxial cells. The measurement of strain in triaxial cells is usually done through displacement transducers affixed to the sample through high stiffness endcaps or a soft membrane with known properties (Figure 3.10(d)). Optical, strain and non-contact displacement sensors can also be adapted to triaxial testing.

The measurement of adsorption-induced strains in triaxial tests is usually performed under constant total stress conditions, where increases of pore pressure during adsorbate injection cause a decrease of effective stress (i.e. poroelastic expansion) and simultaneous swelling of the coal matrix. Such a process can be analyzed through an extension of equation [3.25] separating the fluid stress function $s(P) = s^a(P) + \alpha P$:

$$\varepsilon_{vol} = \frac{1}{K}[\sigma_m + s^a(P) + \alpha P] \qquad [3.31]$$

Moreover, using the definition of the Biot coefficient for helium $\alpha = 1 - K/K_s$ permits rewriting equation [3.31] as:

$$\varepsilon_{vol} = \frac{1}{K}(\sigma_m + \alpha P) + \frac{1}{K_s}\left[\frac{s^a(P)}{1-\alpha}\right] \qquad [3.32]$$

where the first term on the right-hand-side quantifies strain caused by variation of the Biot effective stress (solely for poroelastic deformation excluding adsorption) and the second term quantifies the adsorption-induced strain. Figure 3.10(d) shows an example for this type of measurement. The symbols denote the total measured strain for various experiments on the same coal holding a final total stress at least 1 MPa higher than the target fluid pressure. The lines show a double porosity model interpretation that accounts for both poroelastic deformation and adsorption stress, similar to equation [3.32].

Triaxial testing permits applying more realistic initial stress and pressure conditions to coal samples as well as imposed changes. However, the challenge of this type of test is to separate the poroelastic and adsorption effects. A semi-direct measurement of adsorption-induced strain, as in a dilatometer test (equation [3.28]), can be done by applying alternating changes of pore pressure (adsorbate injection) and stress such that $\Delta P = -\Delta\sigma_m$. In fact, the first term on the right-hand-side of equation [3.32] can cancel out for a change of total stress $\Delta\sigma_m = -\alpha\Delta P$ following an adsorbate injection of magnitude ΔP (Guerrero and Espinoza 2022).

3.5. Adsorption-induced stresses

Triaxial tests can impose constant total stress conditions for changes of fluid pressure that replicate coal adsorption and desorption. However, these are not the most realistic conditions to replicate subsurface stress paths. Typically, coal seams are thin and horizontally long. A simplification of local conditions for such geometry is constant total vertical stress and zero-horizontal strain, also known as the uniaxial-strain condition (Segall 1989; Dudley et al. 2016). For example, permeability increase in coal seams upon methane desorption is not caused by coal seam horizontal shrinkage, but rather coal matrix horizontal shrinkage and horizontal stress relaxation (Espinoza et al. 2015). The following sections explore the direct measurement of "bulk" adsorption stress under isochoric (constant volume) and uniaxial-strain conditions. Section 3.5 summarizes applications of these measurements and highlights the need for adsorption stress measurement to predict coal seam horizontal stress and permeability.

3.5.1. *Isochoric tests*

The isochoric (constant volume) test consists of subjecting a coal sample to an initial stress and injecting a gas at a prescribed target pressure without letting the porous sample deform, that is, no displacement of the exterior boundaries of the sample. Hence, the stress developed by coal subjected to an adsorption process restrained from swelling ($\varepsilon_{vol} = 0$) predicted by equation [3.25] combined with the adsorption stress function where $s(P) = s^a(P) + \alpha P$ is:

$$\sigma_m = -\alpha P - s^a(P) \qquad [3.33]$$

where the first term on the right-hand-side quantifies the total stress caused by changing pore pressure (solely due to fluid pressure change excluding adsorption) and the second term quantifies the bulk adsorption-induced stress. Both P and s^a are positive for coal and N_2, CO_2 and CH_4. Hence, the result of a process of adsorption with restrained sample swelling is more compression, larger than what expected for a porous medium that does not develop adsorption-induced strains and stress. Likewise, the result of a process of de-sorption is less compression.

The constant volume condition can be achieved experimentally with (1) an isochoric – constant volume – cell or (2) a triaxial cell with PID-controlled displacements/strains. The first type requires a cell stiff enough such that the change of stress imparted by adsorption does not compromise the constant volume conditions (Espinoza et al. 2016). The second type requires a strain control through radial and axial stresses precise enough such that momentary changes of strains remain significantly lower than the value expected without enforcing constant volume conditions (Espinoza et al. 2014). Bulk adsorption stresses in coal cores can be in the order of $\sim 10^{+1}$ MPa, hence the rigidity of the sample cell and pressure controllers to apply external stresses must be rated to withstand stresses of this order of magnitude.

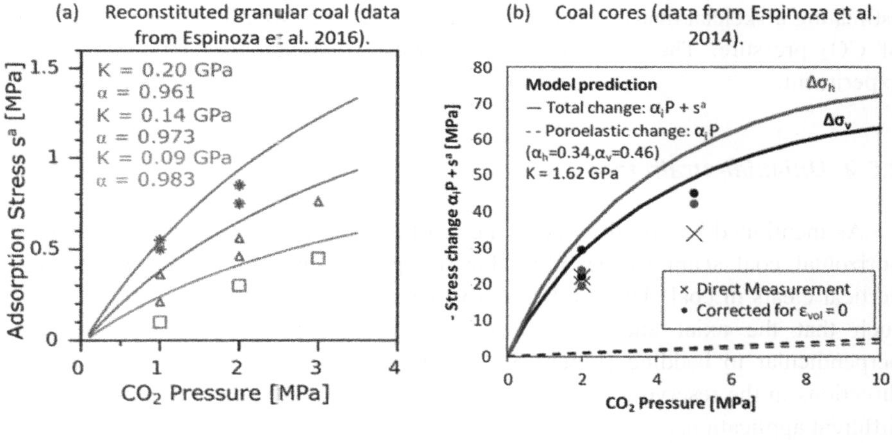

Figure 3.11. *Examples of adsorption-induced stress measurement. (a) Adsorption stress measurement for reconstituted granular coal in an isochoric cell. (b) Adsorption stress measurement for coal cores in a triaxial cell with increasing stress to cancel temporary poroelastic and adsorption-induced deformations. The symbols represent experimental data while lines show model predictions. α_h and α_v are the horizontal and vertical Biot coefficients, respectively*

Figure 3.11(a) shows an example of measurement of bulk adsorption-induced stress in reconstituted granular coal. The bulk adsorption stress is directly proportional to the bulk volumetric modulus K. Hence, the higher K, the higher s^a. Compaction permits achieving varying values of bulk modulus K in reconstituted coal. Figure 3.11(b) shows an example of measurement of bulk adsorption-induced stress in coal cores in horizontal and vertical directions. This is an anisotropic material, that is, the adsorption stresses are different depending on the direction with respect to the planes of material symmetry. In particular, for this sample, an independent set of measurements obtained E_h = 2.74 GPa (horizontal), E_v = 2.55 GPa (vertical) and K = 1.62 GPa. The experiment shown in Figure 3.11(b) included a permeability measurement.

Permeability decreased more than 100 times with the increase of compression stress of ~21 MPa due to CO_2 pressure increase from 0 to 2 MPa, mostly due to adsorption-induced stresses. Both experimental results highlight the relationship between the magnitude of bulk adsorption stress and bulk volumetric modulus K. According to equation [3.28], the skeletal adsorption stress is $s^a/(1-\alpha)$, where the Biot coefficient α decreases as K approaches K_s. The results of Figure 3.11 permit estimating a skeletal/matrix adsorption stress of ~20–55 MPa for this coal at 2 MPa of CO_2 pressure. The variability reflects what is assumed as "matrix" in each experiment.

3.5.2. *Uniaxial-strain test*

As mentioned before, this test is the most appropriate test to predict changes of horizontal coal seam permeability due to opening or closure of predominantly vertical cleats in coal. The test takes advantage of the directionality of coal seams, such that the axial and radial directions of cylindrical coal samples drilled perpendicular to bedding (Figure 3.1(a)) are aligned with vertical and horizontal directions in the reservoir. The test can be performed in two modes, which serve to different applications:

– decreasing adsorbate pressure from an initial state: simulation of reservoir gas depletion;

– increasing adsorbate pressure from an initial state: simulation of gas injection and storage.

The main result is the variation of total horizontal stress σ_h with pressure. This type of experiment aims to represent field conditions for which $|\sigma_v| > |\sigma_h|$. Figure 3.12a shows an example of total horizontal stress variation with decreasing CH_4 pressure (Mitra et al. 2012). This test is conducted in steps where each data point represents an equilibrium state, that is, the measured bulk fluid pressure is in equilibrium with the adsorbate in micropores. Total horizontal stress reduces 230% more with CH_4 than expected for a non-sorbing fluid. Figure 3.12(b) shows an example of total horizontal stress variation with decreasing CO_2 pressure (Espinoza et al. 2015). This test is conducted with a single and rapid pressure change. Hence, pressure (measured outside triaxial cell) decreases quickly and desorption follows with ensuing slow changes of adsorption stress. Total horizontal stress reduces 280% more with CO_2 than expected for a non-sorbing fluid.

(a) CH_4 desorption experiment with multiple pressure steps (data from Mitra et al. 2012)

(b) CO_2 desorption experiment with a single pressure step (data from Espinoza et al. 2015)

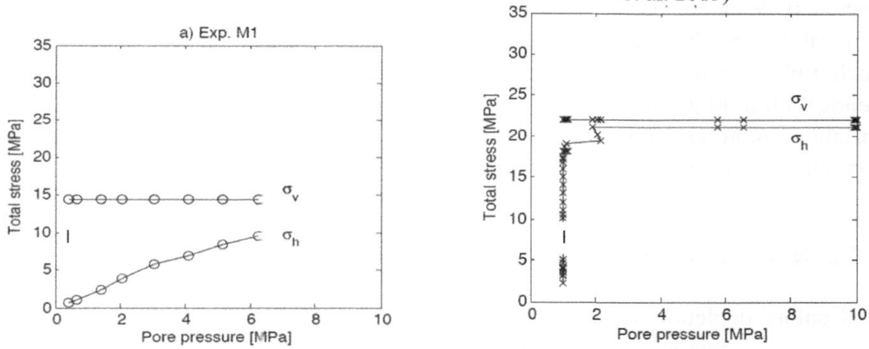

Figure 3.12. *Examples of desorption-induced stress under uniaxial-strain condition. (a) Uniaxial-strain test for CH_4 desorption from P = 6.2 MPa to 0.4 MPa in seven steps. (b) Uniaxial-strain test for CO_2 desorption from P = 10 to 1 MPa in a single step. The pressure is measured externally; hence it is not in thermodynamic equilibrium with the adsorbent at all times in the CO_2 experiment*

3.6. Reservoir scale implications

The following subsections summarize the implications of adsorption-induced strains and stresses in coalbed reservoir geomechanics. Additional information for each process is available in other chapters of this book (see, for instance, Chapters 6 and 7 for modeling case studies of carbon geological storage and gas production, respectively).

The unique geomechanical response of coal can be predicted and anticipated from a combination of laboratory experiments, analytical constitutive models and numerical simulation.

3.6.1. *Natural gas production: stress-dependent permeability*

The reduction of pore pressure during depletion of a conventional natural gas reservoir causes increased effective stresses (Segall 1989). This increase of effective stress is directly linked to the decreases in reservoir permeability typically caused by grain crushing and fracture closing (propped and unpropped) (Shovkun and Espinoza 2017; Sun et al. 2020). However, desorption processes, induced coal shrinkage and horizontal stress relaxation may cause a decrease of horizontal effective stress in coal seams during gas production (Espinoza et al. 2014, 2015). The decrease in horizontal effective stress is particularly impactful for coal seams since cleats (coal natural fractures) tend to be oriented in vertical direction. A 100-times increase of permeability has been observed in the field in some cases (Pan and Connell 2012). Such a phenomenon is very advantageous to gas production and favors production trends with mild production decline trends. Accurate prediction of this permeability variation requires triaxial desorption testing under uniaxial-strain condition combined with simultaneous permeability monitoring (section 3.5.2).

3.6.2. *Natural gas production: desorption-induced shear failure*

Coalbed depletion not only may cause decreases of effective horizontal stress but also causes increases of effective vertical stress as in conventional reservoirs. The reduction of compressive effective horizontal stress (minimum principal effective stress σ_3) and increase of compressive effective vertical stress (maximum principal effective stress σ_1) may result in a state of stress that exceeds the shear failure criterion $\sigma_1 = UCS + \sigma_3\ (1+\sin\phi)/(1-\sin\phi)$, where the unconfined compression strength UCS is usually negligible for cleated coal masses and the friction coefficient is $\phi = 20°–40°$ (Chen et al. 2021).

Despite potential short-term increases of permeability due to desorption-induced shear failure, this is not a desirable scenario because shear failure produces fines that plug cleats, screens, pumps and surface facilities that require workovers to resume production. The minimum bottom-hole production pressure that does not induce shear-failure can be predicted and calibrated with the help of uniaxial-strain desorption tests (section 3.5.2) (Espinoza et al. 2015).

3.6.3. *Carbon geological storage: coal swelling, fracture closing and injectivity decrease*

CO_2 injection into coal seams is not the exact opposite to CH_4 depletion. The main reason is that adsorption capacity of coal is roughly twice the sorption capacity of CH_4 and so are the adsorption-induced strains and stresses. As a result, the compressive effective and total horizontal stresses increase quickly when CO_2 adsorption takes place, even with small increases of reservoir pressure during CO_2 injection. The increase of compressive effective horizontal stress closes cleats and lowers reservoir permeability, therefore, decreasing injectivity. Achieving initial injection rates may require higher injection pressure than initially. Furthermore, the increase of effective horizontal stress could be so high that coal layers might present local compression maxima of σ_{hmin} at coal strata and divert hydraulic fractures to adjacent layers if injection pressure exceeds the fracturing pressure.

The same phenomenon that makes CH_4 production favorable in coalbed seams also makes CO_2 storage difficult because of loss of injectivity (Pekot and Reeves 2002). Injection of N_2-CO_2 mixtures is an alternative to mitigate coal swelling during injection of CO_2 in depleted coal seams (Kovscek et al. 2005; Bustin et al. 2016). This apparently negative implication for carbon geological storage in coal seams might be advantageous in shale caprocks. Clay-rich and organic-matter-rich shales also develop adsorption stress; hence, CO_2 adsorption could prompt increased compressive horizontal stress in caprocks that close potentially existing or new fluid-driven fractures (Busch et al. 2016; Shovkun and Espinoza 2017). Adsorption-induced stress experiments are useful to quantify the expected change of horizontal stress and permeability during a process of CO_2 adsorption in coal and shales.

3.7. Conclusions

Coalbed methane is an alternative to coal mining with much less carbon footprint and surface environmental impact. The physical properties of coal make coalbed OGIP estimations, production trends and potential enhanced-recovery unique and highly dependent on adsorption processes. Hence, field predictions need to be informed by laboratory measurements and appropriate analytical models and numerical tools.

Estimation of OGIP needs to account for adsorption amount that can be measured in the laboratory through a variety of methods. The manometric and gravimetric methods are best suited to measure excess adsorption. The volumetric method can measure total uptake.

Adsorption in coal brings together a mechanical response which manifests as (1) strains in dilatometer (unjacketed) tests and triaxial tests under constant total stress (confinement), (2) adsorption stress in isochoric experiments that restrict strains and (3) a combination of both for mixed mechanical conditions. The dilatometer test is the easiest to perform but is limited to the coal matrix and zero Terzaghi's effective stress. The isochoric test is challenging to perform but permits a direct measurement of "bulk" adsorption stress. The uniaxial-strain test (constant total vertical stress and no horizontal strain) is the closest condition to coal seams and the most representative test to predict variations of horizontal stress as well as coal seam permeability.

Induced strain and stresses increase permeability during gas depletion, a phenomenon unique to CBM, and are very advantageous for production. However, the same phenomenon (1) can be detrimental to CH_4 production with significantly low bottomhole pressures that induce shear failure and fine production, and (2) makes the injection of CO_2 for geological storage purposes in coal seams challenging.

3.8. References

Brochard, L., Vandamme, M., Pellenq, R.J.M., Fen-Chong, T. (2012). Adsorption-induced deformation of microporous materials: Coal swelling induced by CO_2–CH_4 competitive adsorption. *Langmuir*, 28(5), 2659–2670.

Busch, A. and Gensterblum, Y. (2011). CBM and CO_2-ECBM related sorption processes in coal: A review. *International Journal of Coal Geology*, 87(2), 49–71.

Busch, A., Bertier, P., Gensterblum, Y., Rother, G., Spiers, C.J., Zhang, M., Wentinck, H.M. (2016). On sorption and swelling of CO_2 in clays. *Geomechanics and Geophysics for Geo-Energy and Geo-Resources*, 2(2), 111–130.

Bustin, A.M.M., Bustin, R.M., Chikatamarla, L., Downey, R., Mansoori, J. (2016). Learnings from a failed nitrogen enhanced coalbed methane pilot: Piceance Basin, Colorado. *International Journal of Coal Geology*, 165, 64–75.

Chen, Z., Mondal, S., Leonardi, C.R., Flottmann, T. (2021). A new approach for measuring the shear strength of cleated coal. In *55th US Rock Mechanics/Geomechanics Symposium*. OnePetro, Richardson.

Di Giovanni, O., Dörfler, W., Mazzotti, M., Morbidelli, M. (2001). Adsorption of supercritical carbon dioxide on silica. *Langmuir*, 17(14), 4316–4321.

Diamond, W.P. and Schatzel, S.J. (1998). Measuring the gas content of coal: A review. *International Journal of Coal Geology*, 35(1–4), 311–331.

Diamond, W.P., LaScola, J.C., Hyman, D.M. (1986). *Results of Direct-method Determination of the Gas Content of US Coalbeds (Volume 9067)*. US Department of the Interior, Bureau of Mines.

Dudley, J.W., Brignoli, M. Crawford, B.R., Ewy, R.T., Love, D.K., McLennan, J.D., Ramos, G.G., Shafer, J.L., Sharf-Aldin, M.H., Siebrits, E. et al. (2016). ISRM suggested method for uniaxial-strain compressibility testing for reservoir geomechanics. *Rock Mechanics and Rock Engineering*, 49(10), 4153–4178.

Espinoza, D.N. (2021). Introduction to Energy Geomechanics [Online]. Available at: https://dnicolasespinoza.github.io/.

Espinoza, D.N., Vandamme, M., Dangla, P., Pereira, J.M., Vidal-Gilbert, S. (2013). A transverse isotropic model for microporous solids: Application to coal matrix adsorption and swelling. *Journal of Geophysical Research: Solid Earth*, 118(12), 6113–6123.

Espinoza, D.N., Vandamme, M., Pereira, J.M., Dangla, P., Vidal-Gilbert, S. (2014). Measurement and modeling of adsorptive–poromechanical properties of bituminous coal cores exposed to CO_2: Adsorption, swelling strains, swelling stresses and impact on fracture permeability. *International Journal of Coal Geology*, 134, 80–95.

Espinoza, D.N., Pereira, J.M., Vandamme, M., Dangla, P., Vidal-Gilbert, S. (2015). Desorption-induced shear failure of coal bed seams during gas depletion. *International Journal of Coal Geology*, 137, 142–151.

Espinoza, D.N., Vandamme, M., Dangla, P., Pereira, J.M., Vidal-Gilbert, S. (2016). Adsorptive-mechanical properties of reconstituted granular coal: Experimental characterization and poromechanical modeling. *International Journal of Coal Geology*, 162, 158–168.

Gensterblum, Y., Van Hemert, P., Billemont, P., Busch, A., Charriére, D., Li, D., Krooss, B., De Weireld, G., Prinz, D., Wolf, K.-H. (2009). European inter-laboratory comparison of high pressure CO_2 sorption isotherms. I: Activated carbon. *Carbon*, 47, 2958–2969.

Gensterblum, Y., Van Hemert, P., Billemont, P., Battistutta, E., Busch, A., Krooss, B.M., de Weireld, G., Prinz, D., Wolf, K.H. (2010). European inter-laboratory comparison of high pressure CO_2 sorption isotherms II: Natural coals. *International Journal of Coal Geology*, 84(2), 115–124.

Goodman, A., Busch, A., Duffy, G., Fitzgerald, J., Gasem, K., Gensterblum, Y., Krooss, B., Levy, J., Ozdemir, E., Pan, Z. (2004). An inter-laboratory comparison of CO_2 isotherms measured on Argonne premium coal samples. *Energy Fuel*, 18, 1175–1182.

Gor, G.Y., Huber, P., Bernstein, N. (2017). Adsorption-induced deformation of nanoporous materials: A review. *Applied Physics Reviews*, 4(1), 011303.

Gueguen, Y. and Bouteca, M. (1999). Mechanical properties of rocks: Pore pressure and scale effects. *Oil & Gas Science and Technology*, 54(6), 703–714.

Guerrero, J.O. and Espinoza, D.N. (2022). Poroelastic and adsorptive properties of activated carbon. In *56th US Rock Mechanics/Geomechanics Symposium*. OnePetro, Richardson.

Harpalani, S. and Mitra, A. (2010). Impact of CO_2 injection on flow behavior of coalbed methane reservoirs. *Transport in Porous Media*, 82(1), 141–156.

Hol, S. and Spiers, C.J. (2012). Competition between adsorption-induced swelling and elastic compression of coal at CO_2 pressures up to 100 MPa. *Journal of the Mechanics and Physics of Solids*, 60(11), 1862–1882.

Keller, J.U. and Staudt, R. (2005). *Gas Adsorption Equilibria: Experimental Methods and Adsorptive Isotherms*. Springer, Berlin.

Khosrokhavar, R., Wolf, K.H., Bruining, H. (2014). Sorption of CH_4 and CO_2 on a carboniferous shale from Belgium using a manometric setup. *International Journal of Coal Geology*, 128–129, 153–161.

Kovscek, A.R., Tang, G.Q., Jessen, K. (2005). Laboratory and simulation investigation of enhanced coalbed methane recovery by gas injection. In *SPE Annual Technical Conference and Exhibition*. OnePetro, Richardson.

Mitra, A., Harpalani, S., Liu, S. (2012). Laboratory measurement and modeling of coal permeability with continued methane production: Part 1 – Laboratory results. *Fuel*, 94, 110–116.

Moore, T.A. (2012). Coalbed methane: A review. *International Journal of Coal Geology*, 101, 36–81.

Moore, R.L., Loftin, D., Palmer, I. (2011). History matching and permeability increases of mature coalbed methane wells in San Juan Basin. In *SPE Asia Pacific Oil and Gas Conference and Exhibition*. OnePetro, Richardson.

Neimark, A.V. (2017). Reconciliation of Gibbs excess adsorption thermodynamics and poromechanics of nanoporous materials. In *Poromechanics VI*. American Society of Civil Engineers, Reston, Virginia.

Ottiger, S., Pini, R., Storti, G., Mazzotti, M. (2008). Competitive adsorption equilibria of CO_2 and CH_4 on a dry coal. *Adsorption*, 14(4–5), 539–556.

Palmer, I. and Mansoori, J. (1996). How permeability depends on stress and pore pressure in coalbeds: A new model. In *SPE Annual Technical Conference and Exhibition*. OnePetro, Richardson.

Pan, Z. and Connell, L.D. (2012). Modelling permeability for coal reservoirs: A review of analytical models and testing data. *International Journal of Coal Geology*, 92, 1–44.

Pekot, L.J. and Reeves, S.R. (2002). *Modeling Coal Matrix Shrinkage and Differential Swelling with CO_2 Injection for Enhanced Coalbed Methane Recovery and Carbon Sequestration Applications*. Advanced Resources International, Arlington.

Ravikovitch, P.I. and Neimark, A.V. (2006). Density functional theory model of adsorption deformation. *Langmuir*, 22(26), 10,864–10,868. doi: 10.1021/la061092u.

Segall, P. (1989). Earthquakes triggered by fluid extraction. *Geology*, 17(10), 942–946.

Shovkun, I. and Espinoza, D.N. (2017). Coupled fluid flow-geomechanics simulation in stress-sensitive coal and shale reservoirs: Impact of desorption-induced stresses, shear failure, and fines migration. *Fuel*, 195, 260–272.

Sun, Z., Tang, H., Espinoza, D.N., Balhoff, M.T., Killough, J.E. (2020). Grain-to reservoir-scale modeling of depletion-induced compaction and implications on production rate. *SPE Journal*, 25(03), 1543–1556.

Vandamme, M., Brochard, L., Lecampion, B., Coussy, O. (2010). Adsorption and strain: The CO_2-induced swelling of coal. *Journal of the Mechanics and Physics of Solids*, 58(10), 1489–1505.

4

Poromechanical Modeling of Coal Material

Matthieu VANDAMME[1], Jean-Michel PEREIRA[1] and David GRÉGOIRE[2,3]

[1]*Navier, ENPC, CNRS, Institut Polytechnique de Paris, École nationale des ponts et chaussées, Université Gustave Eiffel, Marne-la-Vallée, France*

[2]*LFCR, E2S UPPA, CNRS, Université de Pau et des Pays de l'Adour, Anglet, France*

[3]*Institut Universitaire de France, Paris, France*

This chapter focuses on the modeling of coal behavior at the material scale. Why is there a need for such modeling? Modeling involves establishing a mathematical model – an analogue – of a real system. This model can be used to analyze experimental data obtained from the actual system being modeled or to predict how this system will respond to various stimuli. For example, in the context of coal seams, one might want to predict changes in permeability, temperature or internal stresses within the reservoir during fluid injection or production. Additionally, it may be desirable to predict the injection flow rate over time for a given pressure history.

Various types of modeling exist at different scales. Chapter 5 provides a detailed presentation of molecular modeling methods, which are relevant for simulating phenomena originating at the finest scales of matter. Chapter 6 discusses finite element and finite volume modeling, which is pertinent to continuous media and geological structures. This method of modeling is particularly powerful for conducting reservoir-scale simulations. In this chapter, however, we focus on modeling at the material scale. As we will see later, material-scale modeling is a

For a color version of all of the figures in this chapter, see www.iste.co.uk/vandamme/coal.zip.

Geomechanics of Coal Seams,
coordinated by Matthieu VANDAMME and Jean-Michel PEREIRA. © ISTE Ltd 2025.

necessary prerequisite for reservoir-scale numerical modeling using the finite element/volume method.

Modeling and experimentation are in constant interaction. Experimentation initially allows for the calibration (or identification) of the parameters on which the model depends. Beyond this calibration, experimentation is necessary to validate a model. Independent validation experiments (those not used for model calibration) are crucial for confirming the model's predictive capability. Every model has a certain domain of validity, which is the range of conditions (mechanical stresses, temperature, etc.) over which the model is valid and beyond this range, the model may no longer be valid: it is through experimentation that this domain of validity is identified. A validated model can then be used to explore scenarios that have not been experimentally tested due to time or resource constraints, provided that these scenarios fall within the model's domain of validity.

Consider a physical property dependent on a single parameter (such as temperature). From measurements of this property at various temperatures, one can assess the quality of a trendline calibrated against these data. However, it is not possible to know a priori whether extrapolating this model to more extreme temperatures is reliable. To accurately assess this, additional measurements at temperatures beyond those initially considered are necessary.

The limitations of a model, as revealed through experimentation, inspire the development of more accurate models or those with a broader domain of validity. For instance, in this chapter, we will examine how the swelling of coal in the presence of fluid is an experimentally observed phenomenon that was not predicted by some classical theories, thus motivating the development of more advanced models. Conversely, modeling also informs experimentation by identifying the critical parameters of a phenomenon, thereby facilitating the design of more efficient calibration or validation experiments.

Different modeling philosophies exist. When the goal is to represent the system using a set of mathematical equations, we refer to these as mathematical models. Mathematical models are sometimes called phenomenological when the equations are designed solely to fit experimental data, without necessarily providing a physical understanding of the underlying phenomena. Conversely, when the equations are formulated to incorporate the physical principles of the problem, the models are referred to as physics based. In this chapter, we focus on physics-based mathematical modeling. The domain of validity for a phenomenological model is typically confined to the range of conditions covered by the experiments that inspired it or were used to calibrate it. In contrast, with a physics-based model, there is often an expectation that the model's validity can extend beyond the experimentally tested range.

The distinction between mathematical-based and physics-based models can sometimes be blurred, particularly in the intent behind the model's creation. For example, consider Darcy's law [4.1], which governs the advective flow of fluid in a porous medium. Darcy (1856) derived the law empirically, based on experimental observations, leading one to classify it as phenomenological. However, a century later, Hall (1956) and Hubbert (1956) independently derived this law by integrating the Navier–Stokes equations governing fluid flow in porous media, suggesting that Darcy's law is also physics based. This theoretical derivation provides valuable insights into the meaning of permeability (a parameter in Darcy's law) and the physical significance of the equation itself.

Mathematical models vary in their complexity and the number of parameters they incorporate. Generally, as a mathematical model becomes more complex, it is expected to better align with experimental data; otherwise, there would be no justification for the added complexity. However, increasing a model's complexity typically introduces more parameters and makes calibration more challenging. The physical meaning of the parameters in a complex model is often less intuitive than in a simpler model. Moreover, complexity at the material scale can lead to longer computational times in reservoir-scale simulations. Therefore, depending on the study's objectives, a balance must be struck in the complexity of the models employed.

To develop a relevant physics-based model, it is crucial to have a deep understanding of the underlying physics. This physics is described in section 4.2, following a discussion in section 4.1 on the relevant scale for the "material"-type modeling considered here. Sections 4.3–4.5 address the modeling of the material's hydro-mechanical behavior, focusing on the subtle interplay between hydraulic phenomena (such as flow and pressure) and mechanical phenomena (such as stress and deformation). These sections progressively detail the modeling of increasingly refined physical phenomena. Finally, section 4.6 discusses the limitations (particularly in relation to experimental findings) and future directions of the presented models.

4.1. Relevant scale for material modeling

We consider the material at a scale where it can be treated as a continuous medium, enabling the use of the formalism of continuum mechanics. At this scale, the state of the material, characterized by state variables, is defined at the material point level. This concept of a material point is based on the existence of a representative elementary volume (REV). A REV is a volume of material that is large enough so that the properties of interest (mechanical, hydraulic, etc.) calculated within it are independent of its position, yet small enough relative to the scale of the intended application (especially considering stress and pressure gradients) for these properties to be regarded as point values. For further details, readers may refer to standard textbooks on continuum mechanics, such as Coussy (2004) and Salençon (2005).

At this material point, we can define the state variables that describe the system under study as well as their associated variables. The objective is to determine the relationships between these variables, that is, to establish constitutive models. These models introduce parameters that also depend on the definition of the REV associated with the material point. These parameters, or material properties, can be determined experimentally, theoretically or numerically provided that the microstructure of the material and the relevant physics at a scale smaller than the REV are known. These methods enable the homogenization of materials with varying degrees of microstructural complexity to obtain effective properties applicable at larger scales. Chapters 5 and 7 provide examples of such approaches, where modeling at very fine scales (particularly molecular modeling, see Chapter 5) or multi-scale approaches (see Chapter 7) yield relevant information at the macroscopic scale.

Once these properties are determined, they theoretically represent the material's behavior at larger scales relevant to the intended application. In the case of coal seams, the coal matrix is discontinuous due to the presence of a cleat system that spans several spatial scales. As a result, the value of the measured properties depends on the scale at which they are measured (as discussed in Chapters 2 and 3). For instance, the permeability measured on a laboratory specimen may be representative of material containing fractures with spacing much smaller than the specimen size, but it would not represent the coal seam containing fractures with much larger spacing. Therefore, modeling flow at the scale of the coal seam must explicitly account for these larger fractures. The same consideration applies to mechanical properties and other properties of interest.

Although upscaling (such as to the reservoir scale) may require adjustments to the values of material properties obtained from the calibration of a constitutive model using laboratory test data (adjustments that can be made if in situ monitoring data are available), the development of these "macroscopic" models and the determination of their dependent parameters are crucial for large-scale modeling, such as that of coal seams. The following sections will explain the physics that such models need to describe and the theoretical framework within which they are situated.

4.2. Physics of the phenomena involved

To achieve modeling that is as accurate as possible, a thorough understanding of the underlying physics is essential. This section aims to describe the physical phenomena involved in the injection or production of fluids in coal seams. The discussion first addresses the fluids contained in the seam, then focuses on the solid matrix (i.e. the coal matrix), the interactions between the coal and the fluids it contains and finally introduces the concept of coupling (i.e. the mutual influence of different physical phenomena on each other).

4.2.1. *Fluids*

Coal seams naturally contain hydrocarbons (alkanes), with methane (CH_4) typically being the primary component (White et al. 2005). These seams can also naturally contain carbon dioxide (CO_2), sometimes in very large proportions, up to 99% (Rice 1993). CO_2 can also be injected into the seams (for storage or enhanced methane recovery, see Chapter 1) along with other fluids. Additionally, coal seams often contain water. The quantities of these various fluids vary significantly depending on the type of coal, its rank and its depth or temperature (Rice 1993). During the coalification process, it is estimated that 100–300 cm^3 of methane[1] are produced per gram of coal (Rice 1993).

The state of these fluids in the seam depends on the temperature and pressure conditions. The equation that relates the density of a fluid to its pressure and temperature is known as its equation of state. Despite strong local disparities, the geothermal temperature gradient is approximately 25°C/km, and the hydrostatic pressure gradient is around 10 MPa/km. Regardless of the reservoir depth, the water present in the reservoir remains in liquid form. CO_2 (with a critical temperature of 31.0°C and a critical pressure of 7.38 MPa (Lemmon et al. 2020)) is primarily in a gaseous state in shallow seams (i.e. less than about 800 m deep) and in a supercritical state in deep seams (i.e. more than about 800 m deep). Similarly, CH_4 (with a critical temperature of −82.6°C and a critical pressure of 4.61 MPa (Lemmon et al. 2020)) is mainly gaseous in shallow seams (i.e. less than about 500 m deep) and supercritical in deeper seams (i.e. more than about 500 m deep). In the

1 The gas volumes are provided under standard pressure and temperature conditions, specifically 1,013.25 hPa and 15°C.

supercritical state, reached when both pressure and temperature exceed their critical values, the distinction between liquid and gas phases disappears, eliminating the concept of interfacial tension between these phases. The properties of a supercritical fluid are generally intermediate between those of a gas and a liquid.

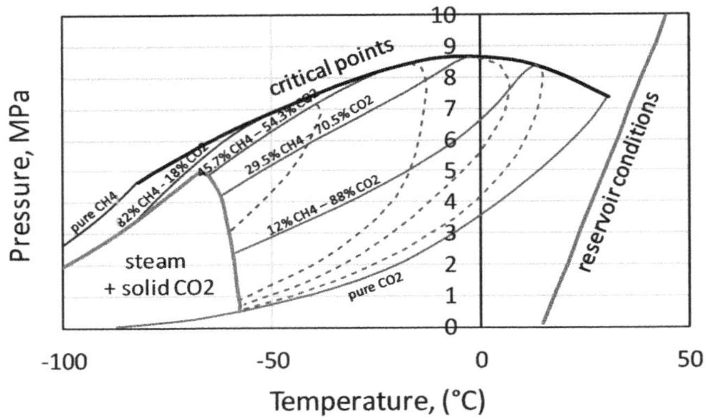

Figure 4.1. *Phase diagram of CH_4-CO_2 systems, overlaid with the typical pressure and temperature curve encountered underground (green line). For each mixture (with the indicated fraction being a molar fraction), the solid fine line represents the boiling curve and the dotted fine line represents the dew curve. The figure is adapted from Donnelly and Katz (1954)*

At all temperatures and pressures relevant to our application, according to the CH_4-CO_2 phase diagram (see Figure 4.1), CH_4 and CO_2 are miscible and form either a gaseous mixture (at shallow depths) or a supercritical mixture (at greater depths). At these same relevant temperatures and pressures, CH_4 and CO_2 are not miscible with water (which, as a reminder, is present in liquid form), but they are soluble in water; Sander (2015) compiled Henry's constants characterizing this solubility under ambient conditions for dilute solutions[2]. The solubility of CH_4 and CO_2 depends on the composition of the water (which may contain salts), increases with pressure and varies with temperature (Wiebe and Gaddy 1940; Culberson and McKetta 1951; White et al. 2005). Up to 40 MPa and 100°C, the solubility of CO_2 decreases with temperature (Wiebe and Gaddy 1940). To provide a concrete example, at a pressure of 10 MPa and a temperature of 50°C, the solubility of CO_2 in pure water is approximately 5 g per 100 g of water (Wiebe and Gaddy 1940), while that of CH_4 is around 0.1 g per 100 g of water (Culberson and McKetta 1951). Data on the

2 The Henry's law constants compiled by Sander are also available on his website at the following address: http://www.henrys-law.org.

solubility of CO_2 in water show deviations from Henry's law, particularly at high pressure and low temperature (Wiebe and Gaddy 1940), indicating the non-ideal behavior of the mixture under these conditions. These solubilities can be modeled using equations calibrated on experimental datasets (e.g. Li and Nghiem 1986; Enick and Klara 1990), or by using numerical codes (e.g. Geochemist's Workbench (GWB) or PHREEQC; White et al. 2005).

The fluid mixture is contained within the fractures and porosity of the coal matrix. The transport properties of the seam are governed by the fractures. Fluid transport along these fractures is primarily Darcian advective, governed by the pressure gradient, as described by Darcy's law (1856), which reads, for isotropic materials:

$$\underline{q} = -k\underline{\nabla}p \tag{4.1}$$

where $\underline{q}$ (in m.s^{-1}) is the volumetric flux vector of the fluid, p (in Pa) is the pressure in the cleats, and k (in m^2·Pa^{-1}·s^{-1}) is the cleat's permeability. This permeability verifies:

$$k = \kappa/\eta \tag{4.2}$$

where η is the dynamic viscosity of the fluid (in Pa.s) and κ is the intrinsic permeability of the material (expressed in m^2 or Darcy, with 1 D = 10^{-12} m^2). The order of magnitude of this intrinsic permeability is provided in Table 4.1. Due to the structure of coal (see Chapter 2), coal permeability is anisotropic: it is lower perpendicular to the bedding plane than within the bedding plane (Yan et al. 2019). Even within the bedding plane, permeability is anisotropic (Yan et al. 2019).

Physical parameter	Unit	Order of magnitude
Intrinsic permeability κ	m^2	10^{-16}–10^{-13} (White et al. 2015)
	mD	0.1–100 (White et al. 2015)
Diffusivity D	m^2.s^{-1}	10^{-9} (Nazarova et al. 2014)
Incompressibility modulus K	GPa	4–12 (Morcote et al. 2010)

Table 4.1. *Orders of magnitude of the physical parameters of the reservoir*

Fluid transport through the coal matrix is significantly slower than through fractures. This transport is generally considered diffusive, governed by the fluid concentration gradient in the matrix. In its simplest form, this transport is governed by Fick's law (Coussy 2004):

$$\underline{w} = -D\underline{\nabla}c \qquad [4.3]$$

where $\underline{w}$ is the mass flux vector of the fluid, c is the fluid concentration in the matrix and D is the diffusivity of the fluid in the matrix.

In fractures, in addition to the Darcian transport of the fluid mixture, if one considers each chemical species in the mixture, each species diffuses, so the transport of each species through the seam via the fractures can be described as being governed by an advection-diffusion equation (Coussy 2004).

4.2.2. *Coal in the absence of fluid*

Coal is a solid material that deforms when subjected to mechanical stress. For sufficiently low levels of stress, these deformations are reversible, and the mechanical behavior of the material is described as elastic. This elasticity may be non-linear if the relationship between stress and deformation is nonlinear. Due to the presence of fractures that are preferentially oriented perpendicular to the bedding plane (see Chapters 1 and 2), the elastic behavior of coal is anisotropic; the deformations induced by stress depend on the orientation of the material relative to the stress field.

If doubling the stress results in a doubling of the deformation, the material's elastic behavior is considered linear. In this case, the relationship between the stress tensor σ_{ij} and the strain tensor ε_{kl} is expressed as:

$$\sigma_{ij} = C_{ijkl}\varepsilon_{kl} \qquad [4.4]$$

where C_{ijkl} is referred to as the elasticity tensor of the material. This relationship, like all the relationships in this chapter, uses Einstein's summation convention, meaning that the summation must be carried out over all repeated indices for all their possible values. Therefore, this equation should be understood as $\sigma_{ij} = \sum_{k=1..3}\sum_{l=1..3} C_{ijkl}\varepsilon_{kl}$.

If the material is isotropic, its elasticity tensor (and therefore its three-dimensional elasticity) is fully defined by two constants, which can be selected from the following quantities: Young's modulus E (in Pa), Poisson's ratio ν (dimensionless), bulk modulus K (in Pa), or shear modulus G (in Pa). Each of these constants can be mathematically expressed as a function of any other pair of constants[3]. However, if the material is anisotropic, more parameters are required to fully define the elasticity tensor. In the case of a generic anisotropic material, its elasticity tensor depends on 21 independent coefficients (Salençon 2005). Due to the increased complexity of considering the material's anisotropy (and the experimental challenges in measuring a greater number of elasticity coefficients), petroleum engineers often resort to isotropic modeling of the material in practice.

Even for an anisotropic material, the bulk modulus K, which relates a change in isotropic stress $\Delta\sigma$ to a change in volumetric strain $\Delta\varepsilon$, can be uniquely defined: $K = \Delta\sigma/\Delta\varepsilon$. For coal samples with a characteristic size of a few centimeters, this modulus ranges from approximately 4–12 GPa (Morcote et al. 2010). It depends notably on the coal rank (Morcote et al. 2010). The modulus of coal also varies with the scale considered, as the behavior at the reservoir scale significantly depends on the fracture network (Hol et al. 2014), raising the question of the applicability of data obtained at the laboratory scale (centimeter or decimeter samples) to the reservoir scale.

The anisotropy of the material can be characterized by Thomsen's parameter ε^* (Thomsen 1986), defined as $\varepsilon^* = (C_{1111} - C_{3333})/(2C_{3333})$. When taking direction 3 as the direction perpendicular to the bedding plane, this parameter has been measured as $\varepsilon^* = 0.075$ (Morcote et al. 2010) or estimated at $\varepsilon^* \approx 0.1$ (Espinoza et al. 2013). This positive value indicates that in these cases, coal is stiffer in the direction parallel to the bedding plane than in the direction perpendicular to it.

Fractures, therefore, induce anisotropy in the mechanical behavior of coal and may also cause the elastic properties to depend on the scale considered. Moreover, fractures introduce non-linearity in the elastic behavior: compressing a coal sample closes its fractures, increasing the material's stiffness (i.e. its modulus of elasticity). Experimentally, this dependence of the modulus on the confining stress is observed up to a stress level of about 5 MPa (Morcote et al. 2010) to 10 MPa (Espinoza et al. 2014) (see Figure 4.2). When the stresses (or more precisely, the effective stresses, see section 4.3.3) exceed 5 or 10 MPa – typically observed at depths of 500 m or 1000 m, respectively – the dependence of the moduli on stress can be neglected, and the material's elasticity can be considered linear. However, for laboratory experiments under low stress levels, the dependence of the modulus on this stress can be significant, as illustrated in Figure 4.2.

---

3 An exhaustive list of the relationships between elasticity quantities can be found on the Wikipedia page dedicated to the "Elastic modulus" (https://en.wikipedia.org/wiki/Elastic_modulus).

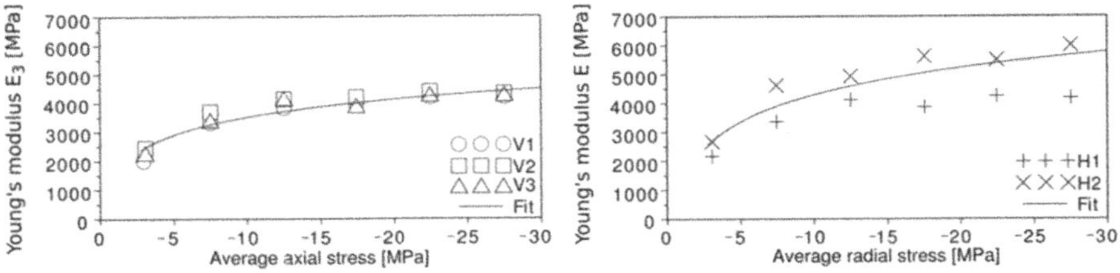

Figure 4.2. *Dependence of Young's moduli E_3 (perpendicular to the bedding plane) and E (parallel to the bedding plane) on axial stress (perpendicular to the bedding plane) or radial stress (within the bedding plane). Adapted from Espinoza et al. (2014)*

When shear stresses are sufficiently increased, coal may rupture. This rupture is classified as brittle because it occurs suddenly, without any clear manifestation of irreversible deformations (plasticity) beforehand. Upon rupture, the coal disintegrates (see Figure 4.3), producing fines that can potentially clog the production well (Espinoza et al. 2015). An empirical failure criterion commonly used for rocks is the Hoek–Brown criterion (Hoek and Brown 1980). According to this criterion, the shear stress that coal can withstand increases with the confining stress to which it is subjected.

Figure 4.3. *Photograph of a cylindrical coal sample broken under simple compression, after the sample has been removed from the membrane. Fragments are arranged from left to right according to their size. Adapted from Espinoza et al. (2015)*

In this section, we have considered *dry* coal, that is, coal in the absence of fluid. However, in practice, coal is always saturated with various fluids that interact with it. These interactions are the focus of the next section.

4.2.3. *Interactions between coal and pore fluids*

Underground, coal is saturated with a fluid mixture of complex composition (see section 4.2.1), which may primarily consist of water (H_2O), methane (CH_4), or carbon dioxide (CO_2). These pore fluids interact with the solid coal matrix.

First, the solid matrix influences the state of the fluid. Due to the presence of mesopores and micropores within the coal matrix, as well as its large specific surface area, adsorption phenomena (also known as sorption) are significant. The fluid within the micropores (i.e. pores smaller than 2 nm) and mesopores (i.e. pores between 2 nm and 50 nm) of the coal matrix exists in an adsorbed state, meaning it interacts intermolecularly with the atoms of the solid matrix. This fluid is referred to as confined. The confined fluid behaves differently from fluid in a macroscopic reservoir: at a given chemical potential, their densities differ. The amount of fluid present in the sample is dependent on the fluid pressure. The relationship between the amount of fluid in the sample and the fluid pressure is called the sorption isotherm (see Chapter 3). If the focus is on the total amount of fluid, it is referred to as the total sorption isotherm. If the focus is on the amount of fluid in excess of what would be contained in the sample if the fluid filled the pore volume at the density it would have in a macroscopic reservoir, it is referred to as the excess sorption isotherm. Due to coal's higher affinity for CO_2 compared to CH_4, when a coal sample is exposed to a reservoir containing a mixture of CH_4 and CO_2, the mole fraction of CO_2 in the fluid mixture within the coal matrix porosity is higher than in the reservoir. Sorption phenomena, arising from intermolecular interactions between the fluid and the solid matrix, are discussed in detail in Chapter 5, which focuses on molecular simulation methods.

Pore fluid causes coal to deform. Immersing coal in CH_4 or CO_2 causes it to swell, as illustrated in Figure 4.4(a). At a given fluid pressure, this swelling is more pronounced with CO_2 than with CH_4 (see Figure 4.4(b)), primarily because coal adsorbs more CO_2 than CH_4 (see Figures 4.4(c) and (d)). Swelling increases with fluid pressure[4], although at high pressures (above approximately 10 MPa), the

4 Strictly speaking, at very low fluid pressures, a slight contraction of coals is observed (Gor et al. 2017); however, this contraction occurs at fluid pressures too low to be relevant for the intended geomechanical application and can therefore be neglected.

material tends to contract again (Hol and Spiers 2012). Due to the wide range of pore sizes, the physical mechanisms by which the fluid deforms the material are diverse.

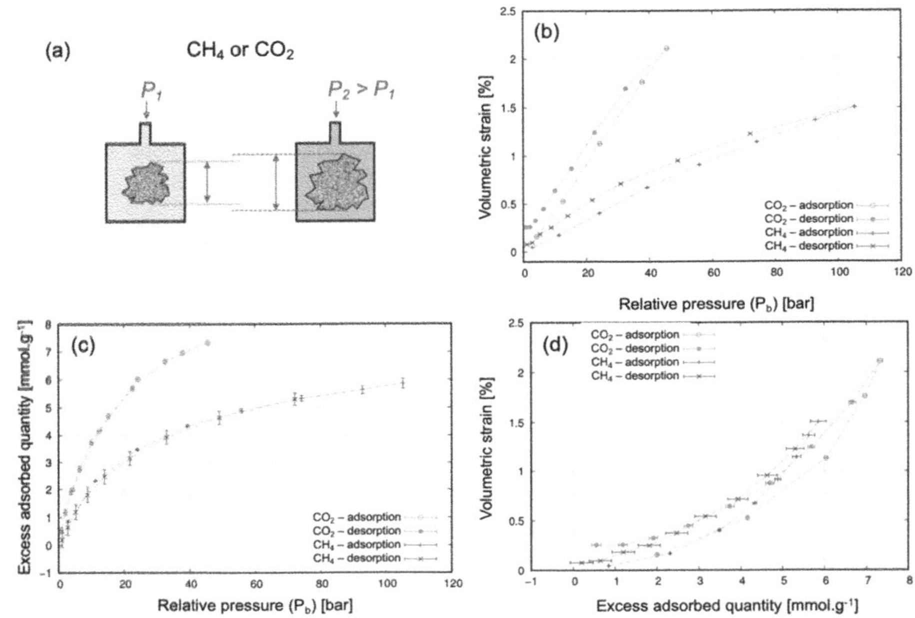

Figure 4.4. *Behavior of a coal sample immersed in CH_4 or CO_2. (a) Diagram indicating coal swelling in the presence of CH_4 or CO_2. Swelling data (b and d) and sorption data (c) as a function of fluid pressure (b and c) or excess adsorbed amount (d). These data, obtained from activated carbon, are adapted from Perrier et al. (2017)*

In fractures, which are macropores (i.e. pores larger than 50 nm by definition), the fluid deforms the material through its mechanical pressure by "pushing" against the walls. Increasing the fluid pressure in the macropores of an unconstrained material causes the fractures to dilate, much like inflating a balloon by blowing into it.

If a macroporous material is immersed in a fluid and the pressure is increased, this pressure is exerted not only within the pores but also on the edges of the sample. Consequently, the macroporous material contracts as the fluid pressure increases, which does not align with experimental observations for coal (see Figure 4.4). This discrepancy suggests that coal contains more than just macropores.

In mesopores and micropores, sorption phenomena prevail, and the deformation of the material is not only due to the pressure of the fluid, but also the intermolecular interactions between molecules of the fluid and atoms of the solid skeleton. It is observed that, to a first approximation, the swelling of the coal matrix is directly proportional to the total amount of adsorbed fluid (see Figure 4.5).

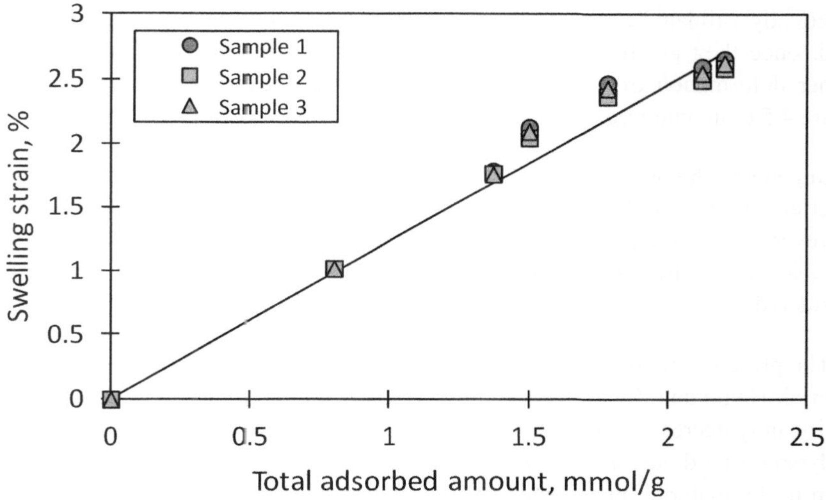

Figure 4.5. *Swelling of a coal sample immersed in CO_2 (adapted from Hol and Spiers (2012)). The swelling is obtained by correcting the total deformation for the elastic deformation of the sample caused by compression from the pressurized fluid*

In mesopores, adsorption is a surface phenomenon. The adsorption of fluid on the surface of mesopores reduces the surface tension[5] at the mesopore surface (following the Gibbs isotherm (1928)): as the surface tension decreases, the

5 More specifically, adsorption reduces surface stress, which is related to surface tension by the Shuttleworth's (1950) formula. However, for disordered materials like coal, Gor and Bernstein (2016) demonstrated that the surface stress variations caused by sorption are equal to the surface tension variations. For a detailed discussion of the relationship between sorption, surface stress variations and surface tension variations, see Gor et al. (2017).

mesopores swell. Empirically, Bangham and Razouk (1938) demonstrated (specifically for coals) that the swelling of an unconstrained material is proportional to the change in surface tension induced by adsorption, as predicted by the Gibbs isotherm[6].

In micropores, adsorption is no longer a surface phenomenon but rather a volumetric phenomenon. Due to the small size of micropores, discussing surface tension at the pore surface, or even simply the pore surface, can be ambiguous. However, the relationship proposed by Bangham and Razouk (1938) has been empirically validated for coals containing micropores. For micropores, on the other hand, once they are filled, any adsorption of any additional molecules results in further deformation of the micropore, making the linear relationship depicted in Figure 4.5 quite intuitive.

In general, the presence of fluid mixtures in a material's porosity can deform the material through capillary effects when the fluids are immiscible (Coussy 2010). However, in the case of coals, we have seen (in section 4.2.1) that CH_4 and CO_2 are miscible, so in the absence of water, such capillary effects do not need to be considered.

The presence of pore fluid can also directly alter the mechanical properties of the material. Hagin and Zoback (2010) observed that saturating a coal sample with CO_2 significantly decreases the static compressibility modulus of the material, though not its dynamic modulus. The same authors observed that CO_2 saturation can induce creep in the material, that is, deformation under constant load that progresses over time. Regarding strength, Czaplinski and Holda (1982) observed that CO_2 sorption reduces the material's strength.

Conversely, Espinoza et al. (2015) found that the failure envelope of the material is only slightly altered by the same sorption, particularly when considered in terms of effective stresses[7].

Thus, the fluid mixture within coal porosity can contribute to its deformation. It is also evident that coal, like any solid material, can be deformed by the direct application of mechanical stresses. In terms of modeling, a theoretical framework is required to predict material deformations based on the applied mechanical stresses and the state of the pore fluid. The development of such a theory is the focus of section 4.3.

---

6 For a discussion on the relationship between the proportionality factor and the material's microstructure, refer to Vandamme et al. (2010).

7 See section 4.3.3 for the concept of effective stresses.

4.2.4. *Impact of temperature*

When a fluid colder than the reservoir is injected, it alters the reservoir's temperature. Any temperature variation induces mechanical effects. It is well known that when a solid is heated, it generally expands: this is referred to as thermal expansion. If the solid is constrained or heterogeneous, heating induces internal stresses within the material. Stresses are also generated if the heating is uneven.

Thus, during the injection of a cold fluid into the reservoir, the area near the injection well cools and tends to contract, while the regions farther from the well experience less cooling. This thermal field heterogeneity results in tensile stresses near the well. The generation of thermally induced stresses and deformations is modeled within the framework of thermo-mechanics.

Temperature variations affect not only the coal but also the pore fluid. The changes in fluid density resulting from temperature variations can be accounted for using the fluid's equation of state. Moreover, temperature alters the interactions between the fluid and the solid matrix. This is why the sorption isotherm (see section 4.2.3) is temperature dependent.

For instance, Sakurovs et al. (2008) demonstrated that the sorption capacity of nitrogen or carbon dioxide in coal decreases significantly with increasing temperature. If temperature variations cause the solid matrix and pore fluid to expand differently, these variations lead to changes in fluid pressure and stress within the solid matrix, a phenomenon known as thermal pressurization.

The sorption process itself can influence temperature since adsorption is an exothermic process, characterized by sorption enthalpy. It should be noted that although the sorption of a pure fluid is exothermic, competitive sorption (when two fluids compete for adsorption, as is common in coal seams) can be either exothermic or endothermic, as the adsorption of one fluid may induce the desorption of another. Another phenomenon, more extreme and not discussed in this chapter but mentioned in Chapter 1, that leads to temperature variations is the underground combustion of coal, a process used in underground coal gasification (Su et al. 2017; Gao et al. 2022).

Despite the importance of accounting for temperature variations in certain applications, the remainder of this chapter will focus on isothermal modeling.

4.3. Classical poromechanical modeling of a saturated material in the absence of sorption

In the previous section, we established that coal, like any porous material, can be deformed by a pore fluid and by mechanical stresses directly applied to the material. In this section, we focus on modeling the mechanical behavior of a porous material saturated with a single fluid. We limit the modeling to a macroporous material, which implies neglecting the effects of sorption. The modeling of these effects will be extended in section 4.4. Although this section is oriented toward modeling reservoir rocks, it also serves as an introduction to poromechanics, that is, the mechanics of porous materials. Readers interested in a more in-depth understanding of the topic are directed to Coussy's reference works (Coussy 2004, 2010).

4.3.1. *System under consideration*

It is important to recall that the following description applies to the scale of the REV of coal (see section 4.1). The system to be modeled is therefore a two-phase system, consisting of a solid porous matrix filled with a fluid. At the considered scale, these two phases can be regarded as continuous media. Typically, the solution to this problem is addressed by assuming that, at every point in space, the continuous medium hypothesis holds: the REV is assumed to be infinitesimal. In this case, at every material point, it is assumed that two material particles (one solid and the other fluid) are superimposed at each instant. From a mechanical perspective, the focus is on the solid particle, allowing the deformations of the porous medium to be deduced. Naturally, these two particles can interact with each other; this is central to defining the relationships that describe the poromechanical behavior of a saturated porous material.

Moreover, a reference configuration must be considered. This is a configuration assumed to be undeformed, generally corresponding to an initial state, which can be relatively easily defined by the modeler before the system's state is disturbed by an external action (mechanical loading, pressure change, etc.). A particular feature of geomechanical problems is that the systems of interest typically experience non-zero[8] initial stresses and fluid pressures (i.e. in the reference configuration).

The solution to this poromechanical problem involves solving a system of two field equations defined at the reservoir scale. These equations are balance equations, each associated with a primary variable. The resolution of these equations will be

8 At the reservoir scale, it is necessary to account for hydrostatic and geostatic gradients in the initial state.

discussed in Chapter 6. However, these equations also involve the behavior laws of each of the phases. The introduction of these laws is what "closes" the problem, enabling its resolution.

4.3.2. *Energy balances and state variables*

Adopting an energy-based approach allows for the rigorous definition of variables that describe the state of the system, along with their associated variables and the relationships linking these different variables.

Consider an infinitesimal volume of a porous medium, including its pore fluids, as the system. In the reference configuration (to which the quantities considered here are related), this volume is denoted as $d\Omega_0$. Applying the first law of thermodynamics to this system gives:

$$dE = \delta W + \delta Q + \mu dN \tag{4.5}$$

where E is the internal energy of the infinitesimal volume, N is the molar quantity of fluid in the volume, μ is the chemical potential of the fluid, δW is the work done on the system and δQ is the heat supplied to the system. The term μdN arises because the system is open, meaning it can exchange fluid with the surroundings. In three-dimensional mechanics, $\delta W = d\Omega_0(\sigma_{ij} d\epsilon_{ij})$, where σ_{ij} is the stress tensor and ϵ_{ij} is the strain tensor.

The second law of thermodynamics states that $\delta Q < TdS$, where T is the temperature and S is the entropy of the system. By introducing the internal energy $e = E/d\Omega_0$ per unit volume of the system, its entropy $s = S/d\Omega_0$ per unit volume, and the fluid quantity $n = N/d\Omega_0$ per unit volume, equation [4.5] becomes:

$$de \leq \sigma_{ij} d\epsilon_{ij} + Tds + \mu dn \tag{4.6}$$

The difference between the right-hand side and the left-hand side is dissipation. In this chapter, we restrict ourselves to reversible processes, meaning no dissipation occurs, and thus the inequality becomes an equality:

$$de = \sigma_{ij} d\epsilon_{ij} + Tds + \mu dn \tag{4.7}$$

We now introduce the internal energy e_s of the solid skeleton and its entropy s_s (both per unit volume of the porous material). The internal energy and entropy of the pore fluid are denoted by e_f and s_f, respectively. By additivity, we have $e = e_s + e_f$ and $s = s_s + s_f$. Knowing the internal energy of the fluid (whose behavior must be described independently by state equations), given by $e_f = n\mu + Ts_f - p\Phi$, and using the Gibbs–Duhem relation $nd\mu + s_f dT - \Phi dp = 0$, equation [4.7] can be rewritten as:

$$de_s = \sigma_{ij} d\epsilon_{ij} + pd\Phi + Tds_s \tag{4.8}$$

This equation can be interpreted as an energy balance for a system composed solely of the solid skeleton, that is, without the pore fluids. By definition, the solid skeleton is a closed system (i.e. it does not exchange matter with the surroundings), hence the absence of the term μdn. Work can be supplied to this system either by deforming it under stress (term "$\sigma_{ij} d\epsilon_{ij}$") or by changing its porosity under fluid pressure (term "$pd\Phi$").

We now introduce a free energy $\psi_s = e_s - p\Phi - s_s T$. The stresses σ_{ij} are decomposed into the volumetric stress $\sigma = (\sigma_{11} + \sigma_{22} + \sigma_{33})/3$ and the deviatoric stresses s_{ij}, such that $\sigma_{ij} = \sigma\delta_{ij} + s_{ij}$ (where δ_{ij} is the Kronecker delta). Similarly, the strains ϵ_{ij} are decomposed into the volumetric strain $\epsilon = \epsilon_{11} + \epsilon_{22} + \epsilon_{33}$ and the shear strains e_{ij}, such that $\epsilon_{ij} = \epsilon\delta_{ij} + e_{ij}$.

Under isothermal conditions (i.e. $dT = 0$), the energy ψ_s satisfies:

$$d\psi_s = \sigma_{ij} d\epsilon_{ij} - \Phi dp = \sigma d\epsilon + s_{ij} de_{ij} - \Phi dp \tag{4.9}$$

The fact that this total differential can be written confirms that the free energy ψ_s depends on the state variables of strain and pressure: $\psi_s = \psi_s(\epsilon, e_{ij}, p)$. Since, according to equation [4.9], the stresses and porosity are partial derivatives of this same energy ψ_s, it follows that the stresses and porosity are also functions of these state variables: $\sigma = \sigma(\epsilon, e_{ij}, p)$, $s_{ij} = s_{ij}(\epsilon, e_{ij}, p)$, $\Phi = \Phi(\epsilon, e_{ij}, p)$.

4.3.3. *Constitutive equations for an isotropic poroelastic material*

Let us consider an isotropic material. For such a material, the symmetry conditions allow us to state that the volumetric state parameters depend only on the volumetric state parameters, and the deviatoric state parameters depend only on the deviatoric state parameters: $\sigma = \sigma(\epsilon, p)$, $s_{ij} = s_{ij}(e_{ij})$, $\Phi = \Phi(\epsilon, p)$.

Finally, let us consider the simplest case of linear poroelasticity. In this case, the state parameters are related in a linear manner. In the generic case of non-zero stresses in the reference state (i.e. in the reference state $\epsilon = 0$ and $\Phi = \Phi_0$ and

$\sigma = \sigma_0$, the stresses are $s_{ij} = s_{ij,0}$ and the pressure is $p = p_0$), the final constitutive equations are:

$$\sigma - \sigma_0 = K\epsilon - b(p - p_0) \quad [4.10]$$

$$\varphi = \Phi - \Phi_0 = b\epsilon + (p - p_0)/N \quad [4.11]$$

$$s_{ij} - s_{ij,0} = 2Ge_{ij} \quad [4.12]$$

where K is the bulk modulus, G is the shear modulus, N is the Biot modulus and b is the Biot coefficient. It should be noted that the presence of the Biot coefficient in the first two equations is due to one of Maxwell's symmetry relations, which originates from thermodynamics: indeed, since $\partial^2\psi_s/(\partial\epsilon\partial p) = \partial^2\psi_s/(\partial p\partial\epsilon)$, it is necessary that $\partial\sigma/\partial p|_\epsilon = \partial(-\Phi)/\partial\epsilon|_p$.

Due to the assumptions introduced during the derivation, it should be noted that these constitutive equations are valid only for macroporous isotropic linear poroelastic materials.

The first constitutive equation indicates that the volumetric deformation ultimately depends on $\sigma + bp$: this stress is referred to as the effective Biot stress. It can be demonstrated that if the solid skeleton is homogeneous with an incompressibility modulus K_s, the Biot coefficient b is given by $b = 1 - (K/K_s) > 0$. Since the porous material cannot be stiffer than the material composing its solid skeleton, it follows that: $b \leq 1$.

The energy-based approach leading to the equations presented above can be extended to anisotropic materials: for such materials, the symmetry considerations discussed at the beginning of section 4.3.2 must be adapted accordingly. The approach can also be extended to irreversible processes, taking into account poro-viscoelasticity, damage, or poro-plasticity (Coussy 2004, 2010).

For such processes, dissipation is non-zero, making inequality [4.6] strict. This chapter does not explore these extensions to irreversible processes in detail; however, anisotropic behaviors are addressed in section 4.5.2.

Although the constitutive equations introduced in this section are applicable to a wide range of porous materials and rocks, they are of limited use for modeling the poromechanical behavior of coals. As will be discussed in the following section, modeling coal as a macroporous material (i.e. by neglecting adsorption effects) leads to predictions that are inconsistent with experimental observations.

4.4. Poromechanical modeling of microporous materials undergoing sorption

In section 4.3, we introduced poromechanical modeling, focusing on a macroporous material and therefore neglecting sorption effects. However, as discussed in section 4.2.3, these effects are significant due to the presence of micropores and mesopores within the coal matrix. Consequently, it is necessary to incorporate these effects into the modeling, which is the subject of this section.

Indeed, applying the constitutive equations [4.10]–[4.12] reveals that the modeling presented in section 4.3 is inconsistent with the experimental results shown in Figure 4.4. Specifically, when the coal sample is immersed in a fluid at pressure p, the fluid exerts a stress $\sigma = -p$ on the external walls of the sample. Applied to this immersion experiment, equation [4.10] predicts a deformation $\varepsilon = -(1 - b)p/K$. The model for macroporous solids presented in section 4.3 thus predicts that (1) the sample should contract (since the Biot coefficient b is always less than 1), (2) the deformation should depend linearly on the fluid pressure and (3) this deformation should be independent of the fluid. These three predictions of the model contradict the experimental observations, demonstrating the inapplicability of the model to coal in the presence of CO_2 or CH_4. Nevertheless, this linear poroelastic model remains valid when sorption phenomena are negligible. This applies to a wide range of macroporous rocks (Coussy 2010) or to microporous materials when in contact with a fluid that adsorbs negligibly, such as helium in the case of coal. However, for coals in the presence of CO_2 or CH_4, the model must be adapted, as its neglect of adsorption effects is prohibitive.

Conceptual reasons also contribute to the inapplicability of the model presented in section 4.3 to coals. The model relies on the concept of porosity, but for a microporous material, how can a porous volume be uniquely defined? On an atomic scale, where does a pore end and the solid wall begin? While the inability to define the pore wall position to the nearest angstrom results in negligible uncertainty for macropores, it introduces significant uncertainty for micropores. One could attempt to define a porous volume experimentally, perhaps by defining it as the volume of fluid in the sample when immersed in that fluid. For instance, when a microporous sample is immersed in CO_2, the CO_2 enters the micropores, assigning the sample a finite pore volume. However, if the sample is immersed in a fluid with large molecules that are too big to penetrate the micropores, a porosity measurement with this fluid would indicate zero porosity for the sample! This demonstrates that, with such an experimentally based definition, porosity would depend on the probe fluid and therefore not be an intrinsic property of the material. Similarly, is it reasonable to speak of mechanical fluid pressure within a micropore when the pore contains only a few fluid molecules, meaning the fluid no longer forms a continuous phase?

For these reasons, the approach presented in section 4.3, which is valid for macroporous materials, must be adapted to be applicable to coals in the presence of a generic fluid. In the following two sections, we present two adaptations with slightly different philosophies. The model in section 4.4.1 is based on the concept of apparent magnitudes (such as introducing a *corrected* fluid pressure) but has the advantage of not requiring an adjustment parameter throughout the loading path. The adsorption-deformation coupling is intrinsic to the model, and the evolution of poromechanical parameters is directly predicted from their initial values.

The model in section 4.4.2, on the other hand, only considers unambiguously defined state variables but introduces a coupling coefficient that must be calibrated to represent the adsorption–deformation coupling throughout the loading path. Despite differences in philosophy and resulting equations, Gor et al. (2017) showed that both approaches ultimately yield identical results!

4.4.1. *Approach based on the concept of apparent magnitudes (confinement coefficient)*

Under isothermal and saturated conditions, for a homogeneous and isotropic porous material with a single porosity, the classical poromechanics equations presented in section 4.3.3 can also be expressed incrementally as follows:

$$d\sigma = K d\epsilon - b dp \quad [4.13]$$

$$d\varphi = d\phi = b d\epsilon + dp/N \quad [4.14]$$

$$ds_{ij} = 2G de_{ij} \quad [4.15]$$

These equations are valid in the absence of adsorption effects, applicable to porous media containing pores large enough for adsorption effects to be negligible or to media with smaller pores when saturated by a fluid that does not adsorb on the substrate, as is typically the case with helium. This non-adsorptive property can be used to experimentally measure the rigidity of the porous medium's skeleton.

The methodology, for example, is detailed by Perrier et al. (2018) for activated carbon and by Perrier (2015) for mine coal. This involves saturating the material with helium at different pressures and recording the induced contraction deformations. In the absence of external mechanical constraints, with $d\sigma = -dp$, equation [4.13] simplifies to $d\varepsilon = -(1-b)dp/K = -dp/K_s$, allowing for the experimental identification of the skeleton's bulk modulus K_s. The obtained values are $K_s = 6.0 \pm 0.6$ GPa for activated carbon and $K_s = 7.7 \pm 0.8$ GPa for mine coal (see Figure 4.6).

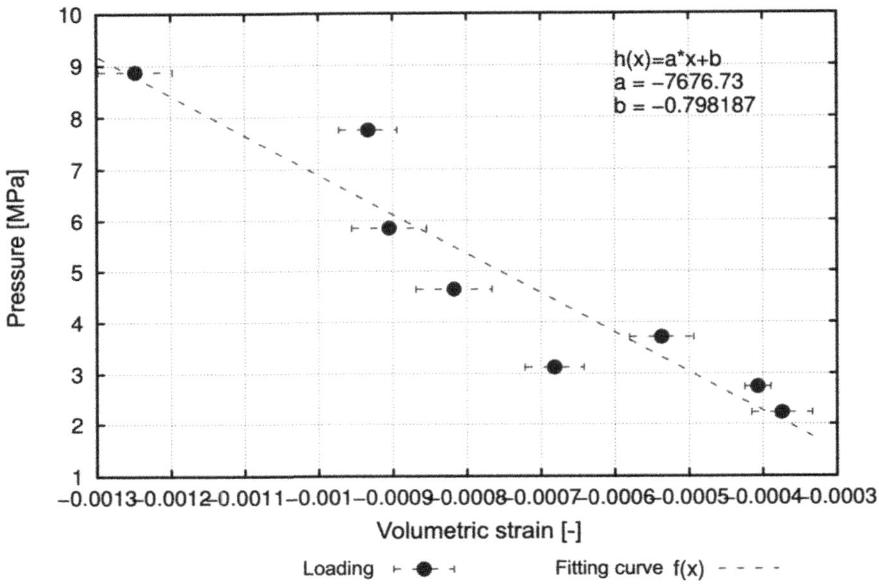

Figure 4.6. *Measurement of contraction deformations induced on a coal mine saturated with helium in the absence of adsorption. The value obtained is $K_s = 7.7 \pm 0.8$ GPa. Adapted from Perrier (2015)*

In the presence of adsorption, equation [4.13] is no longer valid, but a similar equation can be derived by considering the fluid saturating the porous medium in chemical equilibrium with the surrounding fluid. Thermodynamic equilibrium requires the chemical potentials of the surrounding fluid at pressure P_b and the fluid within the porous medium to be equal. While the dependency of porosity on the probe fluid is a real issue, it can be resolved by considering the pore volume as that measured using helium in the absence of adsorption. The fluid pressure within this porosity is also harder to define when pore sizes are small, typically comparable to the size of the adsorbed molecules. In such cases, pressure is no longer a scalar because the normal pressure to the wall differs from the tangential pressure. However, it can be estimated through molecular scale modeling, as detailed in Chapter 5. For example, Figure 4.7 shows the evolution of methane pressure confined within different graphitic micropores as a function of pore size and surrounding pressure, obtained using non-local density functional theory (NLDFT) coupled with statistical associating fluid theory (SAFT) as developed by Grégoire et al. (2018). It is observed that after significant oscillations for ultramicropores, the pressure approaches the surrounding pressure for pore sizes around 2 nm. However, given the complex nanostructure of coals, the very concepts of pore size and fluid-solid interface become questionable at these scales. Similarly, the pressure values obtained in perfectly shaped pores may not be representative of the actual mechanical pressures exerted on micropore walls. This is not problematic, as the extended poromechanical formulation for adsorption does not rely on these concepts and instead uses the overall fluid content to represent the confinement effect on the adsorbed fluid within the smallest pores.

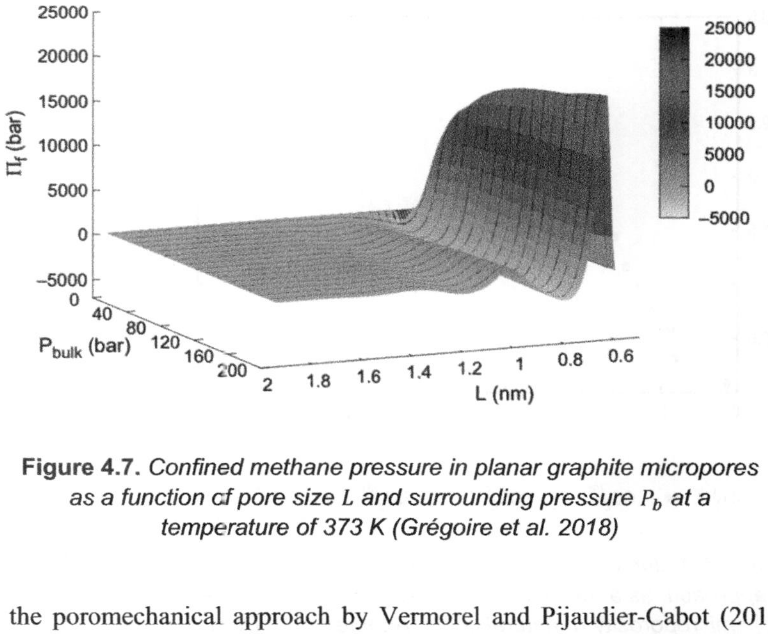

Figure 4.7. *Confined methane pressure in planar graphite micropores as a function of pore size L and surrounding pressure P_b at a temperature of 373 K (Grégoire et al. 2018)*

In the poromechanical approach by Vermorel and Pijaudier-Cabot (2014), the coupling between adsorption and induced deformations is accounted for by introducing a confinement coefficient that measures the deviation of the properties of the interstitial fluid present in the micropores from those of the surrounding fluid with which it is in thermodynamic equilibrium and shares the same chemical potential. Thus, for an isotropic microporous solid under reversible and isothermal conditions, the macroscopic stresses are defined by equations [4.16]–[4.18], which introduce fluid confinement through the parameter $\chi = 1 - \frac{\rho_b}{\rho_f}$, where densities ρ_b and ρ_f represent the surrounding fluid and the interstitial fluid adsorbed in the micropores, respectively. The main advantage of introducing this parameter is that it can be directly expressed as a function of the excess adsorbed matter n_{ex} in the micropores, as detailed in the following equations:

$$\sigma = K(\phi)\epsilon - b(\phi) \int_0^{P_b} \frac{1}{1-\chi(\phi)} dp \qquad [4.16]$$

$$\phi = \phi_0 + b(\phi)\epsilon + \frac{1}{N(\phi)} \int_0^{P_b} \frac{1}{1-\chi(\phi)} dp \qquad [4.17]$$

$$\chi(\phi) = 1 - \frac{\rho_b}{\rho_f} = \frac{n_{ex}}{n_{tot}} = \frac{n_{ex}}{n_{ex} + \frac{\rho_b m_s}{M\, \rho_s} \frac{\phi}{1-\phi}} \qquad [4.18]$$

with $b(\phi) = 1 - \frac{K(\phi)}{K_s}$ and $N(\phi) = \frac{K_s}{b(\phi) - \phi}$.

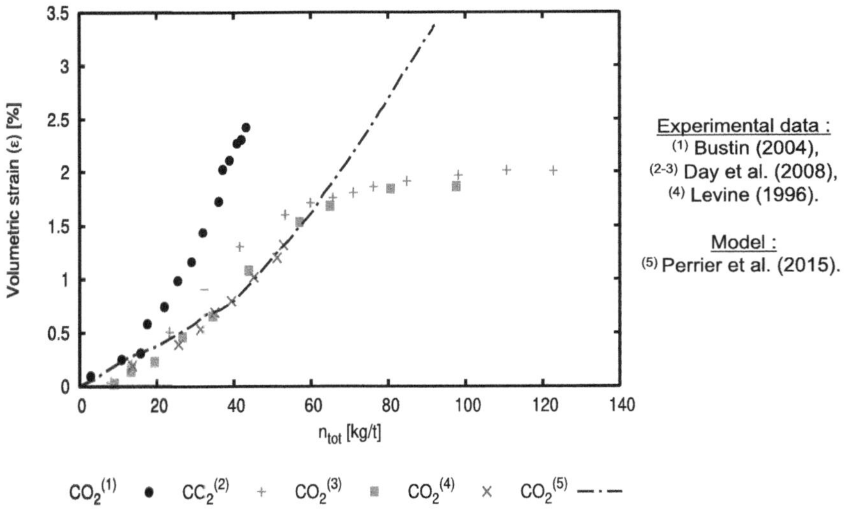

Figure 4.8. *Volume deformations measured and calculated using the described approach, as a function of the total amount of material adsorbed in the microporosity of natural coals. Adapted from Perrier et al. (2015)*

In equation [4.16], the incompressibility modulus $K(\phi)$ is determined through iterative homogenization based on the type of porosity considered (see Perrier et al. (2015) for details). In this approach, the pressure considered is the pressure of the surrounding fluid with which the interstitial fluid in the micropores is in equilibrium. The initial porosity is defined as the ratio between the pore volume measured with helium in the absence of adsorption and the volume of the sample. Thus, the material is fully characterized prior to the test, and the complete evolution of poromechanical parameters, deformations and porosity is directly estimated by the model; the only input data required is the amount of excess material. This excess material can be measured experimentally using manometric or gravimetric techniques (see Chapter 3) or calculated directly, for example, using NLDFT/SAFT theory (see Chapter 5). This approach thus enables the direct calculation of deformations induced by adsorption without requiring any adjustment parameters.

Since the quantity of excess material directly affects the updated porosity due to deformation and confinement effects, the coupling is comprehensive, and the effects of deformation on the quantities adsorbed are intrinsically accounted for. Perrier et al. (2015) demonstrated the necessity of accounting for these effects by updating the porosity during the adsorption-deformation coupling to accurately reproduce the experiments conducted by Ottiger et al. (2008) on natural coal.

Figure 4.8 compares the adsorption-induced deformations obtained using this approach with various experimental results from the literature on coals, which include measurements of total adsorbed quantities.

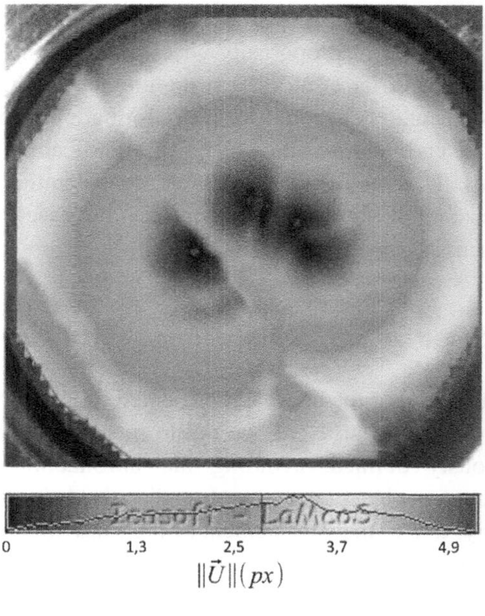

Figure 4.9. *Map of the displacement magnitudes obtained by digital image correlation and induced by CO_2 adsorption on a coal sample ($P_b = 5.47$ MPa and $T = 45°C$). Adapted from Perrier (2015)*

Unfortunately, experimental data in the literature often lack simultaneous measurements of deformation and adsorbed quantities. Sometimes deformations are estimated on one sample and adsorbed quantities on another sample, which may not always come from the same coal mine, complicating comparisons with models.

To address this, Perrier et al. (2017) proposed an original experimental setup that allows for the simultaneous measurement of adsorbed quantities using manometric techniques and deformations using digital image correlation, thereby facilitating the comparison between theory and experiment. Measuring deformation via digital image correlation also provides access to complete and localized displacement and deformation fields, rather than limiting the measurement to the overall volume deformation. For instance, Figure 4.9 presents a map of displacement magnitudes induced by the adsorption of 1.6 mmol/g of carbon dioxide at $P_b = 5.34$ MPa and $T = 45$°C on a natural coal sample with a significant macroscopic natural fracture. The obtained displacement field is complex and can be used to measure the reclosure of the natural fracture induced by the swelling of the microporous matrix.

The availability of various experimental databases in the literature now enables model validation. However, as with many natural and synthetic porous media, coal contains multiple scales of porosity: microporosity where the fluid is trapped in an adsorbed form, and meso- or macroporosity necessary for transporting the fluid to and from the nanopores. Perrier et al. (2018) highlighted the need to consider this dual porosity scale to accurately account for the deformations induced by adsorption on synthetic activated carbons, or otherwise risk significantly overestimating the induced deformations. It is likely that the same consideration applies to natural coals, and the extension of the proposed approach to cases involving different porosity scales is discussed in section 4.5.1.

4.4.2. *Approach based on univocally defined state variables*

The philosophy underlying the approach presented in this section slightly differs from that in the previous section. Here, we opt to model using only state variables that are unambiguously defined, avoiding the introduction of variables with ambiguous definitions, such as porosity (as discussed in the introduction to section 4.4). Similarly, due to the confinement of the fluid, the mechanical effects of the fluid differ from its pressure. Consequently, we will refer only to the thermodynamic pressure P_b of the fluid, which is clearly defined as the pressure prevailing in a macroscopic reservoir containing fluid under the same thermodynamic conditions (i.e. the same molar chemical potential μ and the same temperature T). This thermodynamic pressure satisfies the Gibbs–Duhem relationship $d\mu - \underline{V}_b dP_b + \underline{S}_b dT = 0$ (Coussy 2010), where $\underline{V}_b$ is the molar volume of the unconfined fluid (i.e., in its macroscopic reservoir) and $\underline{S}_b$ is its molar entropy.

To define the state of the fluid in the REV, we choose to use the total quantity n of fluid per unit volume of the reference configuration. Even though not all the fluid present in the sample is adsorbed (since coal also contains macropores), we will refer to this amount as the total adsorbed amount, or the total sorption isotherm. The definition of this quantity n is clear, although this does not imply that it is simple to measure: while the triaxial equipment presented in section 3.2 enables the determination of the total quantity of fluid injected into the sample, and thus allows for the measurement of n, a conventional sorption measurement experiment using manometry or volumetry can only measure the *excess* adsorbed quantity (see section 3.3), not the total adsorbed quantity.

The dual quantity to the total adsorbed amount n is the molar chemical potential μ of the fluid. The internal energy conservation condition [4.5] at the scale of the porous material and the fluid it contains remains valid even for a microporous material. However, without introducing porosity, it is no longer possible to move to the free energy conservation balance [4.8] at the scale of the solid skeleton. Nonetheless, we can work with the balance [4.5]. If we consider only a system subjected to volumetric stresses (for which $\sigma_{ij}\, d\epsilon_{ij} = \sigma d\epsilon$), the fact that the expression $\sigma d\epsilon - n d\mu$ is a total differential (since it is equal to $d(\psi - n\mu)$) allows us to (1) assert that, in the generic case, the volumetric stress σ and the total adsorbed quantity n depend on the volumetric deformation ϵ and the molar chemical potential μ of the fluid, that is, $n = n(\mu, \epsilon)$; (2) write the Maxwell's symmetry relationship $\partial\sigma/\partial\mu|_\epsilon = -\partial n/\partial\epsilon|_\mu$.

After integrating with respect to μ, this equation gives $\sigma = \sigma_{dry}(\epsilon) - \int_{\mu \to -\infty}^{\mu} \partial n/\partial\epsilon|_\mu d\mu'$, where $\sigma_{dry}(\epsilon)$ represents the behavior of the dry material, which, in the case of an isotropic linear elastic material, is $\sigma_{dry}(\epsilon) = K\epsilon$. To express this relationship as a function of the thermodynamic pressure P_b of the fluid rather than its molar chemical potential μ, we can use the Gibbs–Duhem equation given above, leading to the following constitutive equations:

$$\sigma = K\epsilon - s^a(P_b) \qquad [4.19]$$

where $s^a(P_b) = \int_{\mu \to -\infty}^{\mu} \frac{\partial n}{\partial \epsilon}|_\mu d\mu' = \int_{p=0}^{P_b} \frac{\partial n}{\partial \epsilon}|_{P_b} \underline{V}_b dp$ is an adsorption-induced stress, and where:

$$n = n(\epsilon, \mu) = n(\epsilon, P_b) \qquad [4.20]$$

These two equations replace the constitutive equations [4.10] and [4.11] for a microporous material such as coal. Equation [4.12], governing shear behavior, remains unchanged.

These new constitutive equations, which are valid for microporous materials, lead to several observations. First, it is evident that, in the generic case, the sorption isotherm $n(\epsilon, \mu)$ or $n(\epsilon, P_b)$ depends not only on the thermodynamic pressure P_b of the fluid (or its molar chemical potential μ), but also on the deformation ϵ of the coal (the adsorbent). While the dependence of the adsorbed quantity on fluid pressure is obvious (as it is by varying this pressure that the sorption isotherm is traversed, see section 4.2.3), the dependence on the deformation of the adsorbent is much less so. This is particularly due to the fact that, in a conventional sorption experiment (see section 3.3), the stress applied to the adsorbent, and thus its deformation ϵ, is not controlled independently of the fluid pressure p. In a triaxial cell, or in any other equipment that allows independent control of the fluid pressure and the stress (or deformation) of the adsorbent, the dependence of the sorption isotherm on the state of the adsorbent can be demonstrated. For instance, Hol et al. (2011) and Liu et al. (2016) have shown both experimentally and thermodynamically that compressing a coal sample desorbs part of the CO_2 it contains. This dependence of the sorption isotherm on the deformation or stress of the adsorbent is not, of course, limited to coal, as proposed by Grosman and Ortega (2008). Equation [4.19] shows that it is this dependence, via the term $\partial n/\partial \epsilon|_{P_b}$, that causes the fluid to deform the solid.

Equation [4.19] is generic and valid for any porous material whose dry behavior is isotropic linear and whose pore space is saturated with a pure fluid. If we assume the material is macroporous, we can confirm that equations [4.10] and [4.11] are recovered. In the case of a material with generic porosity, all the complexity of poromechanical behavior is actually contained within the complexity of the sorption isotherm $n(\epsilon, P_b)$). The form of this isotherm is not known a priori, but Espinoza et al. (2014) have shown that the model can be calibrated using sorption and swelling experiments. For coals, where deformations do not exceed a few percent, it is reasonable to consider that $n(\epsilon, P_b) \approx n_0(P_b) * (1 + C(P_b)\epsilon)$, where $n_0(P_b)$ is the isotherm that would be measured on a *rigid* material and $C(P_b)$ is a coupling coefficient.

Once calibrated, such a model can be used to predict the evolution of stresses and deformations in the REV under any conditions. For example, Espinoza et al. (2014) verified the validity of the model through isochoric experiments: CO_2 is injected into the sample, while measuring the compressive stresses required to prevent the sample from swelling. The experiment shows that these stresses can be significantly greater than the fluid pressure. By introducing an "apparent Biot coefficient" b^{app}, defined as $b^{app} = -\partial\sigma/\partial P_b|_\epsilon$, an apparent Biot coefficient that can exceed 1 (reaching up to approximately 10 in the presence of CO_2 at low pressure) is found. Such a high apparent Biot coefficient is a direct consequence of sorption phenomena, which amplify the mechanical effects of the fluid. The classical poromechanical model presented in section 4.3 cannot explain the experimental observation of a coefficient $-\partial\sigma/\partial P_b|_\epsilon > 1$, while such an observation can be accurately modeled using the approach derived in this section.

4.5. Examples of possible model enhancements

In sections 4.3 and 4.4, we focused on modeling an isotropic, linear elastic porous material. In section 4.3, the model was limited to a macroporous material (i.e. without considering sorption effects), and in section 4.4, we demonstrated the necessity of extending poromechanical modeling to account for sorption effects, even if only to explain the most basic experimental observations on coals (see Figure 4.4). However, the models derived in section 4.4 are valid only for a linear poroelastic material saturated with a pure fluid, with all material points in local thermodynamic equilibrium. As shown in Chapters 2 and 3, coal is a much more complex material than a simple linear poroelastic medium, and it is subject to additional complexities, such as mixed-fluid stresses. In this section, we provide an overview of how to address some of these complexities in material modeling and incorporate further information regarding the material's microstructure.

4.5.1. *Multi-scale porosity of coal*

4.5.1.1. *Incorporating dual porosity for transport and adsorption*

It is well known that the pore network in coals exhibits certain specific characteristics (see Chapters 1 and 2). Specifically, coal contains a network of natural fractures, which govern the permeability of the reservoir, with the micro- and mesoporous coal matrix situated between these fractures. Since the fractures function as macropores, sorption effects in them are negligible. It may be necessary to incorporate this information into the model, particularly regarding the duality of the porosity system (macroporous fractures and the micro- and mesoporous coal matrix subject to adsorption phenomena).

For instance, this duality must be considered when predicting variations in fracture porosity (and therefore fracture aperture) to estimate changes in reservoir permeability during its exploitation (as described in the aperture/permeability relationships in Chapter 2).

Incorporating this dual porosity into the model can be achieved by building on the thermodynamic approach employed in sections 4.3 and 4.4.2. Let us return to the internal energy balance equation [4.7], which remains valid, and integrate the fact that we can define a Lagrangian porosity, Φ_c, associated with fractures. Since the fluid in the fractures is unconfined (because the fractures function as macropores), it is possible to modify this free energy conservation balance. After a few thermodynamic manipulations, under isothermal conditions (for a detailed derivation, see Nikoosokhan et al. (2012)), we obtain:

$$\sigma_{ij}d\varepsilon_{ij} + p_c d\Phi_c + \mu_m dn_m - d\psi_m = 0 \quad [4.21]$$

where p_c is the pressure of the fluid in the fractures, μ_m is the chemical potential of the fluid in the coal matrix, n_m is the total quantity adsorbed in the matrix (per unit volume of coal, including fractures) and ψ_m is the free energy of the coal matrix (per unit volume of coal, including fractures). This energy balance shows that energy can be supplied to the coal matrix in three ways: by deforming it through mechanical stress (term "$\sigma_{ij}d\varepsilon_{ij}$"), by deforming it through the pressure of the fluid in the fractures (term "$p_c d\Phi_c$"), or by supplying it with fluid (term "$\mu_m dn_m$"). By once again applying Maxwell's symmetry relations, we ultimately arrive at the following constitutive equations:

$$\sigma = K\varepsilon - bp_c - (1-b)s^a(p_m) \quad [4.22]$$

$$\Phi_c - \Phi_{c,0} = b\epsilon + (p_c - s^a(p_m))/N \quad [4.23]$$

where $s^a(p_m) = \int_0^{p_m} \partial n_m/\partial\epsilon|_{p_m} \underline{V_b} dp'_m$ represents the stress induced by adsorption in the coal matrix, p_m is the thermodynamic pressure of the fluid in the coal matrix and b and N are the Biot coefficient and Biot modulus associated with the fracture network.

Equation [4.22] replaces equation [4.19], while equation [4.23] explicitly calculates the variations in porosity associated with fractures and, consequently, their aperture. In these equations, the pressures p_c of the fluid in the fractures and p_m of the fluid in the matrix blocks are distinct, enabling the identification of each term's contribution to the deformations and stresses. The fluid in the fractures directly influences the mechanics through its thermodynamic pressure p_c, as the fluid in the fractures is not confined.

Conversely, the fluid in the coal matrix affects the system via the stress $s^a(p_m)$ induced by adsorption due to the confinement. If the assumption of local thermodynamic equilibrium holds, the thermodynamic pressure of the fluid must be uniform within the REV, and thus $p_m = p_c$.

Similarly, the approach presented in section 4.4.1 can be extended to better account for the different porosity scales present in coal. By considering that coal comprises an adsorption porosity (micro- and mesoporosity with sizes less than 50 nm) where the fluid is trapped in adsorbed form, and a transport porosity (macroporosity between 50 and 300 nm) that facilitates fluid transport to and from the nanopores, Perrier et al. (2018) proposed an extended poromechanical framework that incorporates the coupling between adsorption and deformation in materials exhibiting dual porosity.

As discussed in the previous chapters, in addition to these two porosity scales, coal may present significant macroscopic fractures that cannot be assimilated into macroporosity and are not considered here, as they cannot be homogenized at the material scale.

Figure 4.10 provides a schematic representation of this type of porous medium. Assuming that the fluid saturating the porous medium within each of the two porosities is in chemical equilibrium with the surrounding fluid, the condition of thermodynamic equilibrium imposes the equality of the chemical potentials of the surrounding fluid at pressure P_b, the fluid within the transport porosity ϕ_M at pressure P_M and the fluid within the adsorption porosity ϕ_μ at pressure P_μ.

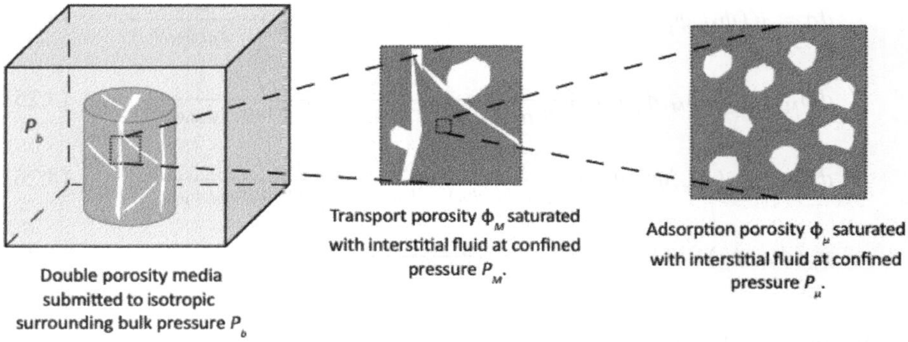

Figure 4.10. *Schematic of a dual-porosity medium with transport porosity ϕ_M and adsorption porosity ϕ_μ (Perrier et al. 2018)*

Both transport and adsorption porosities can be experimentally measured by mercury porosimetry and gas porosimetry, respectively, with the total pore volume measured using helium in the absence of adsorption. Table 4.2 presents typical results for synthetic activated carbon and natural coal. It is evident that synthetic materials, particularly those designed for gas purification, have significantly higher porosities, presenting a greater challenge for modeling.

Material	Transport porosity	Adsorption porosity	Source
Activated carbon	$\phi_M = 41.6 \pm 0.2\%$	$\phi_\mu = 32.2 \pm 0.2\%$	(Perrier et al. 2018)
Natural coal	$\phi_M = 3.7 \pm 0.01\%$	$\phi_\mu = 0.43 \pm 0.01\%$	(Perrier 2015)

Table 4.2. *Adsorption and transport porosities for activated carbon and natural coal*

Since the adsorption effects are minimal in the transport porosity, confinement effects can be neglected $\chi_M \sim 0$, and the pressure of the interstitial fluid is equal to the surrounding pressure: $P_M = P_b$. However, in the adsorption porosity, confinement effects are significant and are accounted for through a confinement coefficient $\chi_\mu = 1 - \rho_b/\rho_\mu$, as developed in section 4.4.1 within the context of a porous medium with single adsorption porosity. The densities ρ_b and ρ_μ represent the densities of the surrounding fluid and the interstitial fluid adsorbed in the adsorption microporosity, respectively. In incremental form, equations [4.13]–[4.14] become[9]:

$$d\sigma = K(\phi_M, \phi_\mu)d\epsilon - b_M(\phi_M, \phi_\mu)dP_b - b_\mu(\phi_M, \phi_\mu)\frac{dP_b}{1-\chi_\mu(\phi_M,\phi_\mu)} \quad [4.24]$$

$$d\phi_M = b_M(\phi_M, \phi_\mu)d\epsilon + \frac{dP_b}{N_{MM}(\phi_M,\phi_\mu)} - \frac{dP_b}{N_{M\mu}(\phi_M,\phi_\mu)(1-\chi_\mu(\phi_M,\phi_\mu))} \quad [4.25]$$

$$d\phi_\mu = b_\mu(\phi_M, \phi_\mu)d\epsilon - \frac{dP_b}{N_{\mu M}(\phi_M,\phi_\mu)} + \frac{dP_b}{N_{\mu\mu}(\phi_M,\phi_\mu)(1-\chi_\mu(\phi_M,\phi_\mu))} \quad [4.26]$$

$$\chi_\mu(\phi_M, \phi_\mu) = 1 - \frac{\rho_b}{\rho_\mu} = \frac{n_{ex}}{n_{tot}} = \frac{n_{ex}}{n_{ex}+\frac{\rho_b m_s}{M\,\rho_s}\frac{\phi_\mu}{1-\phi_\mu-\phi_M}} \quad [4.27]$$

In equations [4.24]–[4.27], the poromechanical properties $(K, b_M, b_\mu, N_{MM}, N_{M\mu}, N_{\mu M}, N_{\mu\mu})$ can be explicitly expressed as a function of (ϕ_M, ϕ_μ, K_s), and (m_s, M, ρ_s) represent, respectively, the mass of the adsorbent, the molar mass of the fluid under consideration and the density of the solid.

9 The reader is referred to Perrier et al. (2018) for the full thermodynamic derivation.

For given values of ambient fluid pressure P_b and ambient fluid density ρ_b, the amount of excess adsorbed material n_{ex} can either be measured experimentally (Perrier et al. 2017) or determined using NLDFT (see Chapter 5 or Grégoire et al. 2018).

Thus, the proposed approach allows direct determination – without any fitting parameters – of both the deformations induced by adsorption and the variations in the different porosities and corresponding poromechanical parameters.

Figure 4.11 shows a comparison of the swelling induced by adsorption of carbon dioxide and methane on activated carbon based on experimental measurements (Perrier et al. 2017) and estimations using a single porosity model (Perrier et al. 2015) or a double porosity model (Perrier et al. 2018).

For this type of material, a single porosity model greatly overestimates the swelling induced by adsorption, as the transport porosity is substantial and undergoes deformation due to the confinement induced in the adsorption porosity.

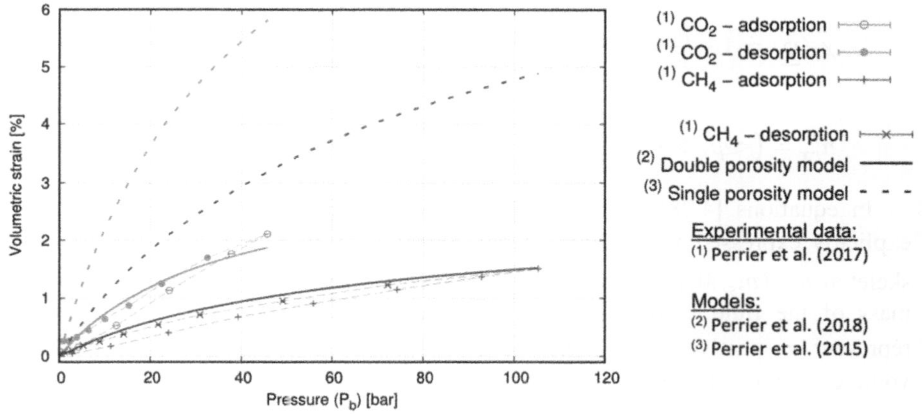

Figure 4.11. *Comparison of swelling results induced by adsorption on activated carbon with carbon dioxide at T = 318 K and methane at T = 303 K in terms of: (1) experimental measurements (Perrier et al. 2017), (2) estimates using a double porosity model (Perrier et al. 2018) and (3) estimates using a single porosity model (Perrier et al. 2015). Adapted from Perrier et al. (2018)*

4.5.1.2. *Consideration of the complete pore size distribution of the material*

In the strategies proposed in section 4.5.1.1, the quantities of adsorbed material are defined globally across the microporosity and mesoporosity of the material, which allows for the definition of only a single adsorption porosity. However, the confinements induced within mesopores and micropores differ significantly. In the case of a complex and heterogeneous material, where different types of adsorption sites are present or if one seeks to account for the full pore size distribution of the material, it becomes necessary to estimate directly the quantities of material adsorbed within each pore size to account for the distinct confinements experienced by the fluid in the various porosities.

This can be achieved through the NLDFT presented in Chapter 5. Moreover, equations [4.24]–[4.27] can be generalized to apply to a porous material exhibiting N distinct porosities $\underline{\phi} = (\phi_i)_{i\in\{1,\ldots,N\}}$, each characterized by different adsorption capacities and inducing distinct confinement effects:

$$d\sigma = K(\underline{\phi})d\epsilon - \sum_{i=1}^{N} b_i(\underline{\phi}) \frac{dP_b}{1-\chi_i(\underline{\phi})} \quad [4.28]$$

$$d\phi_i = b_i(\underline{\phi})d\epsilon + \sum_{j=1}^{N} \frac{dP_b}{N_{ji}(\underline{\phi})(1-\chi_j(\underline{\phi}))} \quad [4.29]$$

$$\chi_j(\underline{\phi}) = 1 - \frac{\rho_b}{\rho_j} = \frac{n_{ex}^j}{n_{tot}^j} = \frac{n_{ex}^j}{n_{ex}^j + \frac{\rho_b m_s}{M \rho_s} \frac{\phi_j}{1-\sum_{i=1}^{N}\phi_i}} \quad [4.30]$$

$$n_{ex}^j = (<\rho_j> - \rho_b)V_j \quad [4.31]$$

In equations [4.28]–[4.31], the poromechanical properties (K, b_i, N_{ji}) can be explicitly expressed as functions of the porosities $\underline{\phi}$ and the stiffness of the solid skeleton K_s, (m_s, M, ρ_s) denote, respectively, the mass of the adsorbent, the molar mass of the fluid in question and the density of the solid, while $(<\rho_j>, V_j)$ represent, respectively, the mean densities calculated via NLDFT and the pore volumes obtained through gas porosimetry (see Chapter 5 for further details).

Once again, this approach allows for the direct determination of the deformations induced by adsorption, as well as the variations in the different porosities and poromechanical parameters, without requiring any fitting parameters, while also accounting for the material's pore size distribution.

However, to rigorously validate the performance of this type of multi-scale model that incorporates the complete pore size distribution of the material, natural materials such as coal may not be the most suitable. This is because they often exhibit diffuse pore size distributions, which can smooth out the coupling between adsorption and deformation. In this context, the dialogue between theory and experiment becomes especially meaningful, as hierarchical synthetic materials with controlled porosities offer a better means of testing the robustness of the developed models. Here, it is no longer the theory attempting to explain the experimental results, but rather the experiment and the material are specifically designed to test the validity framework of the theory.

4.5.2. *Anisotropy of the hydro-poromechanical behavior of coal*

Up until now, we have considered an isotropic model of coal. However, as discussed in section 4.2.2, the presence of fractures predominantly perpendicular to the bedding plane induces material anisotropy. As explained in Chapter 2, the aperture and spacing between fractures can vary for each fracture set. Consequently, permeability can exhibit anisotropy and must be described using a permeability tensor.

This permeability tensor itself can also depend on stress in an anisotropic manner: for instance, Espinoza et al. (2014) experimentally demonstrated that permeability perpendicular to the bedding plane was more affected by stresses within the bedding plane than by stresses perpendicular to it. This phenomenon is easily explained by the fact that the fractures are primarily oriented perpendicular to the bedding plane.

This anisotropy is also evident in the material's mechanical behavior, where different elastic moduli are observed in different directions. Wang et al. (2018) investigated the significance of accounting for the anisotropy of elastic moduli and the anisotropy of the stress dependence of permeability to accurately model changes in permeability during hydro-mechanical loading. Their findings indicated that elastic behavior could be modeled isotropically without introducing significant error in the prediction of permeability changes. However, ignoring the anisotropy in the stress dependence of permeability could result in significant errors when predicting permeability changes under certain geomechanical conditions.

Anisotropy may also manifest as anisotropic swelling in an unstressed sample (see Figure 4.12). It may be desirable to model this anisotropy, as well as to investigate its physical origin. The anisotropy in swelling could stem from either the anisotropy of stresses induced by adsorption or from the anisotropy of the coal's elasticity. To determine the cause of this anisotropy, modeling can be employed.

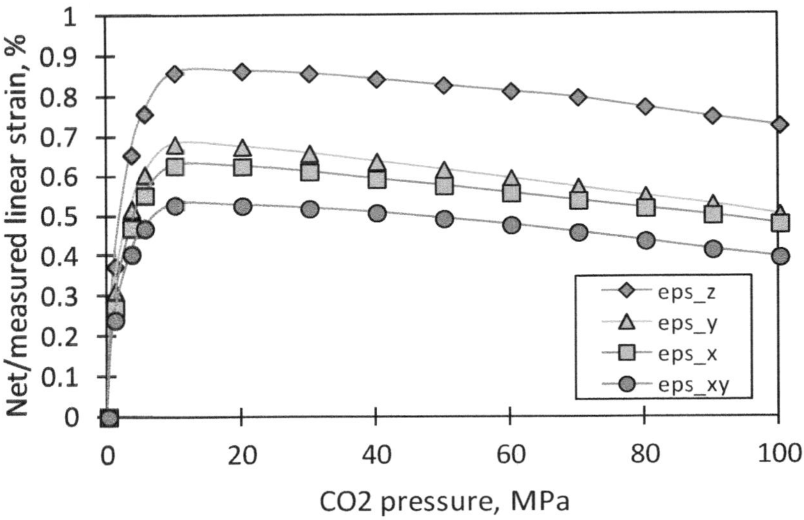

Figure 4.12. *Swelling of a coal sample immersed in CO_2 (adapted from Hol and Spiers (2012)) in three directions. The contraction observed at pressures above 10 MPa is due to the compression of the solid skeleton by the fluid*

To achieve this, using a thermodynamic approach similar to that employed in the previous sections, Espinoza et al. (2013) derived the constitutive equations for a microporous material whose elastic properties, such as the stresses induced by swelling, exhibit transverse isotropy (i.e. the material behaves isotropically within the bedding plane, but differently perpendicular to it).

Once these equations are established, the authors express the deformations of a sample immersed in a fluid at a pressure P_b as follows:

$$\begin{bmatrix} \varepsilon_1 \\ \varepsilon_3 \end{bmatrix} = \begin{bmatrix} \frac{1-\nu}{E} & -\frac{\nu_3}{E_3} \\ -2\frac{\nu_3}{E_3} & \frac{1}{E_3} \end{bmatrix} \begin{bmatrix} s_1^a(P_b) - P_b \\ s_3^a(P_b) - P_b \end{bmatrix} \qquad [4.32]$$

where ε_1 and ε_3 represent the deformations parallel and perpendicular to the bedding plane, E_1 and E_3 are the Young's moduli in these respective directions, s_1^a and s_3^a are the stresses induced by adsorption in these directions, and ν_1 and ν_3 are Poisson's ratios.

With this model incorporating two potential sources of anisotropy, how can we determine whether each source is significant for coal? A promising approach would be to measure the elastic constants E_1, E_3, ν_1 and ν_3. This would not only allow conclusions to be drawn regarding the potential anisotropy of the elastic properties but also facilitate the calculation of the stresses s_1^a and s_3^a from swelling data, thus allowing conclusions about their potential anisotropy as well.

However, such elastic measurements were not available in the work of Hol and Spiers (2012).

Nonetheless, the model shows that if the stresses induced by adsorption are isotropic, the ratio $\varepsilon_1/\varepsilon_3$ should remain constant, independent of the fluid pressure P_b. Espinoza et al. (2013) found that the data from Hol and Spiers (2012) indicate a slight decrease in the $\varepsilon_1/\varepsilon_3$ ratio for pressures below 10 MPa, leading them to conclude a slight anisotropy in the swelling stresses at those pressures. This example illustrates the value of integrating modeling with experimental data to deepen our physical understanding of these phenomena.

Of course, the modeling can be made more complex. For example, Espinoza et al. (2014) developed a poromechanical model of coal that accounts not only for the transverse isotropy discussed in this section but also for the dual aspect of porosity mentioned in the previous section. While such a model remains calibratable, its increased complexity necessitates a more elaborate experimental campaign for proper calibration.

4.5.3. *Transient effects related to transfers between fractures and the matrix*

The assumption of local thermodynamic equilibrium, which posits that the thermodynamic pressure of the fluid within the REV is uniform, is commonly made. However, this hypothesis is not always valid, particularly in laboratory experiments. For example, this is evident in the experiment shown in Figure 4.13, where a cylindrical coal sample under mechanical stress is subjected to CO_2 injection at one of its ends, while the other surfaces of the sample are sealed.

The results demonstrate that the pressure in the fractures equilibrates within approximately 15 min, but the swelling, caused by the diffusion of CO_2 from the fractures into the matrix, continues for several days. Clearly, in this case, the assumption of local thermodynamic equilibrium does not hold. How can this assumption be relaxed in modeling?

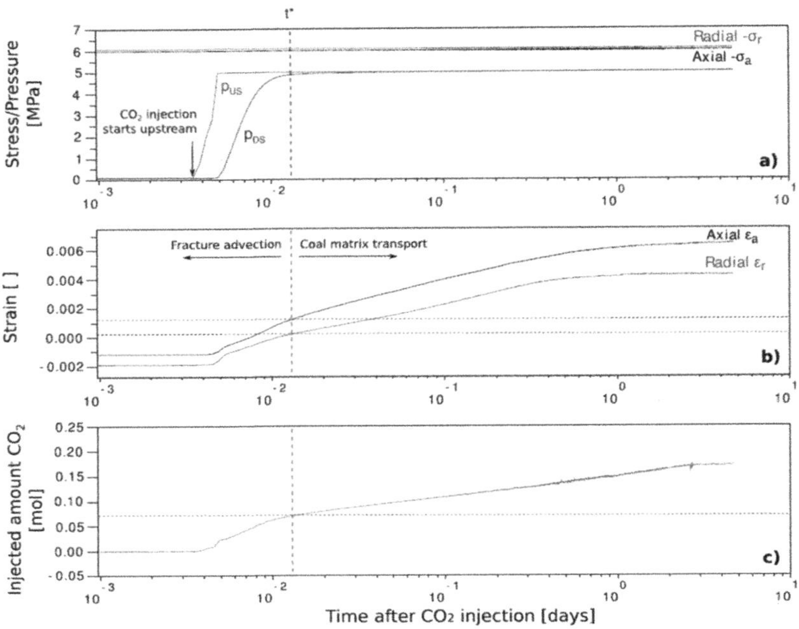

Figure 4.13. *CO_2 injection experiment in a coal sample under constant stress (adapted from Espinoza et al. (2014))*

A first approach might be to differentiate, within the REV, between the thermodynamic pressure p_c of the fluid in the fractures and p_m of the fluid in the matrix, while introducing a fluid transfer kinetics between the fractures and matrix that depends on these pressures. For example, the material flow between the two porous networks could be considered proportional to the pressure difference $p_c - p_m$. This basic principle underpins the double-porosity transport model proposed by Barenblatt et al. (1960). The mechanical consequences of this pressure difference between the two porosities can be captured using the equations derived in section 4.5.1 (see equations [4.22] and [4.23]), which account for a distinct pressure for each porous network. Such a model was employed by Vandamme et al. (2014), as detailed in Chapter 6.

However, even a model of this complexity may not be sufficient to fully capture certain effects of fluid transfer within the REV. Indeed, several studies have demonstrated that these transfers can lead to non-monotonic permeability changes over time (Liu et al. 2011; Qu et al. 2014) and have suggested that some laboratory permeability measurements reported in the literature may have been taken before a steady state was achieved. The principle of these non-monotonic permeability changes is illustrated in Figure 4.14. The injected fluid initially enters the fractures, after which it diffuses from the fractures into the matrix. During this diffusion process, the matrix adjacent to the fracture swells first, leading to fracture closure. Over longer times, as the fluid diffuses throughout the matrix, the entire matrix swells: if the sample is unconstrained, this uniform swelling results in homothetic expansion of the material, leading to fracture reopening and a consequent increase in permeability.

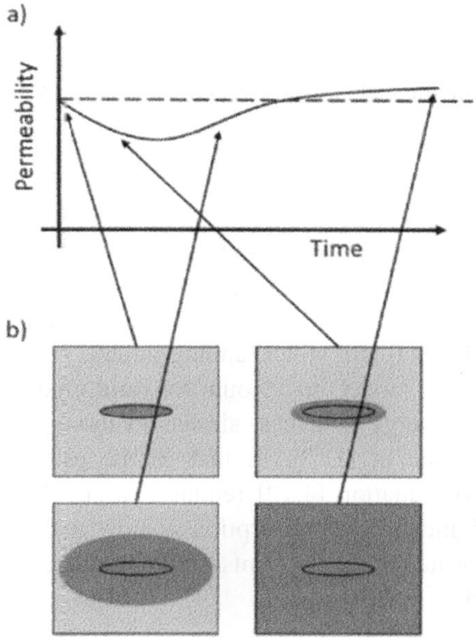

Figure 4.14. *Principle that may lead to non-monotonic changes in permeability over time (adapted from Abouloifa et al. (2021)). The injected fluid is represented in blue, and the ellipsoid indicates a fracture*

Such non-monotonic permeability changes cannot be captured using a model in the spirit of Barenblatt et al. (1960), because that model assumes uniform fluid pressure within the matrix. By combining equations [4.22] and [4.23], it can be demonstrated that for a sample free of external stresses, the porosity Φ_c associated with fractures is an increasing function of both pressures p_c and p_m. Therefore, regardless of the transfer kinetics between fractures and matrix, such a model predicts only a monotonic increase in this porosity, and consequently, in permeability.

To capture the non-monotonic permeability changes illustrated in Figure 4.14, a model must account for the heterogeneity of fluid pressure within the matrix during the transient phase. It is precisely by acknowledging this pressure heterogeneity that the models proposed by Liu et al. (2011), Wei et al. (2019) and Abouloifa et al. (2021) are able to describe these complex changes. Abouloifa et al. (2021) demonstrate that permeability changes are, in fact, dependent on the entire pressure history within the fractures over time. Accurate modeling requires the use of Boltzmann's superposition principle, which adds significant complexity to the model. Further details on this approach are discussed in Chapter 6.

A multi-scale approach, such as the one mentioned in Chapter 7, may also be employed, although it introduces substantial numerical complexity.

4.5.4. *Fluid mixture*

In the models discussed in previous sections, we assumed that the porosity of coal is occupied by a pure fluid. However, as we know (see Chapter 1 and section 4.2.1), in practice, the porosity of coal is typically occupied by a fluid mixture, which may include CH_4, CO_2 and water, among others. This raises the question of how to extend the previous models to account for fluid mixtures. A thermodynamic approach based on free energy balances, similar to those used in earlier sections, demonstrates (Nikoosokhan et al. 2014) that in the presence of a binary fluid mixture, the constitutive equation [4.19] remains valid. The difference lies in the fact that the stress s^a induced by adsorption is now a function of the chemical potentials, μ_1 and μ_2, of the two components of the mixture, rather than the chemical potential of a pure fluid:

$$s^a = s^a(\mu_1, \mu_2)$$

For co-adsorption that is appropriately modeled by a Langmuir binary adsorption model, Corrente et al. (2021) provide the equations needed to calculate the induced stress using sorption and swelling data measured for pure fluids.

The approaches described in sections 4.4.1 and 4.5.1 can also be extended to fluid mixtures. The primary input for the model is the amount of excess adsorbed material, which can be determined experimentally. This requires the addition of a recirculation pump and an infrared analyzer to the experimental setup used for simultaneous adsorption-deformation measurements, as proposed by Perrier et al. (2017). By analyzing the composition of the surrounding supernatant fluid, it is possible to infer the material's selectivity and the adsorbed quantities of each gas present in the mixture. By accounting for the effects of confinement on partial pressures [4.24]–[4.27], the deformations induced by the adsorption of the fluid mixture can be determined.

Similarly, equations [4.28]–[4.31] can be extended to the case of fluid mixtures, since the adsorbed quantities can be directly obtained through the NLDFT.

4.6. Limitations and perspectives of the models

4.6.1. *Experimental limitations: complexity in interpreting experiments in light of models*

Once a model is defined, its calibration requires experimental data, either from laboratory measurements or field observations. In this section, we focus on laboratory tests and the challenges related to determining specific properties, particularly permeability (discussed in Chapter 2) and adsorption properties (discussed in Chapter 3). We deliberately omit issues of sample representativeness, size effects, scale effects and so on.

To measure the permeability of a coal specimen in the laboratory (see Chapter 2), for example, in a triaxial cell to simulate stress conditions representative of in situ states, a pressure gradient must be applied along the sample to generate gas flow. Although the sample is subjected to a homogeneous total stress, it experiences a heterogeneous effective stress, and similarly, the thermodynamic pressure of the injected fluid is also heterogeneous. This pressure heterogeneity induces a non-uniform swelling of the coal matrix, leading to an uneven opening of the fractures in the sample. As a result, the local permeability along the specimen is heterogeneous, which in turn produces a nonlinear pressure gradient. However, these heterogeneities are not directly measurable in practice, and permeability determination is thus based on external parameters imposed or measured from outside the specimen. Another challenge in permeability measurement lies in the fluid mass exchanges between the fracture network and the coal matrix. These diffusive exchanges are slow but can influence permeability measurements, which are primarily dominated by advective transport through the fractures (see section 2.4.2.1). Moreover, these exchanges contribute to delaying the establishment of a steady-state flow of the injected fluid.

The various methods for characterizing the sorption properties of coal have been detailed in Chapter 3 (section 3.3). Sorption measurements are typically performed by immersing the material in the adsorbate, where the material is mechanically stressed only by the fluid pressure, and sorption is measured under zero effective stress. However, in subsurface coal seams, effective stresses are generally not zero. As we have demonstrated in this chapter, there is a coupling between sorption and mechanical stresses or deformations. Therefore, at a given fluid pressure, the quantities of gas adsorbed in a reservoir may differ from those measured in the laboratory.

Hol et al. (2011) experimentally showed that applying compressive stress to recompacted ground coal samples results in the release of some of the CO_2 present in the sample. They estimate that the sorption capacity of coal in situ could be 5–50% lower than what is measured in the laboratory. This reduction, due to the mechanical constraint on the coal in situ, can be modeled using the approaches presented in this chapter, which account for the coupling between adsorption and deformation/stress. Thus, a dialogue between experimental data and models can be established. For instance, models can be used to refine the interpretation of experimental results, as demonstrated by Espinoza et al. (2016), who utilized the dual-porosity models presented in section 5.1. They showed that, in the experiments conducted by Hol et al. (2011), only 10–15% of the released CO_2 originated from the coal grains, with the remainder coming from the pore spaces between the grains.

4.6.2. *Model perspectives*

In this chapter, we have sought to present coal modeling at the material scale, focusing on the underlying philosophy of the chosen approaches and the decisions faced by the modeler.

In section 4.5, we demonstrated that models can be enhanced to capture a wider range of phenomena. From the examples described, it becomes clear that this enhancement often leads to increased model complexity. This added complexity may necessitate a more extensive experimental program for model calibration and may also incur greater computational costs, particularly when these equations are integrated into finite element calculations (see Chapter 6). Thus, a balance must be struck between the necessary model complexity and the intrinsic complexity of the material and the physical phenomena under consideration.

Through the examples provided in sections 4.5.1.2, 4.5.2 and 4.6.1, we hope to have highlighted the importance of fostering a dialogue between modeling and experimentation. Beyond this specific instance, such dialogue is generally essential. Experimental work helps define the limits of a model and identifies areas for improvement. In turn, modeling can guide the design of experimental campaigns and is often crucial for properly analyzing experimental results.

We have seen that the models presented can be complexified indefinitely. Among the necessary areas for further development, we identify the need to account for material nonlinearities, which is essential for analyzing certain low-stress laboratory experiments[10] where nonlinearities are confirmed (see Figure 4.2). Additionally, there is a need to model the poromechanical behavior of coal in the presence of mixtures more complex than binary, particularly when water is present. However, studying ternary mixtures introduces significant experimental challenges, as well as the burden of extensive experimental programs (or molecular simulations, if one seeks to explore the adsorption of ternary mixtures using the methods discussed in Chapter 5).

The poromechanical modeling approaches detailed here, though presented for coal, are applicable to a broader range of materials. They are relevant to all geomaterials (rocks, soils), as well as to cementitious materials, wood and other porous materials. Notably, water sorption phenomena and the resulting deformations are significant in cementitious and clay-based materials, as well as in wood. These adsorption-induced deformations are of interest for a range of applications that extend well beyond coal geomechanics; for example, they are used in the design of sensors and actuators (Gor et al. 2017). Furthermore, as models become more complex, it may be necessary to synthesize specific model materials with controlled porosity to better understand their limitations.

One of the challenges in applying material-scale modeling to practical scenarios lies in scaling up to the reservoir level. Laboratory samples are often too small to represent the rock at the reservoir scale. While the form of the equations may remain valid during this scaling, the material properties used in these equations must change between the laboratory scale and the reservoir rock scale. Field data remains the reference point – provided such data are available.

Another difficulty in scaling arises from the intrinsic heterogeneity of coal. To capture this heterogeneity and predict macroscopic properties from microscopic ones, both analytical (Dormieux et al. 2006) and numerical homogenization methods are available. Numerical methods for performing scale transitions or conducting reservoir-scale simulations of coal are addressed in Chapters 7 and 6, respectively. Chapter 5 focuses on numerical methods for simulating matter at the finest scales, namely, molecular-scale simulations, while connecting these approaches with the material-scale models discussed in this chapter.

10 We are thinking, for example, of the isochoric injection experiments by Espinoza et al. (2014).

4.7. References

Abouloifa, N., Vandamme, M., Dangla, P. (2021). Modeling transient variations of permeability in coal seams at the reservoir scale. *Journal of Natural Gas Science and Engineering*, 88, 103796. doi: 10.1016/j.jngse.2021.103796.

Bangham, D.H. and Razouk, R.I. (1938). The swelling of charcoal. Part V. The saturation and immersion expansions and the heat of wetting. *Proceedings of the Royal Society A: Mathematical, Physical and Engineering Sciences*, 166, 572–586. doi: 10.1098/rspa.1938.0112.

Barenblatt, G.I., Zheltov, I.P., Kochina, I.N. (1960). Basic concepts in the theory of seepage of homogeneous liquids in fissured rocks [strata]. *Journal of Applied Mathematics and Mechanics*, 24, 1286–1303.

Bustin, R.M. (2004). Acid Gas Sorption by British Columbia Coals: Implications for Permanent Disposal of Acid Gas in Deep Coal Seams and Possible Co-production of Methane. Technical Report, The University of British Columbia [Online]. Available at: https://bcogris.ca/documents/scek/Final_Reports/d-ET-Com-UBC-2000-16-Rep.pdf.

Corrente, N.J., Zaręba, K., Neimark, A.V. (2021). Deformation of nanoporous materials in the process of binary adsorption: Methane displacement by carbon dioxide from coal. *The Journal of Physical Chemistry C*, 125(38), 21310–21316.

Coussy, O. (2004). *Poromechanics*. John Wiley & Sons, New York. doi: 10.1002/0470092718.

Coussy, O. (2010). *Mechanics and Physics of Porous Solids*. John Wiley & Sons, New York.

Culberson, O.L. and McKetta, J.J. (1951). Phase equilibria in hydrocarbon-water systems III – The solubility of methane in water at pressures to 10,000 PSIA. *Journal of Petroleum Technology*, 3(08), 223–226.

Czapliński, A. and Holda, S. (1982). Changes in mechanical properties of coal due to sorption of carbon dioxide vapour. *Fuel*, 61, 1281–1282.

Darcy, H. (1856). *Les fontaines publiques de la ville de Dijon*. Victor Dalmont, Paris.

Day, S., Fry, R., Sakurovs, R. (2008). Swelling of Australian coals in supercritical CO_2. *International Journal of Coal Geology*, 74, 41–52. doi: 10.1016/j.coal.2007.09.006.

Donnelly, H.G., Katz, D.L. (1954). Phase equilibria in the carbon dioxide–methane system. *Ind. Eng. Chem.*, 46, 511–517. doi: 10.1021/ie50531a036.

Dormieux, L., Kondo, D., Ulm, F.-J. (2006). *Microporomechanics*. John Wiley & Sons, New York.

Enick, R.M. and Klara, S.M. (1990). CO_2 solubility in water and brine under reservoir conditions. *Chemical Engineering Communications*, 90, 23–33.

Espinoza, D.N., Vandamme, M., Dangla, P., Pereira, J.-M., Vidal-Gilbert, S. (2013). A transverse isotropic model for microporous solids: Application to coal matrix adsorption and swelling. *Journal of Geophysical Research: Solid Earth*, 118, 6113–6123. doi: 10.1002/2013JB010337.

Espinoza, D.N., Vandamme, M., Pereira, J.-M., Dangla, P., Vidal-Gilbert, S. (2014). Measurement and modeling of adsorptive–poromechanical properties of bituminous coal cores exposed to CO_2: Adsorption, swelling strains, swelling stresses and impact on fracture permeability. *International Journal of Coal Geology*, 134–135, 80–95. doi: 10.1016/j.coal.2014.09.010.

Espinoza, D.N., Pereira, J.-M., Vandamme, M., Dangla, P., Vidal-Gilbert, S. (2015). Desorption-induced shear failure of coal bed seams during gas depletion. *International Journal of Coal Geology*, 137, 142–151. doi: 10.1016/j.coal.2014.10.016.

Espinoza, D.N., Vandamme, M., Dangla, P., Pereira, J.-M., Vidal-Gilbert, S. (2016). Adsorptive-mechanical properties of reconstituted granular coal: Experimental characterization and poromechanical modeling. *International Journal of Coal Geology*, 162, 158–168. doi: 10.1016/j.coal.2016.06.003.

Gao, W., Zagorščak, R., Thomas, H.R. (2022). Insights into ground response during underground coal gasification through thermo-mechanical modeling. *International Journal for Numerical and Analytical Methods in Geomechanics,* 46, 3–22. doi: 10.1002/nag.3287.

Gibbs, J.W. (1928). *The Collected Works of J. Willard Gibbs – Volume I: Thermodynamics.* Longmans, Green and Co., New York.

Gor, G.Y. and Bernstein, N. (2016). Revisiting Bangham's law of adsorption-induced deformation: Changes of surface energy and surface stress. *Phys. Chem. Chem. Phys.*, 18, 9788–9798. doi: 10.1039/C6CP00051G.

Gor, G.Y., Huber, P., Bernstein, N. (2017). Adsorption-induced deformation of nanoporous materials–A review. *Applied Physics Reviews*, 4, 011303. doi: 10.1063/1.4975001.

Grégoire, D., Malheiro, C., Miqueu, C. (2018). Estimation of adsorption-induced pore pressure and confinement in a nanoscopic slit pore by a density functional theory. *Continuum Mechanics and Thermodynamics,* 30, 347–363. doi: 10.1007/s00161-017-0602-x.

Grosman, A. and Ortega, C. (2008). Influence of elastic deformation of porous materials in adsorption-desorption process: A thermodynamic approach. *Physical Review B*, 78, 1–12. doi: 10.1103/PhysRevB.78.085433.

Hagin, P. and Zoback, M. (2010). Laboratory studies of the compressibility and permeability of low-rank coal samples from the Powder River basin, Wyoming, USA. In *44th US Rock Mechanics Symposium and 5th US-Canada Rock Mechanics Symposium*. ARMA, Salt Lake City, UT.

Hall, W.A. (1956). An analytical derivation of the Darcy equation. *Transactions American Geophysical Union*, 37, 185–188.

Hoek, E. and Brown, E.T. (1980). Empirical strength criterion for rock masses. *Journal of the Geotechnical Engineering Division*, 106, 1013–1035. doi: 10.1061/AJGEB6.0001029.

Hol, S. and Spiers, C.J. (2012). Competition between adsorption-induced swelling and elastic compression of coal at CO_2 pressures up to 100 MPa. *Journal of the Mechanics and Physics of Solids*, 60, 1862–1882. doi: 10.1016/j.jmps.2012.06.012.

Hol, S., Peach, C.J., Spiers, C.J. (2011). Applied stress reduces the CO_2 sorption capacity of coal. *International Journal of Coal Geology* 85, 128–142. doi: 10.1016/j.coal.2010.10.010.

Hol, S., Gensterblum, Y., Massarotto, P. (2014). Sorption and changes in bulk modulus of coal – Experimental evidence and governing mechanisms for CBM and ECBM applications. *International Journal of Coal Geology*, 128–129, 119–133. doi: 10.1016/j.coal.2014.04.010.

Hubbert, M.K. (1956). Darcy's law and the field equations of the flow of underground fluids. *Transactions of the AIME*, 207, 222–239.

Lemmon, E.W., Bell, I.H., Huber, M.L., McLinden, M.O. (2022). Thermophysical properties of fluid systems. In *NIST Standard Reference Database 69: NIST Chemistry WebBook*, Linstrom, P.J. and Mallard, W.G. (eds). National Institute of Standards and Technology, Gaithersburg, MD.

Levine, J.R. (1996). Model study of the influence of matrix shrinkage on absolute permeability of coal bed reservoirs. *Geological Society, London, Special Publications*, 109, 197–212. doi: 10.1144/GSL.SP.1996.109.01.14.

Li, Y.-K. and Nghiem, L.X. (1986). Phase equilibria of oil, gas and water/brine mixtures from a cubic equation of state and Henry's law. *The Canadian Journal of Chemical Engineering* 64, 486–496. doi: 10.1002/cjce.5450640319.

Liu, J., Wang, J., Chen, Z., Wang, S., Elsworth, D., Jiang, Y. (2011). Impact of transition from local swelling to macro swelling on the evolution of coal permeability. *International Journal of Coal Geology*, 88, 31–40. doi: 10.1016/j.coal.2011.07.008.

Liu, J., Spiers, C.J., Peach, C.J., Vidal-Gilbert, S. (2016). Effect of lithostatic stress on methane sorption by coal: Theory vs. experiment and implications for predicting in-situ coalbed methane content. *International Journal of Coal Geology*, 167, 48–64. doi: 10.1016/j.coal.2016.07.012.

Morcote, A., Mavko, G., Prasad, M. (2010). Dynamic elastic properties of coal. *Geophysics*, 75, E227–E234. doi: 10.1190/1.3508874.

Nikoosokhan, S., Vandamme, M., Dangla, P. (2012). A poromechanical model for coal seams injected with carbon dioxide: From an isotherm of adsorption to a swelling of the reservoir. *Oil & Gas Science and Technology – Revue d'IFP Energies nouvelles*, 67, 777–786. doi: 10.2516/ogst/2012048.

Nikoosokhan, S., Vandamme, M., Dangla, P. (2014). A poromechanical model for coal seams saturated with binary mixtures of CH_4 and CO_2. *Journal of the Mechanics and Physics of Solids*, 71, 97–111. doi: 10.1016/j.jmps.2014.07.002.

Ottiger, S., Pini, R., Storti, G., Mazzotti, M. (2008). Competitive adsorption equilibria of CO_2 and CH_4 on a dry coal. *Adsorption*, 14, 539–556. doi: 10.1007/s10450-008-9114-0.

Perrier, L. (2015). Couplage entre adsorption et déformation en milieux microporeux. PhD Thesis, Université de Pau et des Pays de l'Adour.

Perrier, L., Pijaudier-Cabot, G., Grégoire, D. (2015). Poromechanics of adsorption-induced swelling in microporous materials: A new poromechanical model taking into account strain effects on adsorption. *Continuum Mechanics and Thermodynamics*, 27, 195–209. doi: 10.1007/s00161-014-0359-4.

Perrier, L., Plantier, F., Grégoire, D. (2017). A novel experimental setup for simultaneous adsorption and induced deformation measurements in microporous materials. *Review of Scientific Instruments*, 88, 035104. doi: 10.1063/1.4977595.

Perrier, L., Pijaudier-Cabot, G., Grégoire, D. (2018). Extended poromechanics for adsorption-induced swelling prediction in double porosity media: Modeling and experimental validation on activated carbon. *International Journal of Solids and Structures*, 146, 192–202. doi: 10.1016/j.ijsolstr.2018.03.029.

Qu, H., Liu, J., Pan, Z., Connell, L. (2014). Impact of matrix swelling area propagation on the evolution of coal permeability under coupled multiple processes. *Journal of Natural Gas Science and Engineering*, 18, 451–466. doi: 10.1016/j.jngse.2014.04.007.

Rice, D.D. (1993). Composition and origins of coalbed gas. In *AAPG Studies in Geology No. 38, Hydrocarbons from Coal*, Law, B.E. and Rice, D.D. (eds). The American Association of Petroleum Geologists, Tulsa, OK.

Sakurovs, R., Day, S., Weir, S., Duffy, G. (2008). Temperature dependence of sorption of gases by coals and charcoals. *International Journal of Coal Geology* 73, 250–258.

Salençon, J. (2005). *Mécanique des milieux continus. Tome I – Concepts généraux*. Les Éditions de l'École Polytechnique, Paris.

Sander, R. (2015). Compilation of Henry's law constants (version 4.0) for water as solvent. *Atmospheric Chemistry and Physics*, 15(8), 4399–4981. doi: 10.5194/acp-15-4399-2015.

Shuttleworth, R. (1950). The surface tension of solids. *Proceedings of the Physical Society. Section A*, 63, 444–457.

Su, F., Hamanaka, A., Itakura, K., Deguchi, G., Sato, K., Kodama, J. (2017). Evaluation of coal combustion zone and gas energy recovery for underground coal gasification (UCG) process. *Energy Fuels,* 31, 154–169. doi: 10.1021/acs.energyfuels.6b01922.

Thomsen, L. (1986). Weak elastic anisotropy. *Geophysics*, 51, 1954–1966. doi: 10.1190/1.1442051.

Vandamme, M., Brochard, L., Lecampion, B., Coussy, O. (2010). Adsorption and strain: The CO_2-induced swelling of coal. *Journal of the Mechanics and Physics of Solids*, 58, 1489–1505. doi: 10.1016/j.jmps.2010.07.014.

Vandamme, M., Patrick, D., Nikoosokhan, S., Brochard, L. (2014). Modeling the poromechanical behavior of microporous and mesoporous solids: Application to coal. In *Nonlinear Elasticity and Hysteresis*, Kim, A.H. and Guyer, R.A. (eds). Wiley-VCH Verlag, Weinheim. doi: 10.1002/9783527665068.ch5.

Vermorel, R. and Pijaudier-Cabot, G. (2014). Enhanced continuum poromechanics to account for adsorption induced swelling of saturated isotropic microporous materials. *European Journal of Mechanics-A/Solids*, 44, 148–156. doi: 10.1016/j.euromechsol.2013.10.010.

Wang, L.L., Vandamme, M., Pereira, J.-M., Dangla, P., Espinoza, D.N. (2018). Permeability changes in coal seams: The role of anisotropy. *International Journal of Coal Geology*, 199, 52–64. doi: 10.1016/j.coal.2018.09.014.

Wei, M., Liu, J., Elsworth, D., Li, S., Zhou, F. (2019). Influence of gas adsorption induced non-uniform deformation on the evolution of coal permeability. *International Journal of Rock Mechanics and Mining Sciences*, 114, 71–78. doi: 10.1016/j.ijrmms.2018.12.021.

White, C.M., Smith, D.H., Jones, K.L., Goodman, A.L., Jikich, S.A., LaCount, R.B., DuBose, S.B., Ozdemir, E., Morsi, B.I., Schroeder, K.T. (2005). Sequestration of carbon dioxide in coal with enhanced coalbed methane recovery a review. *Energy & Fuels*, 19, 659–724.

Wiebe, R. and Gaddy, V.L. (1940). The solubility of carbon dioxide in water at various temperatures from 12 to 40° and at pressures to 500 atmospheres. Critical phenomena. *J. Am. Chem. Soc.*, 62, 815–817. doi: 10.1021/ja01861a033.

Yan, Z., Wang, K., Zang, J., Wang, C., Liu, A. (2019). Anisotropic coal permeability and its stress sensitivity. *International Journal of Mining Science and Technology*, 29, 507–511.

5

Molecular Scale Modeling of Coal and Its Behavior

Laurent BROCHARD[1], Christelle MIQUEU[2] and David GRÉGOIRE[2,3]

[1] *Navier, ENPC, CNRS, Institut Polytechnique de Paris, École nationale des ponts et chaussées, Université Gustave Eiffel, Marne-la-Vallée, France*

[2]*LFCR, E2S UPPA, CNRS, Université de Pau et des Pays de l'Adour, Anglet, France*

[3]*Institut Universitaire de France, Paris, France*

5.1. Introduction

Coals are organic geomaterials derived from the transformation of biomass over geological time (see Chapter 1). They belong to the family of source rocks, which consist of rocks that contain a bio-sourced fraction. At depth, due to the effects of pressure and temperature, source rocks undergo maturation, which modifies their chemical composition and structure. Biomass can be approximated as a semi-crystalline polymer analogous to cellulose ($(C_6H_{10}O_5)_n$). As maturation progresses, the organic matter in the source rocks releases water (H_2O), carbon dioxide (CO_2) and hydrocarbons (e.g. CH_4), resulting in a shift in elemental composition toward a predominantly carbon-based and more crosslinked molecular structure. The final stage of maturation is graphite, where the structure contains only sp^2-hybridized carbon. This transformation in composition (C/H/O) can be visualized through a Van Krevelen diagram (Figure 5.1). In this diagram, coals are classified as mature rocks. Specifically, a distinction is made between lignites with moderate maturation (55%–75% C), highly mature bituminous coal (75%–90% C)

For a color version of all of the figures in this chapter, see www.iste.co.uk/vandamme/coal.zip.

Geomechanics of Coal Seams,
coordinated by Matthieu VANDAMME and Jean-Michel PEREIRA.

and very mature anthracites (>90% C). This classification based on carbon content is relevant because many properties, notably calorific value, are correlated with it.

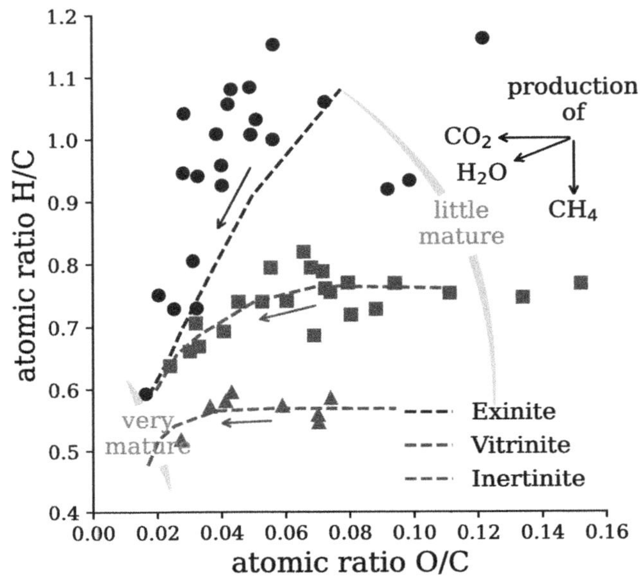

Figure 5.1. *Van Krevelen diagram (adapted from Meyers (1982, Chapter 5))*

The molecules released during maturation primarily migrate out of the source rock and eventually accumulate beneath an impermeable geological layer, thereby forming conventional hydrocarbon reservoirs. This migration of maturation products causes the rock to contract. In coal, this contraction leads to the formation of a network of natural fractures, which plays a crucial role in reservoir permeability (see Chapter 2).

However, a fraction of the maturation products may remain trapped in the source rock if it has low permeability or exhibits a particular *adsorption affinity* for the released molecules. Adsorption is defined as the increase in fluid density at the interface with a solid due to attractive molecular interactions between the fluid and solid molecules (Figure 5.2).

In the case of coal, light hydrocarbons, primarily methane (CH_4), have a strong affinity with the rock and are therefore often present in an adsorbed form. The release of this methane during mine excavation is well known, as it is the cause of the infamous firedamp explosions. This adsorbed gas, also referred to as *coalbed methane* (or commonly coal gas), represents an unconventional hydrocarbon resource.

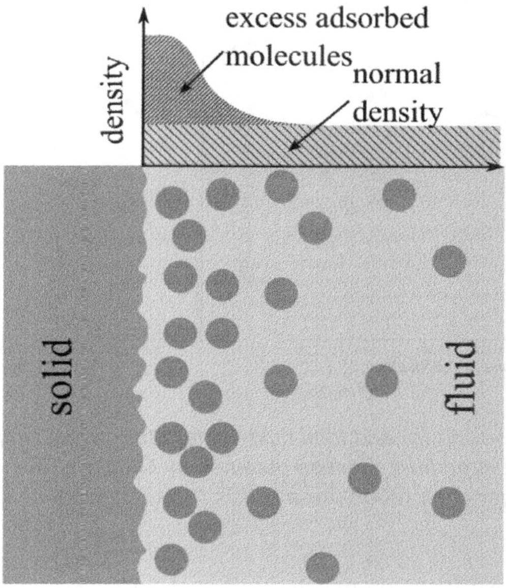

Figure 5.2. *The adsorption phenomenon refers to the accumulation of fluid on the surface of a solid due to favorable atomic interactions between the fluid and the solid*

The significant adsorption of methane in coal is due to its very large *specific surface area*, which exceeds 100 m^2/g. As discussed in section 5.2, this specific surface area results from the porous nanostructure of coal, with pore sizes on the order of a few gas molecules. In its adsorbed form, a fluid exhibits properties that differ significantly from its ordinary state. For example, the transition from a pure to an adsorbed state is an exothermic process that induces temperature changes.

From a mechanical perspective, adsorption on a surface induces *surface stresses*, while adsorption in micropores (< 2 nm) generates *disjoining pressures*. These pore-scale effects manifest at the macroscopic level as abnormal thermo-poro-mechanical couplings, which will be further discussed in section 5.5.

Consequently, in stress-free coals, significant swelling (resulting in volumetric deformations of several percent) is observed following adsorption. This swelling depends on the type of gas and is evidently correlated with adsorption affinity: the higher the affinity, the greater the swelling (Figure 5.3). Under geological conditions, coal is not free to swell, and the abnormal poro-mechanical couplings generate significant stresses that play a crucial role at the reservoir scale, affecting permeability and potentially causing rock damage (Chapters 2 and 4).

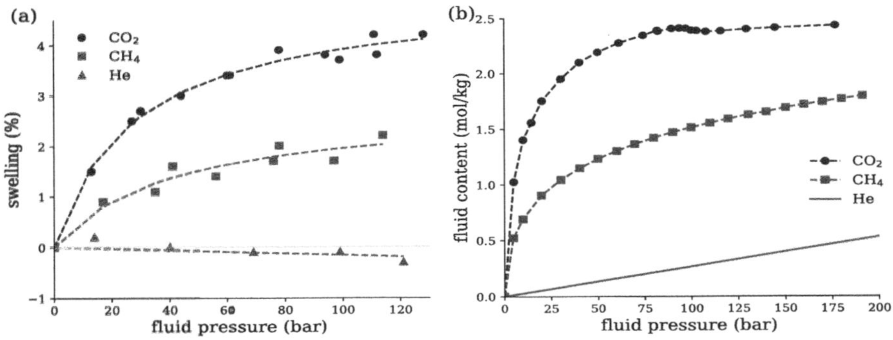

Figure 5.3. *Swelling (a) and total fluid content (b) of coal simply immersed in a pressurized fluid. Results obtained by Ottiger et al. (2006) and Ottiger et al. (2008) for a Sulcis (Sardinia) coal at 45°C*

In this chapter, we focus on the study of coal through *molecular modeling and simulation*. The molecular scale is the appropriate level for gaining a detailed understanding of adsorption mechanisms within nanopores. Experimental investigation of mechanics at this scale is highly limited, and molecular simulation offers a complementary approach to experiments, relying on elementary physical interactions between atoms and molecules. However, it is important to note that we are concerned here with so-called classical simulations, which are based on *empirical interaction potentials* between atoms, as opposed to ab initio simulations. The latter are grounded in solving the Schrödinger equation, that is, based solely on first principles. The size of the systems being studied places them beyond the reach of ab initio simulations. However, such simulations remain valuable in coal research, as they help calibrate the empirical interaction potentials used in classical simulations. Classical simulations are, in turn, limited to studying systems no larger than a few tens of nanometers.

Therefore, the contribution of molecular simulations to coal geomechanics is confined to investigating the microporous matrix (pores < 2 nm), potentially including mesopores (pores > 2 nm and < 50 nm). The scale of a representative elementary volume of rock, with its mineral heterogeneities or potential fissures, is beyond the capability of molecular simulation methods. Thus, molecular simulation is particularly useful for phenomena that originate at the scales accessible to these methods, namely, adsorption.

At the molecular scale, atoms are not static but undergo dynamic changes due to thermal motion. The distribution of atomic states follows *Boltzmann statistics*, and molecular simulations aim to sample atomic configurations according to these statistics to estimate average properties. On a macroscopic scale, fluctuations become negligible and these averages dominate. The three key components of a molecular simulation are (1) the atomic structure of the system being studied, (2) the potential for interaction between the atoms within the system and (3) the method used to sample atomic configurations (atomistic simulation) or establish the thermodynamic state (modeling using density functional theory (DFT)). Section 5.2 introduces atomic structure models and interaction potentials for coals. The choice of simulation method depends on the phenomenon being studied: in this case, adsorption. In section 5.3, we present the study of adsorption in coals using grand canonical Monte Carlo (GCMC) molecular simulations. Section 5.4 is dedicated to modeling adsorption using another thermodynamic formalism introduced for inhomogeneous fluids: DFT. Beyond adsorption, molecular approaches can also be employed to explore abnormal thermo-poro-mechanical couplings induced by adsorption, which will be the focus of section 5.5. Finally, in section 5.6, we summarize the contributions of molecular simulations to coal geomechanics, highlight the connections with larger scales and discuss other areas of interest in coal research where molecular simulation can provide valuable insights.

5.2. Nature of coal at the molecular scale

Understanding the atomic structure of coal is essential for its study through molecular simulation. Since coal is not crystalline, its atomic structure is not precisely known, except in the limiting case of graphite. Furthermore, there is no single atomic structure, as the composition evolves with maturation. Various atomic structure models have been proposed in the literature, aiming to represent the experimentally known characteristics of coals. This process is known as *reconstruction*. The most advanced models also incorporate interaction energies between atoms to ensure structures that are both experimentally accurate and stable from an energetic and thermodynamic perspective. Below, we summarize key elements for constructing molecular models of coal. For a more detailed explanation, the interested reader may consult the dedicated literature, such as Meyers (1982).

– *Elemental composition*: the analysis of the chemical composition of coal differentiates between crystalline minerals from the organic phase (ranging from a few percent to 50% by mass), whose components are called *macerals* (Vassilev et al. 1996, 1997). From an adsorption perspective, the majority of the specific surface area comes from the organic phase, so molecular simulations focus solely on this phase. The organic phase itself is subdivided into a volatile part (water, hydrocarbons), which desorbs upon heating, and a nonvolatile part, which remains fixed. Ultimate combustion analysis allows for the determination of the complete atomic composition of the organic phase (Table 5.1). Coal is composed primarily of carbon, oxygen and hydrogen, with trace amounts of sulfur and nitrogen. Coal maturity primarily influences the volatile and nonvolatile proportions (from 50–50% to 20–80%) as well as the carbon (from 60% to 95%) and oxygen (from 25% to 5%) contents. The hydrogen content remains around 5%.

– *Nanostructure*: unlike the mineral component, the organic matrix of coal is noncrystalline, meaning its atomic arrangement is not precisely known. However, while the diffraction spectrum does not exhibit a sharp peak, oscillations are present, indicating a moderate degree of order, primarily due to the presence of stacked polyaromatic sheets within the structure, at the scale of a few nanometers (Meyers 1982). These aromatic sheets can be directly observed using high-resolution electron microscopy (Fernandez-Alos et al. 2011). Chemically, the presence of *aromaticity* (the proportion of aromatic carbon) correlates with higher carbon content, suggesting a link with coal maturity. Observations confirm that both aromaticity and local structural order increase as coal matures. Another important feature of the organic nanostructure of coal is its significant *nanoporosity*, which leads to a particularly large specific surface area. Typically, the specific surface area ranges from 100 to 300 m^2/g, with the pore size distribution (PSD) showing a predominance of micropores (< 2 nm), except in lower maturity coals (Meyers 1982).

– *Chemistry*: beyond the aromaticity of carbon, the chemical structure of coal also includes *functional groups* associated with heteroatoms, primarily oxygen. Due to their polarity and lower bond energies, these functional groups play a key role in coal reactivity. Additionally, they contribute to the cohesion of the carbon structure through electrostatic interactions and, in the case of linking functional groups, covalent bonds. Oxygen is mainly found in the form of hydroxyl (phenols, alcohols), carbonyl, or carboxyl terminations, as ether or ester bridges between carbon chains, or as substitutions within polyaromatic sheets (Levine et al. 1982). The abundance of different functional groups, as determined by infrared spectroscopy and selective reactivity analysis, reveals that hydroxyls and ethers are the most common, with correlations established between these groups and the coal's elemental composition (Meyers 1982).

	Lignite	Semi-bituminous coal	Bituminous coal/anthracite
Volatile (% by mass)	50	40	20
Nonvolatile (% by mass)	50	60	80
C (% by mass)	60–75	75–85	85–95
O (% by mass)	25	15	5
H (% by mass)	5	5	4
N (% by mass)	1	1	1
S (% by mass)	1	1	1

Table 5.1. *Approximate elemental composition of the organic phase of coals (Vassilev et al. 1996, 1997)*

The reconstruction of molecular models of coal has been extensively studied in the literature. For a comprehensive review, readers may refer to Mathews and Chaffee (2012). In an initial approximation, graphite can be considered as a model for highly mature coal. The sheet-like structure of graphite allows for adsorption in the inter-sheet spaces. When modeling van der Waals molecular interactions using a Lennard–Jones potential (with the parameters σ for characteristic distance and ε for characteristic energy), and assuming a continuous, homogeneous sheet with a numerical density per unit area ρ, the interaction energy between a fluid molecule located at a distance z from the sheet and the entire sheet (per unit thickness) can be analytically determined by integration:

$$U(z) = 2\pi\varepsilon\rho\sigma^2\left[\frac{2}{5}\left(\frac{\sigma}{z}\right)^{10} - \left(\frac{\sigma}{z}\right)^{4}\right] \qquad [5.1]$$

This model enables the simulation of a Lennard–Jones fluid confined within an ideally homogeneous, graphitic planar pore. The most well-known case is the *Steele potential*, which represents a flat pore surrounded by an infinite number of sheets (Steele 1973). This model is the simplest molecular representation for simulating adsorption in coals and is frequently used in conjunction with NLDFT (nonlocal density functional theory) approaches to determine PSDs of micro- to mesoporous coals. These determinations are made by analyzing the adsorption isotherm of a probe gas such as nitrogen or argon at cryogenic temperatures (Landers et al. 2013).

The planar pore model can be refined to account for heterogeneities within the sheets (e.g. heteroatoms and vacancies) (Vishnyakov et al. 1998; Tenney and Lastoskie 2006). For instance, it has been demonstrated (Ravikovitch et al. 2000) that assuming energetically uniform, graphite-like pore walls in coals creates an artificial minimum in PSD around 1 nm. This is associated with transitions between adsorbed monolayers and bilayers when gas porosimetry results are interpreted using NLDFT. Consequently, two approaches have been developed to address the heterogeneity of carbon surfaces in nonlocal DFT models: QSDFT (quenched solid DFT), which accounts for surface roughness (Neimark et al. 2009), and 2D-NLDFT, which incorporates the spatial non-uniformity of the adsorbent surface in two dimensions (Jagiello and Olivier 2013) (see Figure 5.4). The QSDFT and 2D-NLDFT methods were tested on three different types of coal by Puziy et al. (2016), who found similar PSDs across the three materials, despite the differences in the ways the two methods accounted for surface heterogeneity.

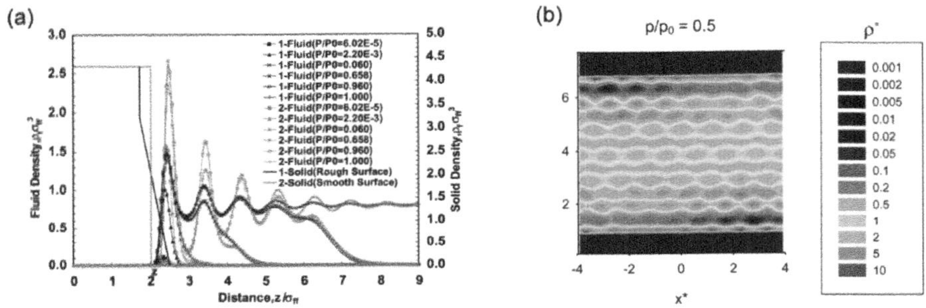

Figure 5.4. *Consideration of surface heterogeneity for the analysis of microporous coals using gas porosimetry with DFT methods: (a) Formation of Ar layers at 87 K on the surface of Cabot BP-280 coal when the surface density of the solid varies linearly (QSDFT approach) (Neimark et al. 2009); (b) nitrogen density distribution at 77 K in a rough graphitic pore calculated by 2D-NLDFT for a reduced pressure of 0.5 (Jagiello and Olivier 2013)*

However, these approaches are limited to specific geometries. More advanced methods aim to reconstruct the molecular structure of coals in 3D. These methods typically involve optimizations under constraints, where the optimality criteria may be in agreement with certain experimental data (e.g. diffraction spectrum) or the interatomic energy, while constraints are properties that define the search space and are deemed essential (e.g. elemental composition, aromaticity and functional groups). Figure 5.5, for example, illustrates a model obtained using the *hybrid reverse Monte Carlo* (HRMC) method. In this method, reconstruction involves optimizing both the interatomic energy and the structure factor through an inverse Monte Carlo approach, similar to simulated annealing, while imposing constraints on elemental composition and density (Jain et al. 2006a). These reconstructions are particularly relevant for studying adsorption: the predicted PSDs and adsorption isotherms align well with experimental data (Jain et al. 2006b). These reconstructions reveal that the molecular structure of coals is microporous and disordered, with pore sizes in the sub-nanometer range, similar to interatomic distances. As a result, the concepts of porosity and specific surface area become ambiguous, challenging the traditional geomechanical view that treats a porous medium as a solid matrix with one or more fluids in the pores. Reconciling molecular and geomechanical perspectives is the subject of section 5.5. Sections 5.3 and 5.4 provide a more detailed discussion of molecular simulation methods and their application to coal studies.

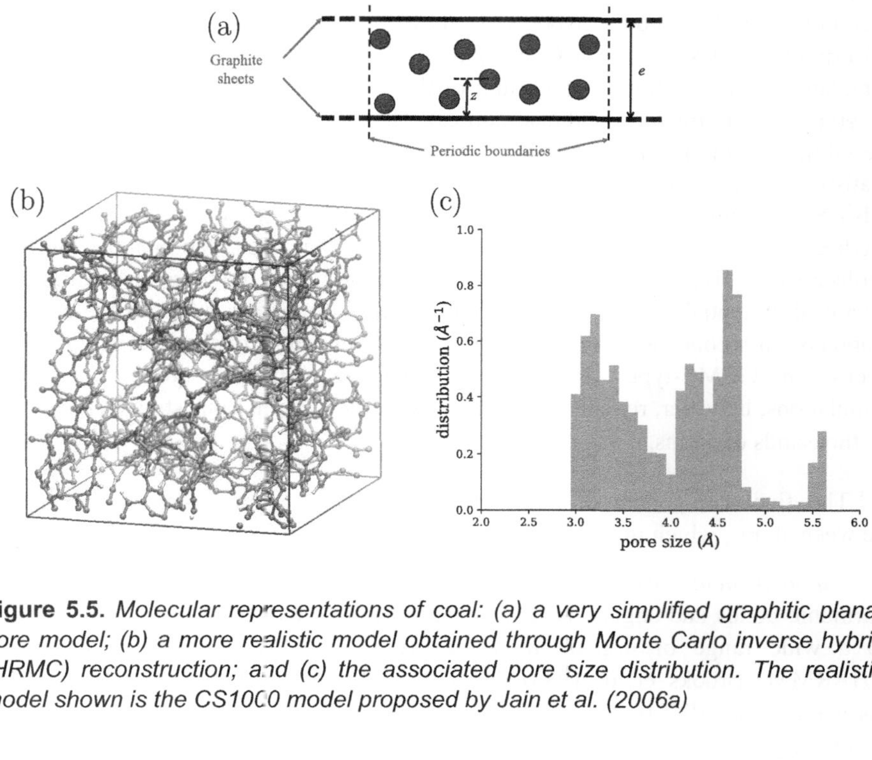

Figure 5.5. *Molecular representations of coal: (a) a very simplified graphitic planar pore model; (b) a more realistic model obtained through Monte Carlo inverse hybrid (HRMC) reconstruction; and (c) the associated pore size distribution. The realistic model shown is the CS1000 model proposed by Jain et al. (2006a)*

5.3. Atomic simulations of adsorption in coals

This section focuses on atomistic simulations, where the elementary particles being simulated are atoms, considered within the framework of classical physics. An atomic system is described by the positions and momenta of all atoms. The temporal evolution of each atom is governed by Newtonian dynamics under the influence of interactions with all other atoms (e.g. electrostatic and van der Waals interactions). At equilibrium, on a macroscopic time scale, the atomic configurations explored are distributed in the phase space in accordance with *Boltzmann statistics*. Thus, for a given macroscopic state, there corresponds a multitude of atomic states, and thermodynamic properties, derived from averaging the properties of all of these atomic states. Atomistic simulation methods aim to sample atomic states according to Boltzmann statistics in order to estimate the thermodynamic properties of interest by averaging over the sampled states. There are two main categories of atomistic simulations: 1) *molecular dynamics* simulations, which simulate the temporal evolution of atoms based on Newtonian dynamics; and 2) *Monte Carlo* simulations, which sample Boltzmann statistics without reference to temporal evolution. According to the ergodic hypothesis, these two approaches are equivalent. For adsorption studies, the GCMC method is often preferred because it allows the simulation of the osmotic equilibrium of the adsorbed fluid with an external fluid reservoir (e.g. in fractures and macropores). This osmotic equilibrium is achieved by permitting the insertion and deletion of fluid molecules in order to satisfy Boltzmann statistics in the grand canonical ensemble (with imposed chemical potential μ of the adsorbate, volume V of the adsorbent and temperature T). Adsorption can also be studied using molecular dynamics; however, in that case, the number of fluid molecules is fixed, and studying osmotic equilibrium requires estimating the chemical potential of the fluid. Unfortunately, these estimates are often very imprecise for condensed systems (e.g. Widom insertion method). Therefore, we focus on GCMC-type simulations to achieve osmotic equilibrium. These simulations, however, remain limited to nanometric systems consisting of a few tens of thousands of atoms.

The final component of a molecular simulation is the interaction potential between atoms, which must reflect elementary atomic interactions, specifically:

– *Ionocovalent bonds* within the coal matrix: these are generally modeled by harmonic bonds and angles. Several universal force fields exist that can be applied to a wide range of organic molecules (e.g. AMBER (Cornell et al. 1995), CHARMM (Brooks et al. 1983) and OPLS (Jorgensen and Tirado-Rives 1988; Jorgensen et al. 1996)). These force fields differentiate atoms based on the type of bonds they are involved in (for instance, an ether oxygen is distinguished from a hydroxyl oxygen). However, a specific challenge with coal is that the configurations encountered in reconstructions, particularly for carbon atoms, cannot always be categorized within the classifications used by universal force fields. A potential alternative is to use reactive force fields (e.g. REBO (Brenner 1990; Brenner et al. 2002) and ReaxFF (van Duin et al. 2001)), which can automatically determine the specific configuration of each atom and interpolate the energy between different configurations. For example, the reconstruction model of Jain et al. (Figure 5.5(b)) is based on the REBO reactive potential. The analytical expressions of both universal and reactive force fields are generally complex, and readers interested in more details can refer to the specialized literature on each potential. Although complex, these various potentials are available in standard molecular simulation tools, and thus their implementation does not pose significant difficulties.

– *Electrostatic interactions* between fluid molecules and polar coal fragments. For point charges q_i and q_j separated by a distance r_{ij}, these interactions follow the standard form: $U = \frac{1}{4\pi\varepsilon_0}\frac{q_i q_j}{r_{ij}}$, where ε_0 is the vacuum permittivity. Instead of using point charges, point dipoles (for polar molecules) or even higher order multipole expansions (e.g. quadrupole moments for carbon dioxide) are sometimes used. The charge distribution within the coal matrix is either fixed in universal force fields or calculated using charge equilibration methods in reactive force fields. In some cases, the force field does not account for charges (e.g. REBO). In such cases, the charge distribution must be determined in advance through ab initio simulations or empirical methods (e.g. PEOE (Gasteiger and Marsili 1980) and QEq (Rappe and Goddard 1991)). For fluid molecules, the charges are typically static. This is particularly true for common models of water (e.g. SPC (Berendsen et al. 1987)) and carbon dioxide (e.g. EPM (Harris and Yung 1995)). As for methane, it possesses a weak octopole moment, which is usually neglected in molecular modeling. Considering electrostatic interactions in molecular simulations presents a particular challenge due to the long-range nature of these interactions, which often exceed the size of the simulated systems. To address this, specific methods are used, with Ewald summation being the most common for periodic systems.

– *Van der Waals interactions* (attractive dispersion forces and steric repulsion): these interactions apply to all atoms and are often modeled by the Lennard–Jones potential (or an equivalent potential), which involves two parameters: (ε characteristic energy and σ characteristic distance): $U = 4\varepsilon\left[\left(\frac{\sigma}{r_{ij}}\right)^{12} - \left(\frac{\sigma}{r_{ij}}\right)^{6}\right]$. Lennard–Jones parameter sets are typically provided for both common fluid models and universal or reactive force fields.

While the formulation of force fields is grounded in atomic physics, their parameterization is often adjusted to reflect the macroscopic thermodynamic properties of materials (e.g. the liquid–vapor coexistence curve). It is important to remember that these force fields are inherently empirical and seek to balance accuracy with computational efficiency. For example, a simple model for methane uses a single Lennard–Jones center, which is a significant simplification compared to an accurate ab initio calculation that accounts for its pyramidal structure (see Figure 5.6). In studies conducted under geological conditions (high pressure, high temperature), the Lennard–Jones parameters are calibrated to reflect the equation of state of methane in supercritical conditions (Kurniawan et al. 2006) and its phase equilibrium in cases of immiscible mixing, such as with water (Errington et al. 1998). Therefore, the choice of a force field for a molecular simulation largely depends on the specific context. For instance, it is common to assume rigid solids when studying adsorption, which allows the omission of a force field for the solid itself. However, in the case of coals, this assumption is debatable since adsorption can induce structural deformations.

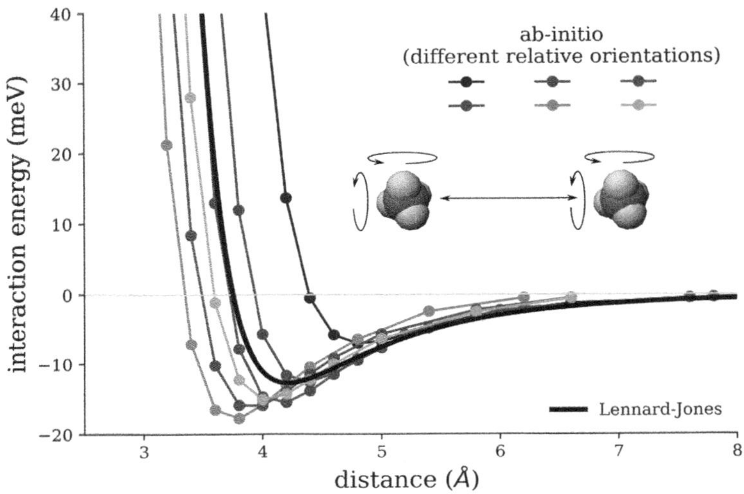

Figure 5.6. *Interaction energy between two methane molecules as calculated using a simple Lennard-Jones potential (Kurniawan et al. 2006), or by ab initio methods for different relative orientations (Tsuzuki et al. 1998)*

Below we present GCMC simulation results for the adsorption of methane and carbon dioxide in the CS1000 molecular model, as shown in Figure 5.5(b), which was obtained via HRMC reconstruction (Jain et al. 2006a). The interaction potential for methane is a Lennard–Jones potential (Figure 5.6), and for carbon dioxide, it is the EPM potential, where each atom is modeled by a Lennard–Jones center and a point charge (Harris and Yung 1995). Initially, the solid is assumed to be rigid. The cross Lennard–Jones parameters for fluid–solid interactions are derived by applying Lorentz–Berthelot mixing rules, with the solid atom parameters taken from the study by Jain et al. (2006b) study on adsorption in CS1000. As for the partial charges in the solid, these are calculated using the empirical PEOE method (Gasteiger and Marsili 1980). More detailed information on the methods and models used in these simulations can be found in Vandamme et al. (2010).

In a GCMC simulation, the *chemical potential* of the fluid is imposed, while the number of fluid molecules fluctuates according to the Boltzmann distribution in the grand canonical ensemble. The adsorption isotherm can then be obtained by estimating the average number of adsorbed molecules as a function of the chemical potential. In practice, experimental adsorption isotherms relate the excess adsorbed quantity (i.e. taking into account the buoyant force) to the pressure of the pure fluid reservoir, rather than to the chemical potential. To achieve this, a series of simulations of the pure fluid must be conducted separately to establish the relationship between chemical potential and the pressure/density of the pure fluid. Methane shows behavior similar to an ideal gas, as its critical temperature is much lower than geological temperatures (–82.6°C). In contrast, carbon dioxide deviates significantly from ideal gas behavior. Though supercritical under geological conditions, its critical temperature (31.3°C) remains close to subsurface temperatures. Once the link between pure fluid pressure and chemical potential is established, adsorption at a given fluid pressure is simulated using GCMC by applying the chemical potential corresponding to the desired fluid pressure. As an illustration, Figure 5.7 presents examples of molecular configurations sampled for adsorption simulations, either on the surface of a mesopore or within micropores.

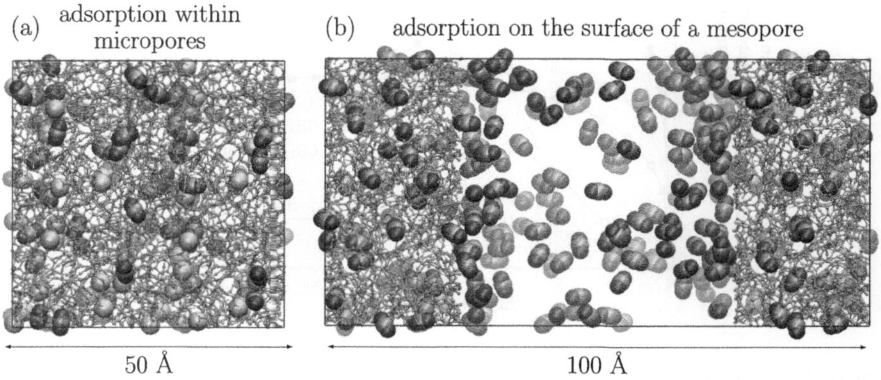

Figure 5.7. *Example of molecular configurations sampled for adsorption simulation within micropores (a) or on the surface of a mesopore (b). The adsorbed molecules are represented as spheres (yellow for methane, blue and red for carbon dioxide), while the solid structure is shown as rods*

These two types of adsorption differ significantly: the interaction energy with the solid is stronger in the micropores than on the mesopore surface, and the volume available for the adsorbed fluid is highly restricted in micropores, whereas the surface of a mesopore allows for the adsorption of a larger fluid volume. This results in very distinct adsorption isotherms (Figure 5.8). Adsorption in micropores follows a type I isotherm, resembling site adsorption, while adsorption on the mesopore surface corresponds to a type II or III isotherm, where adsorption gradually increases with rising reservoir pressures. Both methane and carbon dioxide are supercritical at geological temperatures (here 310 K), meaning there is no saturating vapor pressure limiting the isotherm. However, pure carbon dioxide still exhibits a notable increase in density at pressures near its critical pressure (7.3 MPa), transitioning from a state resembling a gas to one resembling a liquid. This is reflected in the adsorption isotherms as a drop in excess adsorbed quantities between 7 and 10 MPa. Comparison with experimental data (Figure 5.8) aligns with adsorption occurring primarily within the micropores rather than on the mesopore surface. While the presence of mesopores cannot be ruled out, their contribution to the total adsorption isotherm appears to be dominated by micropore adsorption. However, in the presence of capillary condensation (subcritical fluids), mesoporous volume may significantly contribute to the excess adsorbed quantity.

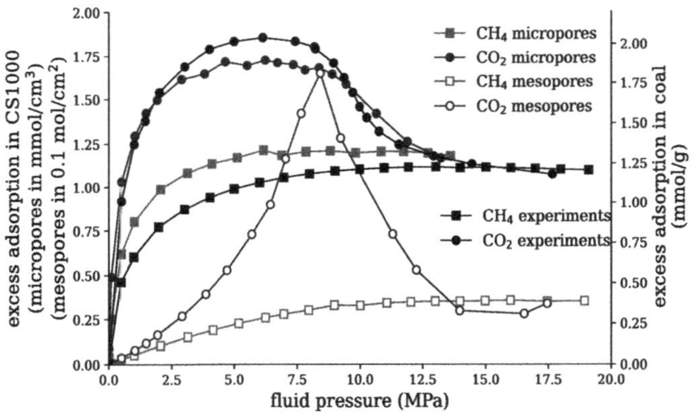

Figure 5.8. *Excess adsorption isotherms for methane and pure carbon dioxide in the micropores of CS1000 and on the surface of CS1000 mesopores. The comparison with experimental data (Ottiger et al. 2006, 2008) indicates that the majority of adsorption in coals occurs within the micropores*

Adsorption is highly sensitive to the nature of the fluid. In this case, for coal, the adsorption affinity is greater for carbon dioxide than for methane. The discrepancy between methane and carbon dioxide appears slightly less pronounced in simulations compared to experiments (Figure 5.8). Setting aside the issue of the representativeness of the coal model (CS1000), two possible explanations can be proposed:

– The flexibility of the solid, which is neglected in these simulations. Carbon dioxide induces more pronounced swelling than methane. Thus, for a flexible matrix, a more significant difference between methane and carbon dioxide can be expected.

– Accessibility to micropores, which is not accounted for in the GCMC method. Osmotic equilibrium is simulated by randomly introducing or removing fluid molecules within the microporous volume. In this process, the entire micro-porosity is considered accessible. However, experimentally, it is likely that methane and carbon dioxide do not have the same access to micro-porosity. Greater accessibility can be expected for carbon dioxide than for methane, as steric repulsion allows carbon dioxide to access narrower pores compared to methane, and because the attractive interactions favor carbon dioxide more.

For geological carbon storage applications, it is necessary to replace methane with carbon dioxide within a coal reservoir while maintaining an approximately constant fluid pressure of around 10 MPa. This process resembles the adsorption of a mixture with a changing composition, which can be simulated by GCMC in the same way as the adsorption of pure substances. The state variables imposed are the chemical potentials of the two fluids, which must first be linked to the (variable) composition and the (constant) pressure of the fluid. Therefore, an initial series of GCMC simulations of the reservoir fluid enables the determination of the chemical potentials of methane and carbon dioxide for mixture compositions ranging from pure methane to pure carbon dioxide. Five temperature/pressure levels were considered, corresponding to the typical conditions at depths of 300–1,500 m. Following this, the mixture adsorption simulations predict the quantities adsorbed in the microporous matrix as a function of reservoir composition at various depths (Figure 5.9). As expected, the preferential affinity for carbon dioxide enhances its adsorption over methane, so that the composition of the adsorbed fluid is consistently richer in carbon dioxide than that of the fluid in the reservoir. A simple indicator reflecting this preferential affinity is the *selectivity* S_{CO_2}, defined as follows: $S_{CO_2} = \left(\frac{x_{CO_2}}{x_{CH_4}}\right)_{adsorbed} \Big/ \left(\frac{x_{CO_2}}{x_{CH_4}}\right)_{bulk}$, where x_i represents the molar fraction of component *i*. The "adsorbed" subscript refers to the fluid adsorbed in the microporous coal, while the "bulk" subscript refers to the fluid in a macroscopic reservoir in equilibrium with the coal, whose behavior is not influenced by adsorption effects. A selectivity greater than 1 indicates a preferential affinity, which is indeed observed for carbon dioxide relative to methane (Figure 5.9). Thus, a 50/50 composition of the adsorbed fluid roughly corresponds to a 25/75 composition in the "bulk", that is, external fluid. It is also noted that the adsorbed quantities increase only slightly with depth, while the reservoir densities change significantly. From an application perspective, this implies that a coal reservoir is an effective carbon storage site even at shallow depths, whereas other geological storage sites must be located deeper than 800 m to benefit from sufficiently high carbon dioxide density.

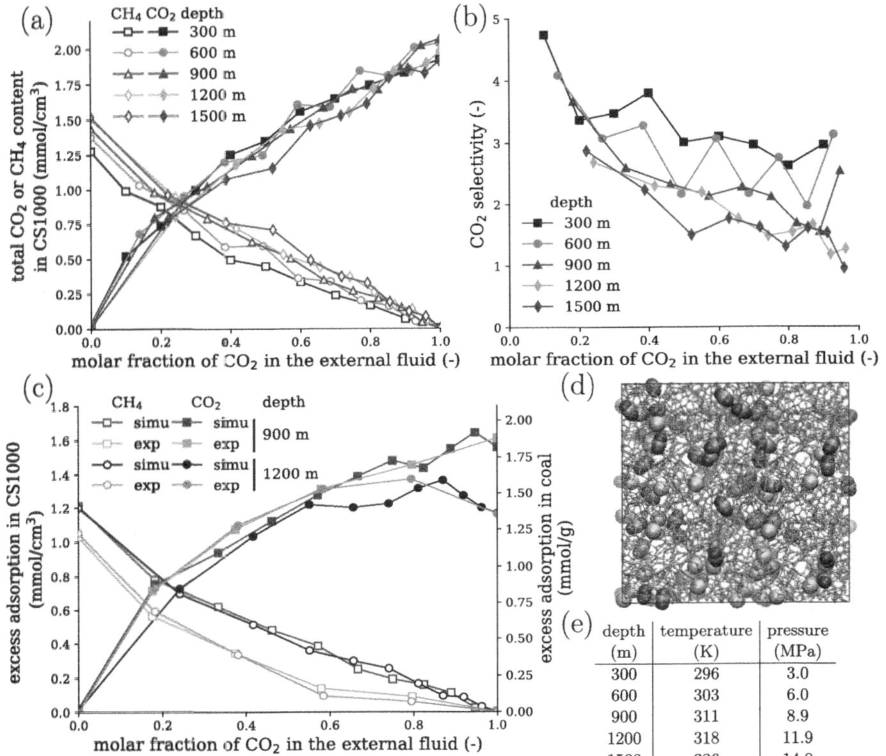

depth (m)	temperature (K)	pressure (MPa)
300	296	3.0
600	303	6.0
900	311	8.9
1200	318	11.9
1500	326	14.8

Figure 5.9. *Total adsorbed quantities (a), CO_2 selectivity (b). Comparison to experimental results for excess quantities (c) during the adsorption of a CO_2/CH_4 mixture in CS1000 (example of molecular configuration in (d)). The different curves show the variation as a function of the CO_2/CH_4 composition of the external fluid maintained at constant pressure and temperature. Five pressure/temperature cases are considered, reflecting five different depths relevant to carbon storage and enhanced coal gas recovery applications (e). The experimental curves are obtained by interpolating the results of Ottiger et al. (2006, 2008) for the two depths at the temperatures closest to the experiment (318 K). See Brochard et al. (2012) for further details*

Another mixture of interest for practical applications is the methane–water mixture. To study this, we consider the MSPC/E water model and the TraPPE methane model, both of which are well suited for simulating methane–water mixtures at high temperature and high pressure (Errington et al. 1998). Unlike the methane-carbon dioxide mixture, this mixture is *biphasic* because water is not in the supercritical phase. The methane–water mixture thus exhibits a gas phase that is predominantly methane and a liquid phase that is predominantly water. These two phases have a coexistence point, and only compositions richer in methane for the gas and richer in water for the liquid are possible.

To determine the *coexistence point*, specific molecular simulation methods must be used. Here, we employ *Gibbs ensemble* Monte Carlo simulations, which involve simulating two distinct systems corresponding to the two phases, while allowing for the exchange of molecules and volume, all while keeping the total number of molecules and total volume constant. This approach simulates the osmotic and mechanical equilibrium of the two phases. This type of simulation ensures the equality of chemical potentials and pressure between the liquid and gas, representing the coexistence point. Unlike pure substances, the pressure and temperature at coexistence are no longer directly related and can be chosen independently. Once the coexistence point is established, GCMC simulations are conducted for each phase to determine the chemical potentials as a function of the possible compositions (from pure methane to coexistence for the gas, and from pure water to coexistence for the liquid). The adsorption of the mixture can then be simulated for various compositions (Figure 5.10). It is observed that the affinity for water is very low in the CS1000 model to the extent that even in contact with liquid water, methane remains the majority of the adsorbed fluid. This is due to the fact that water is a polar molecule that is highly sensitive to the presence of electronegative elements in the solid (mainly oxygen).

However, the CS1000 model represents a highly mature coal structure, where the oxygen content is less than 1% by mass. In contrast, coals of low to moderate maturity exhibit significant oxygen contents, around 5% by mass. To provide insight into the importance of oxygen, Figure 5.10 presents the adsorption of a methane–water mixture in CS1000 doped with 5% oxygen. Only bridging oxygens (which generally have the highest partial charges, making them more favorable for water adsorption) are introduced. Charge calculations using the QEq method (Rappe and Goddard 1991) show that the partial (negative) charges on the oxygens are much greater than on the carbons and hydrogens (ranging from –0.5 to –1 elementary charges).

The presence of these charges has little impact on methane adsorption (non-polar), but a significant effect on water adsorption (polar). For an oxygen content of 5% (representative of moderately mature coal), water adsorption becomes comparable to methane adsorption (Figure 5.10).

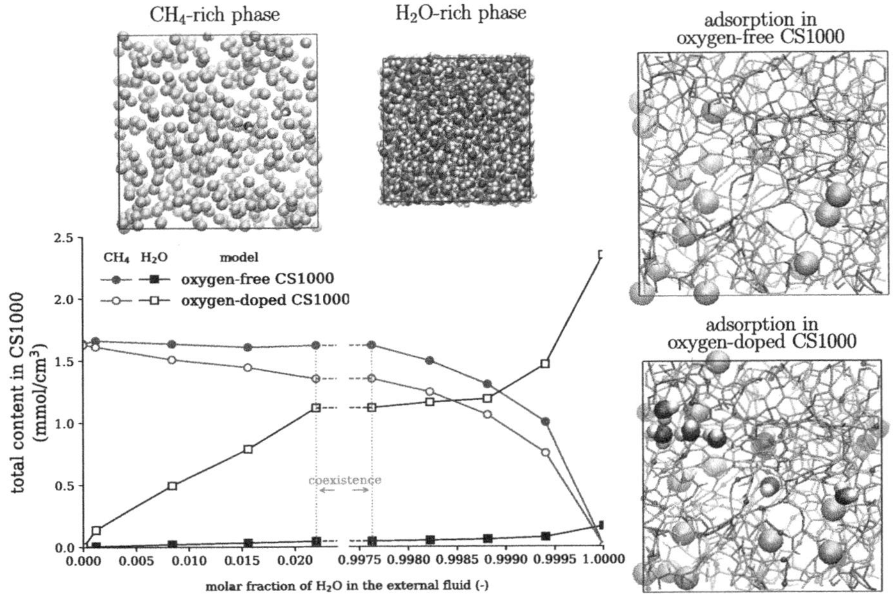

Figure 5.10. *Adsorption of a water-methane mixture in CS1000 and CS1000 doped with 5% oxygen. The water-methane mixture is biphasic under the conditions considered (T = 430 K, P = 30 MPa, corresponding to a depth of about 3,000 m). Prior to the adsorption simulations, the coexistence point is established through Gibbs ensemble simulations, and the relationship between the chemical potentials and the composition of each phase is determined via grand canonical simulations*

5.4. Adsorption modeling using DFT

Among the theoretical approaches developed to describe adsorption, DFT has emerged as one of the most powerful and practical molecular methods. DFT approaches retain detailed information on the microstructure of inhomogeneous fluids, while offering a computational cost that is, on average, two orders of magnitude lower than molecular simulations. Comprehensive descriptions of DFT can be found in the reviews of Evans (1992), Lutsko (2010) or the book by Davis (1996). Here, only the key points relevant to understanding fluid adsorption in pores will be addressed.

In this context, the DFT formalism expresses the grand thermodynamic potential, Ω, of an inhomogeneous fluid subjected to an external potential, V_{ext}, within the grand canonical ensemble (μ, V, T), as a functional of the molecular densities, $\rho_i(\vec{r})$, for each species:

$$\Omega[\rho_i(\vec{r})] = \mathcal{F}[\rho_i(\vec{r})] + \sum_i \int \{V_{ext}^i(\vec{r}) - \mu_i\} \rho_i(\vec{r})\, d^3r \qquad [5.2]$$

where $\mathcal{F}[\rho_i(\vec{r})]$ represents the Helmholtz free energy of the fluid. The equilibrium density profiles are obtained by minimizing the grand potential, which involves solving the corresponding Euler–Lagrange equations:

$$\frac{\delta\Omega[\rho_i(\vec{r})]}{\delta\rho_i(\vec{r})} = \frac{\delta\mathcal{F}[\rho_i(\vec{r})]}{\delta\rho_i(\vec{r})} + V_{ext}^i(\vec{r}) - \mu_i = 0 \qquad [5.3]$$

To compute the amount of fluid adsorbed in a pore, and hence the thermodynamic properties of the adsorbed fluid, three key components are required:

– A reliable thermodynamic model to describe the homogeneous fluid in equilibrium with the adsorbed fluid. For fluids relevant to coal, such as methane, alkanes, CO_2 and particularly water, SAFT (statistical associating fluid theory) equations (Müller and Gubbins 2001) are highly effective for this purpose.

– An external potential that describes the interaction between the fluid and the solid. The Steele (1973) potential, a generalization of the Lennard–Jones potential for a semi-infinite solid, is commonly used to study gases adsorbed in coal.

– An expression for the Helmholtz free energy, $\mathcal{F}[\rho_i(\vec{r})]$, of the inhomogeneous fluid. Finding an appropriate function for this free energy is the central challenge in DFT treatments. Various proposals exist in the literature depending on the fluids and configurations involved. For fluids adsorbed on solid surfaces, only a *nonlocal treatment* (NLDFT) can adequately capture significant density variations. This approach describes short-range interaction contributions using weighted densities that account for the influence of the surroundings at a specific point in space. Two perspectives have been proposed: *coarse-grained* approaches, as initially suggested by Tarazona (1984), and methods based on fundamental measure theory (Roth 2010), which are better suited to high-density systems.

As an illustration, Figure 5.11 presents the density profiles of methane and water, predicted by the coupling of NLDFT with SAFT in micropores and small mesopores with a planar graphitic surface, compared with molecular simulation results (Malheiro et al. 2014a, 2014b).

These results demonstrate the efficacy of this theoretical approach in modeling the adsorption of these compounds within model pores.

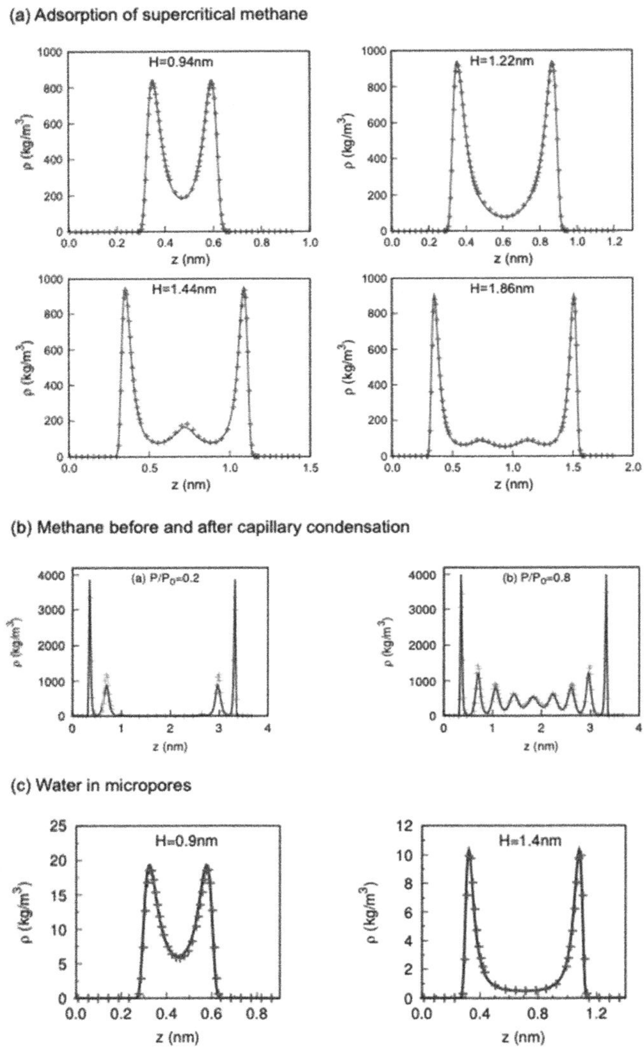

Figure 5.11. *Adsorption of methane or water in micropores: (solid line) by NLDFT/SAFT, (symbols) by grand canonical Monte Carlo molecular simulation. (a) Supercritical methane (T = 353 K, P = 6 MPa), showing the influence of pore size on the number and density ρ of adsorbed layers (Malheiro et al. 2014a). (b) Gaseous methane (T = 120 K) adsorbed in a 3.7 nm pore; density profiles are shown before capillary condensation (P/P_0 = 0.2) and after, when the fluid is highly structured (P/P_0 = 0.8) (Malheiro et al. 2014a). (c) Water in graphitic plane micropores at T = 425 K and P = 0.05 MPa (Malheiro et al. 2014b)*

One advantage of NLDFT is its applicability to mixtures, allowing the calculation of selectivity, which indicates the preferential affinity of the porous medium for a given compound, as explained in section 5.3.

For example, the effect of water on methane adsorption in graphitic pores has been studied (Malheiro 2014), and the impact of various parameters, such as thermodynamic conditions or mixture composition, on selectivity has been evaluated (Figure 5.12).

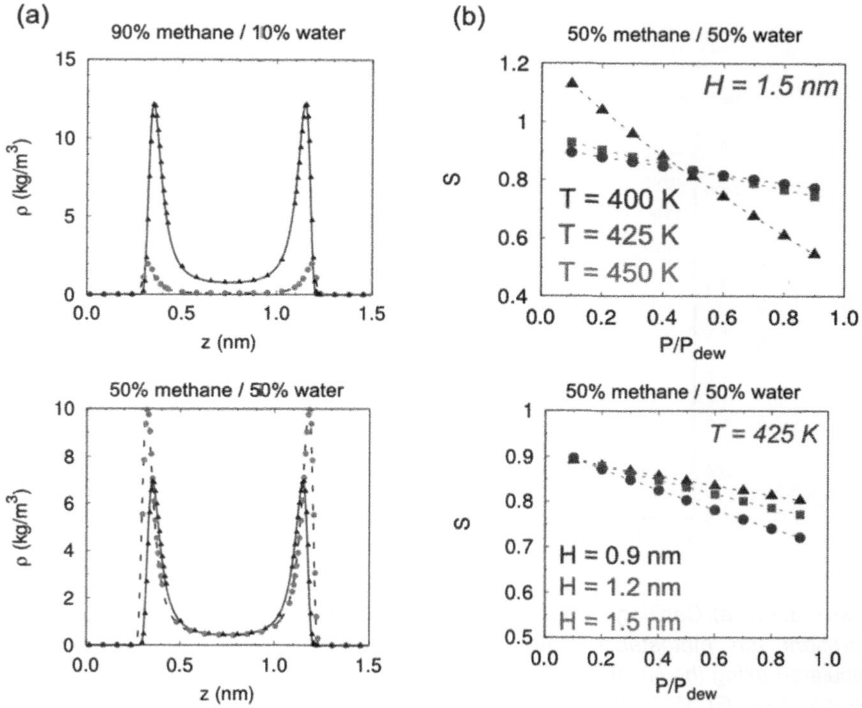

Figure 5.12. *Adsorption of a methane/water mixture in graphitic plane micropores. (a) Density profiles ρ (kg/m³) at T = 425 K and P = 0.1 MPa in a 1.5 nm pore: (solid line) by NLDFT/SAFT, (symbols) by grand canonical Monte Carlo molecular simulation (Malheiro 2014) (triangles: methane, circles: water). (b) Selectivity factor S of methane relative to water for several configurations as a function of relative pressure (P_{dew} being the bulk condensation pressure at the considered temperature) (Malheiro 2014)*

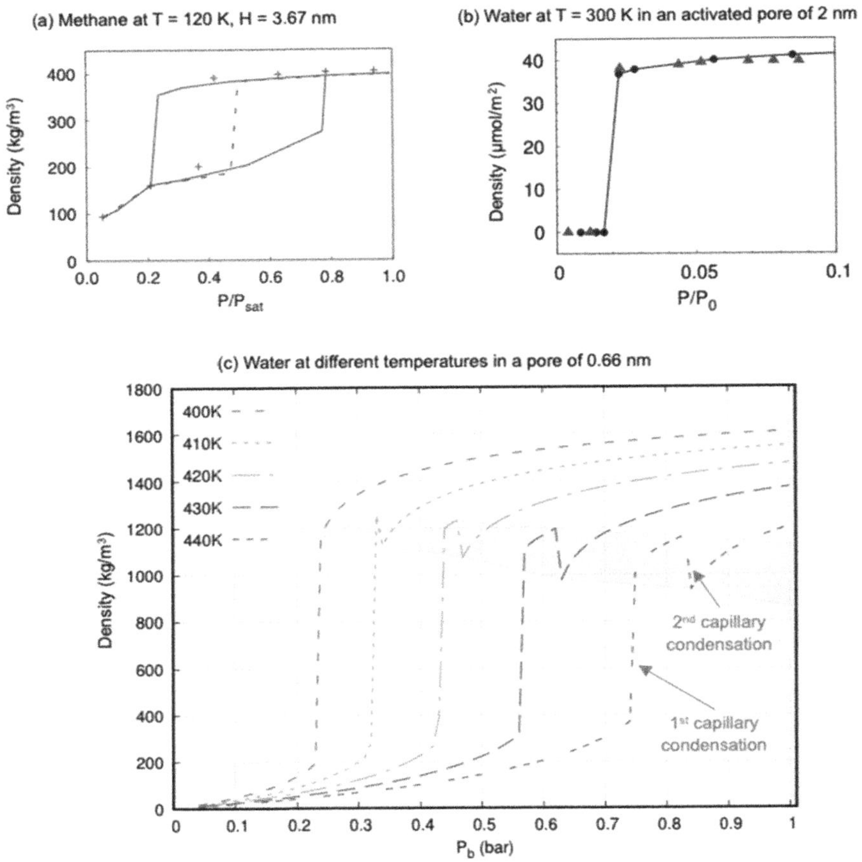

Figure 5.13. *(a) Capillary condensation and evaporation of methane. The solid lines represent the metastable states of capillary condensation and its hysteresis, calculated using the NLDFT/SAFT-VR model. The dotted line (NLDFT/SAFT-VR) and the symbols (GCMC) represent the equilibrium transition of capillary condensation (Malheiro et al. 2014a). (b) Desorption isotherm of water at 300 K in an activated graphitic plane pore with a size of 2 nm and a site density of 1 nm^{-2}. The circles represent the results obtained with the NLDFT/SAFT-VR model (Malheiro et al. 2014b), while the triangles show the molecular simulation results from Muller et al. (1996). (c) Successive capillary condensations of water in a graphitic plane pore of 0.66 nm (data from Miqueu and Grégoire (2020)). The fraction of free hydrogen bonds in the intermediate phase (i.e. between the two transitions) corresponds to an association degree of approximately 50%, characteristic of a phase more organized than liquid, such as a monolayer of ice*

The density profiles of each component obtained using the NLDFT/SAFT-VR predictive model are in excellent agreement with those from molecular simulations, validating this theoretical approach for mixtures. For a binary mixture initially composed of 50% methane and 50% water, the selectivity factor, *S*, is shown for three different temperatures and three pore sizes as a function of reduced pressure in Figure 5.12. It should be noted that the selectivity factor for methane relative to water is less than 1 in most cases (except for the smallest pore at the lowest pressures).

This indicates that the adsorbed phase consists of a higher proportion of water than methane, even though the *bulk* mixture is equimolar. Thus, as shown in Figure 5.10 with molecular simulations, coal can become more selective for water than for methane. In other words, the presence of water can significantly reduce methane adsorption capacity compared to when methane alone is present in the pores under the same thermodynamic conditions.

The density profiles obtained can then be spatially integrated to determine the total quantities of fluid adsorbed in the pores, typically represented as adsorption isotherms. These quantities may vary monotonically with pressure, as observed for supercritical methane, or exhibit abrupt variations, particularly during condensation, as explained in section 5.3. Figure 5.13 illustrates these variations for various configurations of methane and water in graphitic pores.

To determine the adsorption isotherm for the entire material, rather than a single pore size, the contribution of each pore making up the material must be summed. This strategy involves using the PSD, which can be obtained, for example, through gas porosimetry. The PSD indicates the types and quantities of pores present in the material. This approach differs from that in section 5.3, which directly calculates the adsorbed quantities for a representative sample of coal (the CS1000 molecular model, in this case).

The main disadvantage of this method, compared to using a molecular model, is that it neglects the connectivity between pores, treating the porous medium as a collection of disconnected pores, as represented in gas porosimetry. However, its advantage is that the PSD corresponds to the material being studied, not a generic model, and can be determined systematically and quickly for each new sample of coal. To make a meaningful comparison with a measured isotherm, it is necessary to calculate the excess adsorbed amount, as the absolute adsorbed amount is not directly measurable. Excess adsorption is measured using calibrated volumes, often with helium as the calibration gas. The excess adsorbed amount can be calculated via:

$$n_{exc} = \sum_i \langle \rho_i \rangle V_i - \rho_{bulk} V_{He} \qquad [5.4]$$

where V_i is the pore volume for a pore of size H_i (both H_i and V_i are obtained via gas porosimetry), and $<\rho_i>$ is the average density of the confined fluid in the pore of size H_i, calculated using DFT. The density ρ_{bulk} is calculated from the equation of state used as the homogeneous limit of the DFT version employed. The volume accessible to helium can be (i) calculated by applying DFT to helium in the pores considered, determining the volume it occupies, (ii) taken from experimental calibration if available or (iii) obtained from gas porosimetry.

This methodology is illustrated in Figure 5.14, which shows a comparison between experimental results and the NLDFT-SAFT model for methane adsorption on a carbon molecular sieve (Carboxen 1012), whose PSD is typical of activated carbon and is represented on Figure 5.14. This PSD was obtained using nitrogen porosimetry at 77 K (with a Micromeritics ASAP 2020 porosimeter), and the nitrogen isotherm was interpreted using the model developed by Malheiro et al. (2015).

The Steele potential was used to describe the methane/solid interaction, with a parameter of $\epsilon_{ss}/k_B = 35\,K$. The excellent agreement between theoretical and experimental results demonstrates the consistency of the characterization and modeling, and their ability to accurately reproduce methane adsorption isotherms on this carbon molecular sieve.

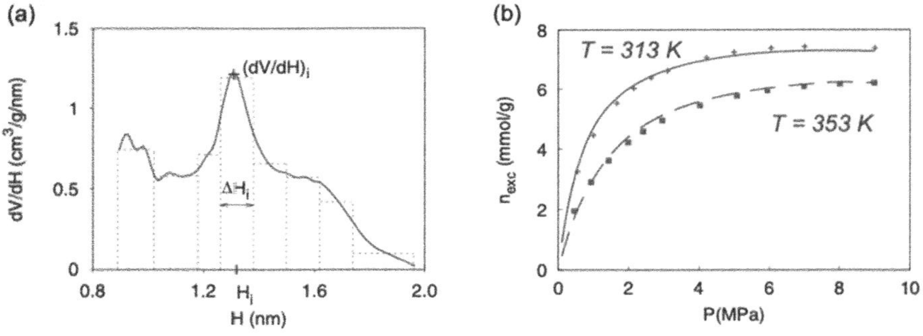

Figure 5.14. *(a) Pore size distribution dV/dH of Carboxen-1012 and its discretization used for the NLDFT calculation; $(dV/dH)_i \times \Delta H_i$ represents the volume assigned to each pore of size i. (b) Methane adsorption isotherms (excess adsorbed amount n_{exc} as a function of pressure P) on Carboxen-1012. The solid and dashed lines represent the results modeled by NLDFT/SAFT-VR at 313 K and 353 K, respectively. The symbols correspond to experimental points obtained using gravimetric techniques (Malheiro et al. 2014a)*

5.5. Mechanics-adsorption coupling

5.5.1. *Thermodynamic analysis of coupling and extended poromechanics*

The interactions between fluids and porous solids lie at the core of modeling challenges in geomechanics. For the majority of geomaterials, the fluid phase occupying the pores exists in its ordinary state, also known as the bulk state, meaning that its law of behavior is not affected by the presence of the solid. The thermodynamic description of a porous medium, therefore, takes the form of two superimposed phases in thermal equilibrium that can exchange volume through variations in porosity. This is the description adopted in classical poromechanics, which we will review below (see also Chapter 4). However, when the fluid becomes adsorbed – whether on the surface or in micropores – its behavior changes due to molecular interactions with the solid. The laws governing bulk fluid behavior are no longer applicable, necessitating a revision of the conventional poromechanical theory. This leads to an "extended" poromechanics, which applies to both adsorbed and bulk fluids. The essential steps of this extended poromechanics will be summarized in this section. At this point, molecular approaches become vital tools for characterizing the behavior of adsorbed fluids at the scale of an individual pore or a small number of pores. Applying extended poromechanics to a disordered porous medium, such as coal, is not straightforward. Different strategies can be employed. For instance, we will see that the inherent disorder allows for simplifying assumptions, resulting in a simplified poromechanical description that is easier to implement. Alternatively, we can rely on the distribution of pore sizes, summing the contributions from each pore size, similar to what is done in DFT approaches for adsorption isotherms (see Figure 5.14).

Let us first review the classical poromechanical theory (Coussy 2010). In the absence of adsorption, a porous medium consists of a solid matrix and a bulk fluid within the pores. For simplicity, we assume here that the solid matrix is elastically isotropic, and we are concerned only with the volumetric mechanical behavior. Furthermore, we limit the discussion to a single fluid phase (a saturated medium). Consider a porous medium maintained under drained, isothermal and isochoric conditions, meaning that the system is at a fixed temperature T, fixed volume V and can exchange fluid with an external bulk fluid reservoir at the same temperature and imposed pressure P. This situation corresponds to a system in thermal and osmotic equilibrium (for the fluid only), where thermodynamic equilibrium is achieved through the minimization of a potential Ω, representing the Helmholtz free energy for the solid and the grand potential for the fluid: $\Omega = E - \sum_{i \in f} \mu_i N_i - TS$, where E is the internal energy, S is the entropy, N_i is the number of molecules of the fluid components i and μ_i is the corresponding chemical potential. The state variables of the system are T, V and μ_i, while the conjugate quantities to these variables are the entropy S, the spherical stress σ and the N_i, leading to variations of Ω in the following form:

$$d\Omega = \sigma dV - \sum_{i\in f} N_i d\mu_i - SdT \qquad [5.5]$$

Poromechanics assumes that the fluid remains in its bulk state. For a bulk fluid, however, the Gibbs–Duhem relationship ($\sum_{i\in f} N_i d\mu_i + S_f dT = V_f dP$) simplifies this equation:

$$d\Omega = \sigma dV - V_f dP - S_s dT \qquad [5.6]$$

The role of the fluid in the system is effectively summarized by its bulk pressure P, and the specifics of its composition do not influence the analysis. Additionally, the quantity conjugate to the pressure is the volume occupied by the fluid V_f, which corresponds to the pore volume. It is also observed that, following this simplification, the temperature is conjugated only to the entropy of the solid $S_s = S - S_f$. Since the problem is extensive with respect to the undeformed volume of the porous matrix $V_0 = V(\sigma = 0, P = 0)$, it is common to use a reduced formulation relative to V_0, which introduces the concepts of Lagrangian volumetric deformation $d\varepsilon = dV/V_0$, Lagrangian porosity $\phi = V_f/V_0$, and Lagrangian entropy $s_s = S_s/V_0$:

$$d\omega = \frac{d\Omega}{V_0} = \sigma d\varepsilon - \phi dP - s_s dT \qquad [5.7]$$

The constitutive equations of poromechanics describe how the conjugate variables (σ, ϕ, s_s) evolve in response to changes in the control variables (ε, P, T). For an isothermal framework ($dT = 0$), these equations are:

$$\begin{cases} d\sigma = K d\varepsilon - b dP \\ d\phi = b d\varepsilon + \frac{dP}{N} \end{cases} \qquad [5.8]$$

Where $K = \left.\frac{\partial\sigma}{\partial\varepsilon}\right|_{P,T}$ is the incompressibility modulus of the porous medium, $b = -\left.\frac{\partial\sigma}{\partial P}\right|_{\varepsilon,T} = \left.\frac{\partial\phi}{\partial\varepsilon}\right|_{P,T}$ is the Biot coefficient, and $\frac{1}{N} = \left.\frac{\partial\phi}{\partial P}\right|_{\varepsilon,T}$ is the Biot modulus. These three moduli are sufficient to fully characterize the poromechanical response of the medium for any isothermal and isotropic change. K and N describe the uncoupled responses: the change in stress with deformation and the change in porosity with pressure. The coefficient b characterizes poromechanical coupling, that is, the effect of pressure on stress and the effect of deformation on porosity. The same coefficient governs these two changes due to the Maxwell relationships arising from the double differentiability of ω: $\left.\frac{\partial^2\omega}{\partial P\partial\varepsilon}\right|_T = \left.\frac{\partial^2\omega}{\partial\varepsilon\partial P}\right|_T$. Additionally, thermodynamic stability requires that ω be convex, meaning that the Hessian matrix associated with it must be positive definite, thus imposing $K > 0$ and $N > 0$. Initially, it may seem that these moduli depend on both the fluid and the solid. However, it turns out that the differential [5.7] of the thermodynamic potential ω is strictly identical to the free energy of the solid alone, adjusted for pressure loading in the porosity: $\eta_s = \frac{E_S - TS_S}{V_0} - \phi P$. This potential η_s is independent of the fluid and depends only on the solid. Consequently, K, b and N are properties of the porous solid alone, independent of the fluid. In the specific case where the solid material is an isotropic elastic material with an incompressibility modulus k_s, an immersion-type loading (where $d\sigma = -dP$) induces a uniform contraction $d\epsilon = -\frac{1-b}{K}dP = -\frac{dP}{k_s}$. Thus, b and K are not independent, leading to: $b = 1 - \frac{K}{k_s}$. Hence, with $0 < K < k_s$, it follows that b must lie between 0 and 1.

It is instructive to examine the immersion of coal in various fluids (Figure 5.3) in the context of classical poromechanical theory. This theory predicts that $d\epsilon = -\frac{1-b}{K}dP < 0$, indicating a contraction that is independent of the fluid's nature and of relatively small magnitude (typically not exceeding several percent). It appears that only immersion in helium aligns with classical poromechanics. This is understandable because, unlike methane and carbon dioxide, helium exhibits negligible adsorption in coal due to very weak molecular interactions between the fluid and the solid. In contrast, the other two gases demonstrate significant adsorption, leading to a pronounced swelling response that is highly nonlinear with respect to pressure and dependent on the fluid type. To understand this swelling phenomenon, we need to revisit poromechanical theory, this time considering that the fluid in the pores is no longer a bulk fluid but an adsorbed fluid (Brochard et al. 2012).

The condition for thermodynamic equilibrium remains the minimization of the hybrid potential ω. However, in this scenario, the Gibbs–Duhem relationship [5.6] used previously is no longer applicable. Instead, we must adopt a more general formulation where the fluid effect cannot be simplified to its bulk pressure alone:

$$d\omega = \sigma d\varepsilon - \sum_{i\in f} n_i d\mu_i - sdT \qquad [5.9]$$

where $n_i = N_i/V_0$ represents the Lagrangian concentration of component i of the fluid. Consequently, the system's description now differentiates between fluids, and two fluids with different properties at the same bulk pressure can produce different poromechanical responses. The control variables are ε, μ_i and T, while the conjugate variables are σ, n_i and s, where s denotes the total Lagrangian entropy (of both solid and fluid). For isothermal changes, the most general form of the constitutive equations is:

$$\begin{cases} d\sigma = \widetilde{K} d\varepsilon - \sum_{i\in f} \widetilde{b_i} d\mu_i \\ dn_i = \widetilde{b_i} d\varepsilon + \sum_{j\in f} \frac{d\mu_j}{\widetilde{N_{ij}}} \end{cases} \quad [5.10]$$

here, $\widetilde{K} = \left.\frac{\partial\sigma}{\partial\varepsilon}\right|_{\varepsilon,\mu_i,T}$ is the incompressibility modulus, $\widetilde{b_i} = -\left.\frac{\partial\sigma}{\partial\mu_i}\right|_{\varepsilon,\mu_{j\neq i},T} = \left.\frac{\partial n_i}{\partial\varepsilon}\right|_{\mu_j,T}$ represents the poromechanical coupling modulus, and $\frac{1}{\widetilde{N_{ij}}} = \left.\frac{\partial n_i}{\partial\mu_j}\right|_{\varepsilon,\mu_{k\neq j},T}$ are coupling moduli related to fluid concentrations. This formulation ensures the Maxwell relations for poromechanical coupling are satisfied. Furthermore, Maxwell's relationships concerning fluid concentrations imply that $\frac{1}{\widetilde{N_{ij}}} = \frac{1}{\widetilde{N_{ji}}}$. The structure of these equations [5.10] is naturally suited to the conjugate variable pairs in the problem. However, it is more illustrative to rewrite these equations using a framework akin to classical poromechanics by introducing the densities ρ_i^b of the bulk fluid components (from the external reservoir, not to be confused with the densities in the porous medium):

$$\begin{cases} d\sigma = K^{eff} d\varepsilon - \sum_{i\in f} b_i^{eff} \rho_i^b d\mu_i \\ d\left(\frac{n_i}{\rho_i^b}\right) = b_i^{eff} d\varepsilon + \sum_{j\in f} \frac{\rho_j^b}{N_{ij}^{eff}} d\mu_j \end{cases} \quad [5.11]$$

The quantity n_i/ρ_i^b represents the effective porosity that component i of the fluid would occupy if it were at bulk density. In this formulation, the notations are intentionally chosen to align with the usual poromechanical modules, but these are now "effective" because they depend on the fluid as well as the solid: the effective incompressibility modulus $K^{eff} = \widetilde{K}$, the Biot coefficients $b_i^{eff} = \widetilde{b_i}/\rho_i^b$ for each fluid component and the cross Biot moduli $\frac{1}{N_{ij}^{eff}} = \frac{1}{\rho_i^b\rho_j^b}\frac{1}{\widetilde{N_{ij}}} - \frac{n_i}{\left(\rho_i^b\right)^2\rho_j^b}\left.\frac{\partial\rho_i^b}{\partial\mu_j}\right|_{\mu_{k\neq j},T}$ between components. It is crucial to understand that this formulation is no longer the natural framework for thermodynamic analysis of the problem, including Maxwell's relationships or stability analysis. In particular, Maxwell's relationships concerning $\widetilde{N_{ij}}$ impose $\left.\frac{\rho_i^b\rho_j^b}{N_{ij}^{eff}} + \frac{n_i}{\rho_i^b}\frac{\partial\rho_i^b}{\partial\mu_j}\right|_{\varepsilon,\mu_{k\neq j},T} = \left.\frac{\rho_i^b\rho_j^b}{N_{ji}^{eff}} + \frac{n_j}{\rho_j^b}\frac{\partial\rho_j^b}{\partial\mu_i}\right|_{\varepsilon,\mu_{k\neq i},T}$. From a practical standpoint, the focus is often on the stress or deformation induced by adsorption, which can be determined by integrating the first constitutive equation.

$$\sigma(\varepsilon,\mu_i)=\sigma_0(\varepsilon)-\frac{\partial}{\partial\varepsilon}\left[\int_{-\infty,-\infty,\dots}^{\mu_1,\mu_2,\dots}\sum_{i\in f}n_i d\mu_i\right]_{\mu_j,T} \qquad [5.12]$$

where $\sigma_0(\varepsilon)=\sigma(\varepsilon,\mu_i\rightarrow-\infty)$ represents the behavior of the "empty" medium, that is, in the absence of fluid. This poromechanical formulation generalizes the conventional formulation by incorporating cases where the fluid in the pores is not necessarily a bulk fluid. Specifically, if the fluid in the pores is a bulk fluid, the Lagrangian concentration of a fluid component is the product of the Lagrangian porosity and the bulk density of that component: $n_i=\phi\rho_i^b$. Consequently, since bulk densities depend solely on μ_i and T (and are independent of ε), the Biot coefficients for the various fluid components are all equal: $b_i^{eff}=\frac{\partial\phi}{\partial\varepsilon}\Big|_{\mu_j,T}=b$, defining a single Biot coefficient. Similarly, by applying the Gibbs–Duhem relationship for the bulk fluid, it can be shown that the Biot moduli are also equal $\frac{1}{N_{ij}^{eff}}=\frac{\partial\phi}{\partial P}\Big|_{\varepsilon,\mu_{k\neq j},T}=\frac{1}{N}$, which defines a single Biot modulus. The pressure increment naturally appears in the constitutive equations ($\sum_{i\in f}\rho_i^b d\mu_i=dP$) so the generalized equations [5.11] reduce to the conventional equations [5.8].

A simpler case is that of a pure fluid (a single constituent). In this scenario, the chemical potential μ can be substituted with the bulk pressure P using the Gibbs–Duhem relationship for the reservoir fluid. It is important to note that the Gibbs-Duhem relationship applies only to the external fluid reservoir, meaning the pressure considered is that of the external reservoir, not the pressure within the pores. This yields a form of the constitutive equations that is analogous to the standard poromechanics, reflecting the thermodynamics of the problem:

$$\begin{cases} d\sigma=K^{eff}d\varepsilon-b^{eff}dP \\ d\phi^{eff}=b^{eff}d\varepsilon+\frac{dP}{N^{eff}} \end{cases} \qquad [5.13]$$

Here, we introduce the effective porosity $\phi^{eff}=n/\rho^b$, which can be interpreted as the porosity that the fluid would occupy if its density were the bulk density. This results in a unique Biot coefficient $b^{eff}=-\frac{\partial\sigma}{\partial P}\Big|_{\varepsilon,T}=\frac{\partial\phi^{eff}}{\partial\varepsilon}\Big|_{P,T}$, and a unique Biot modulus $\frac{1}{N}=\frac{\partial\phi^{eff}}{\partial P}\Big|_{\varepsilon,T}$, both defined from the pressure of the bulk fluid. However, it should be noted that all of these moduli are effective, as they depend on the fluid, and the usual restrictions such as $0<b<1$ do not apply to effective moduli. In particular, values of b^{eff} greater than 1 are possible, allowing for swelling during immersion experiments. In practice, the stress or deformation induced by the adsorption of a pure fluid is obtained by integrating the first constitutive equation:

$$\sigma(\varepsilon,\mu) = \sigma_0(\varepsilon) - \frac{\partial}{\partial\varepsilon}\left[\int_{-\infty}^{\mu} n d\mu\right]_{\mu,T} = \sigma_0(\varepsilon) - \frac{\partial}{\partial\varepsilon}\left[\int_0^P \frac{n}{\rho^b} dP\right]_{P,T} \qquad [5.14]$$

In the elastic framework, the formalism can equivalently be established in terms of imposed stress rather than imposed deformation. In the case of a pure fluid, this yields:

$$\varepsilon(\sigma,P) = \varepsilon_0(\sigma) + \frac{\partial}{\partial\sigma}\left[\int_0^P \frac{n}{\rho^b} dP\right]_{P,T} \qquad [5.15]$$

The mechanical effect of the fluid is directly related to the fluid content n, with a complex relationship that is both integral and differential. Implementing this requires understanding how n depends on both osmotic (μ or P) and mechanical (ε or σ) loads. Typically, adsorption is characterized under immersion conditions where these two loads are not independently controlled ($\sigma = -P$). Additionally, investigations are generally conducted on the scale of the heterogeneous medium. However, understanding the coupling between adsorption and deformation requires examination at the pore scale. Different pore sizes and morphologies can exhibit very different couplings, meaning experiments measure only the average result of heterogeneous behaviors. To address this, molecular approaches are valuable complements to experimental methods, allowing (1) the study of coupling at the adsorption scale (micropores, mesopore surfaces) and (2) a complete characterization of this coupling by independently applying mechanical and osmotic loads.

Before discussing the contribution of molecular simulation to the study of mechanical-adsorption interactions, it is useful to differentiate between two types of mechanical effects of adsorption, depending on whether it occurs at the surface of mesopores or within micropores (see Figure 5.15). For surface adsorption, the mechanical effect manifests as a surface stress. This is because, in the case of surface adsorption, the density at the center of the pore remains at the bulk density (which is independent of the pore thickness e, as shown in Figure 5.15(a)). Consequently, $\frac{\partial}{\partial e}\left[\frac{N}{\rho^b}\right]_{P,T} = 1$, implying that the normal mechanical force at the interface equals the bulk pressure, similar to the situation without adsorption. However, in the tangential direction to the interface, adsorption alters the surface stress, leading to deformation. This mechanical effect of surface adsorption is also referred to as the Bangham effect. Since adsorption in coal predominantly involves micropores, we will not delve into how the extended poromechanical theory applies specifically to surface adsorption here. For detailed discussions on this topic, the reader is referred to Vandamme et al. (2010) and Brochard et al. (2012). When the pore size is sufficiently small, the fluid at the center of the pore is no longer in its bulk state (see Figure 5.15(b)). In such cases, an additional normal mechanical effect, known as disjoining pressure, arises at the interface. The onset of disjoining pressure depends on the range of molecular interactions and thus varies significantly from one system to another.

For methane and carbon dioxide adsorbed in coal, where interactions are primarily van der Waals, it is commonly accepted that disjoining pressure occurs in pores smaller than 1–2 nm. This is illustrated in Figure 5.16, which shows the pressure calculated by NLDFT exerted by methane confined in planar graphitic pores (Grégoire et al. 2018). The data indicate that, after a series of significant oscillations for ultramicropores, the pressure converges to its bulk value for pores around 2 nm in size.

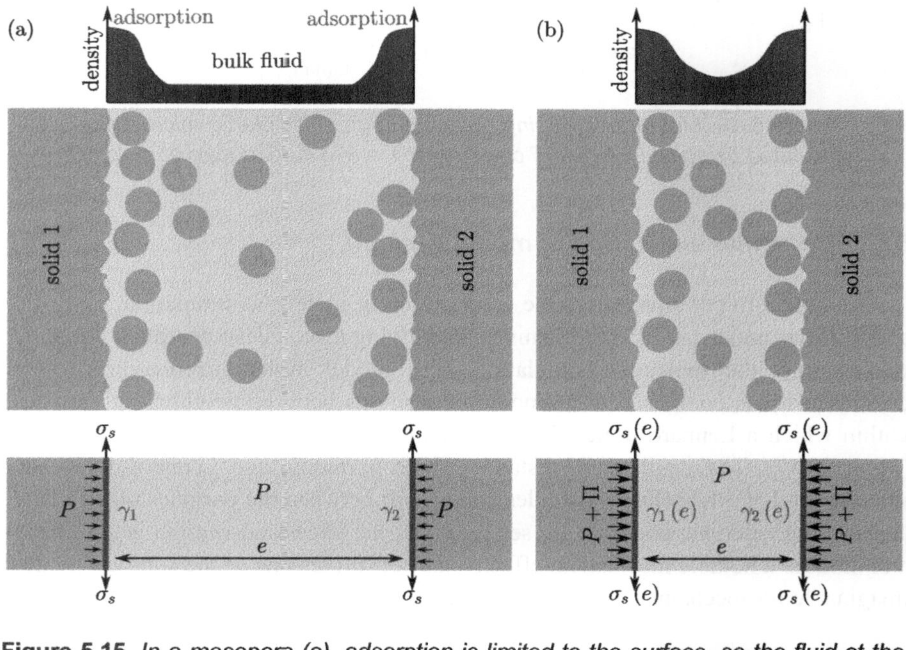

Figure 5.15. *In a mesopore (a), adsorption is limited to the surface, so the fluid at the center of the pore remains in its bulk state. The mechanical effect of adsorption is thus confined to a tangential force on the surface, altering the surface stress at the fluid–solid interface. In a micropore (b), the fluid state deviates from the bulk state at the center of the pore, resulting in an additional normal force, known as disjoining pressure*

Nevertheless, due to the complex nanostructure of coals, concepts such as pore size and fluid–solid interface become problematic. This is not an issue in itself, as the extended poromechanical formulation for adsorption does not rely on these concepts but instead uses the overall fluid content.

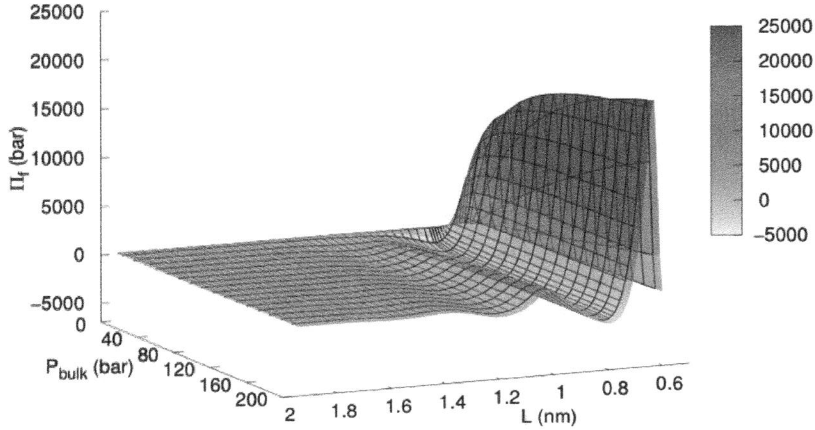

Figure 5.16. *Methane disjoining pressure in graphitic planar micropores, calculated using NLDFT/SAFT coupling at T = 373 K (Grégoire et al. 2018)*

5.5.2. Molecular simulation of model systems

We now turn our attention to the contribution of molecular simulation. To begin, and to better understand the mechanisms underlying mechanics-adsorption coupling, we examine the molecular simulation of a model system: a one-dimensional crystalline chain composed of harmonic interactions between neighboring particles, within which a Lennard–Jones fluid can adsorb between the particles of the chain (Figure 5.17). The equilibrium distances (both harmonic and Lennard–Jones) are selected so that a few fluid molecules can adsorb between the particles of the chain, and the characteristic energies are set such that the excess adsorption is significant and induces a notable mechanical effect. This constitutes one of the simplest models imaginable for mechanics-adsorption coupling, and it proves to be quite instructive.

A series of grand canonical molecular simulations at various deformations and chemical potentials allows for precise characterization of fluid content and the stress induced by the fluid under different osmotic and mechanical loads (Figure 5.18). The fluid content exhibits a fairly typical evolution under osmotic loading (type I isotherm). In contrast, the evolution under mechanical loading is notably complex, including plateau effects and even decreasing fluid content with deformation. As we have seen, theory predicts a mechanical effect that depends on the derivative of fluid content with respect to deformation. Thus, the mechanical effect of the fluid is expected to vary significantly: disjoining when the content increases with deformation and cohesion when the content decreases with deformation. This is indeed what is observed in the stress data (Figure 5.18(b)). Additionally, the relationship [5.14] between fluid content and induced stress is confirmed to apply accurately.

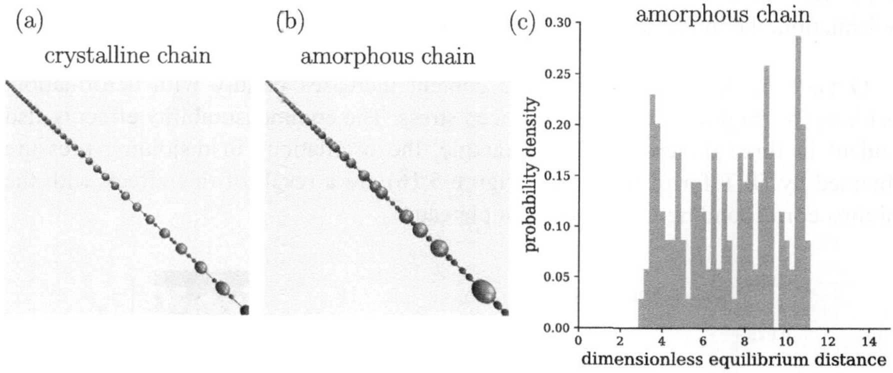

Figure 5.17. *Examples of molecular configurations for the one-dimensional crystalline (a) and amorphous (b) chains. Blue particles represent the atoms of the chain, and red particles represent the adsorbed fluid molecules. Distribution of equilibrium distances between atoms of the amorphous chain without fluid and free of stress (c)*

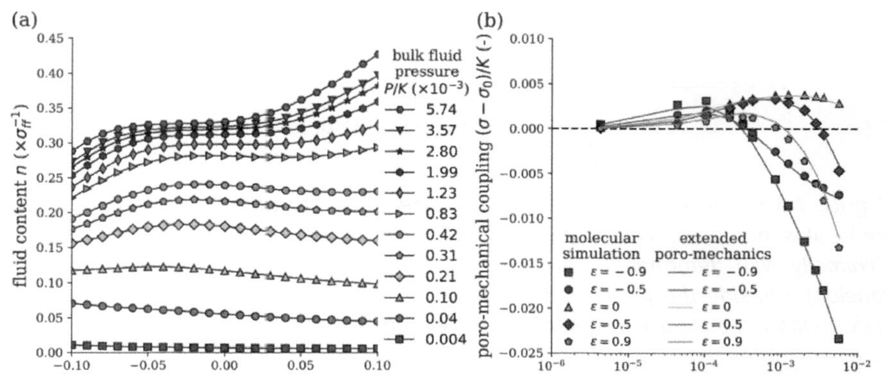

Figure 5.18. *Fluid content n (a) and stress induced by adsorption (b) in the crystalline one-dimensional chain. The quantities are dimensionless with respect to the incompressibility modulus for pressure/stress units and with respect to the fluid-fluid Lennard–Jones parameter for length units*

A detailed analysis helps us to better understand the origins of the complex poro-mechanical behavior observed in this seemingly simple case study. The nonlinear evolution of the fluid content with deformation is actually a consequence of the commensurability of the fluid molecules with the spacing between two particles of the chain (Figure 5.19). The fluid content plateau corresponds precisely to two fluid molecules between the particles of the chain. Since this plateau is associated with a fluid content that changes little or even decreases with deformation, the induced stress tends to be cohesive (tensile).

Outside of this plateau, the fluid content increases rapidly with deformation, resulting in a highly compressive induced stress. The commensurability effect is also evident in three dimensions. For example, the oscillations in disjoining pressure obtained by DFT for planar pores (Figure 5.16) are a result of this effect, with the minima corresponding to fluid content plateaus.

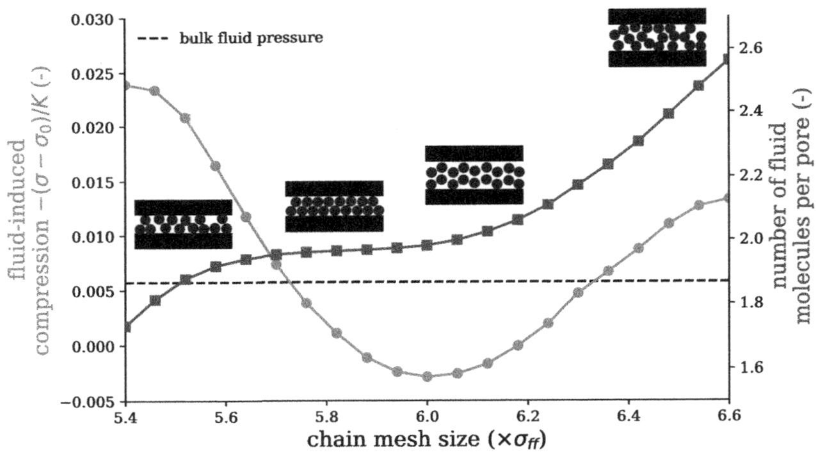

Figure 5.19. *The fluid content plateau corresponds to an integer number of fluid molecules per pore. This plateau is associated with a fluid content that changes minimally with deformation, leading to a cohesive mechanical force (i.e. inducing tension). Outside the plateau, the number of fluid molecules increases rapidly with deformation, resulting in a compressive mechanical force*

As previously noted, coal is not a crystalline material but rather an amorphous one with a wide range of pore sizes. Thus, a more relevant case study for understanding coal is the amorphous one-dimensional chain (Figure 5.17(b)). The parameters for this case are identical to those of the crystalline chain, except that the equilibrium distances between neighboring particles are randomly chosen to represent a varied PSD. Similarly, the poro-mechanical behavior of this amorphous chain is simulated in detail under various osmotic and mechanical loads (Figure 5.20).

The behavior observed is remarkably simple: the fluid content depends almost linearly on deformation, resulting in a compressive mechanical effect of the fluid that is nearly independent of the deformation state. Thus, despite being more complex than the initial crystalline case, this amorphous chain exhibits a much simpler apparent poro-mechanical coupling. As with the crystalline chain, it can be confirmed that the extended poro-mechanical formulation [5.14] applies to the amorphous chain.

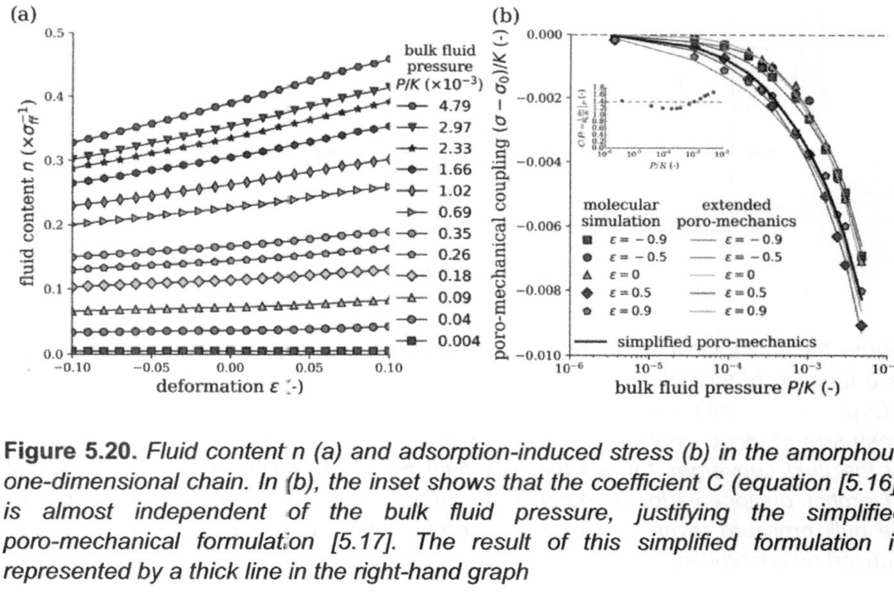

Figure 5.20. *Fluid content n (a) and adsorption-induced stress (b) in the amorphous one-dimensional chain. In (b), the inset shows that the coefficient C (equation [5.16]) is almost independent of the bulk fluid pressure, justifying the simplified poro-mechanical formulation [5.17]. The result of this simplified formulation is represented by a thick line in the right-hand graph*

When examining deformation at the scale of each pore, we observe highly heterogeneous deformations: some pores contract while others expand (Figure 5.21). For a given pore size, the pore deformation can be compared to the stress induced by adsorption in an equivalent crystalline chain (one where all pores are the same size). It is found that the deformation for a given pore size is very precisely correlated with the stress induced by adsorption (Figure 5.21).

Thus, the overall behavior of the amorphous chain results from the superposition of heterogeneous local behaviors at the pore scale. The global behavior is compressive/dilating because local behaviors are more often dilating than contracting, particularly in the smallest pores that accommodate one to two molecules.

This behavior is also applicable in 3D, where it is observed that the first two oscillations of disjoining pressure dominate (Figure 5.16).

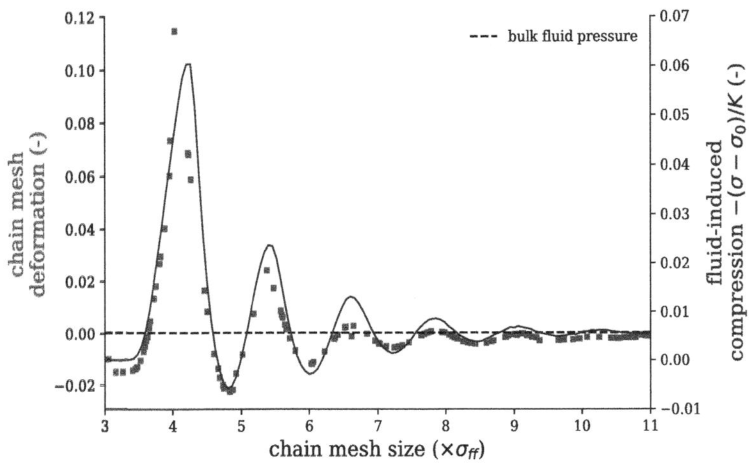

Figure 5.21. *The local deformation at the scale of each cell of the amorphous chain as a function of the cell size is shown in red. In blue, the fluid-induced compression ($\sigma(\varepsilon,\mu) - \sigma_0(\varepsilon)$, with $\sigma_0(\varepsilon) = \sigma(\varepsilon, \mu \to -\infty)$) in a crystalline chain as a function of the mesh size of this crystalline chain. In the large mesh limit, the compression induced by the fluid converges toward the bulk fluid pressure, indicating that the effect of adsorption disappears for large pores. Aligning the origin of the deformation axis with the bulk pressure reveals an excellent correlation between deformation and stress induced by adsorption*

5.5.3. *Simplified poro-mechanical model applied to coal*

The overall behavior of the amorphous chain suggests that, for microporous materials with heterogeneous PSDs (e.g. coal), it is reasonable to consider a linear relationship between fluid content and deformation:

$$n(\varepsilon, \mu) = n_0(\mu)(1 + C\varepsilon) \qquad [5.16]$$

Generally, the coupling coefficient C may depend on the osmotic loading (μ or P). However, this coefficient is found to be nearly independent of the osmotic loading for the amorphous chain (see insert in Figure 5.20(b)). This leads to a significantly simplified poro-mechanical formulation:

$$\sigma(\varepsilon,\mu) = \sigma_0(\varepsilon) - C\int_{-\infty}^{\mu} n_0 d\mu = \sigma_0(\varepsilon) - C\int_0^P \frac{n_0}{\rho^b} dP \qquad [5.17]$$

Applying this simplified formulation to the case of the amorphous chain predicts the poro-mechanical coupling quite well (Figure 5.20(b)). The use of this simplified model requires knowledge of the constant C and the fluid content n_0 as a function of the osmotic loading at a constant mechanical load. In practice, adsorption experiments are typically immersion experiments where osmotic and mechanical loads are not independent ($\sigma = -P$). However, by simultaneously measuring the fluid content n_{im} and the deformation ε_{im} during an immersion experiment, it is possible to deduce n_0 and C from equation [5.16] and by differentiating equation [5.14] with respect to P:

$$n_0 = \frac{n_{im}}{1+C\varepsilon_{im}} \text{ and } C = \frac{1+\left.\frac{d\sigma_0}{d\varepsilon}\right|_{\varepsilon=\varepsilon_{im}} \frac{d\varepsilon_{im}}{dP}}{\frac{n_{im}}{\rho^b} - \varepsilon_{im}\left(1+\left.\frac{d\sigma_0}{d\varepsilon}\right|_{\varepsilon=\varepsilon_{im}} \frac{d\varepsilon_{im}}{dP}\right)} \qquad [5.18]$$

Using the joint adsorption-deformation measurements from Ottiger et al. (2006, 2008) (Figure 5.3), it is observed that the coefficient C is approximately independent of bulk fluid pressure (Figure 5.22(a)), similar to the amorphous chain. This finding is quantitatively confirmed by molecular simulations of methane adsorption in flexible CS1000 (Brochard et al. 2012). However, the coefficient C does depend on the nature of the fluid, with a significantly larger value for carbon dioxide compared to methane. Using these average C values, coal swelling can be predicted from the isotherms estimated by molecular simulation (Figure 5.7) and the simplified model (equation [5.17]). These predictions align well with experimental results (Figure 5.22(b)).

The simplified poro-mechanical formulation [5.17] appears to be well suited for the case of coal. Although this global response is the result of highly heterogeneous local responses, analyzing this effective response remains valuable. The coupling coefficient C is directly related to the effective Biot coefficient: $b^{eff} = \left.\frac{\partial \phi^{eff}}{\partial \varepsilon}\right|_{P,T} = C\phi_0^{eff}$, where $\phi_0^{eff} = \frac{n_0}{\rho^b}$ represents the effective porosity in the undeformed state. In the case of a bulk fluid within the pores, the coefficient C is similar to the ratio between the Biot coefficient and the reference porosity. Often, porosity and the Biot coefficient are correlated, and their ratio is approximately one. However, for coal, the C values (ranging from 6 to 10) are significantly higher.

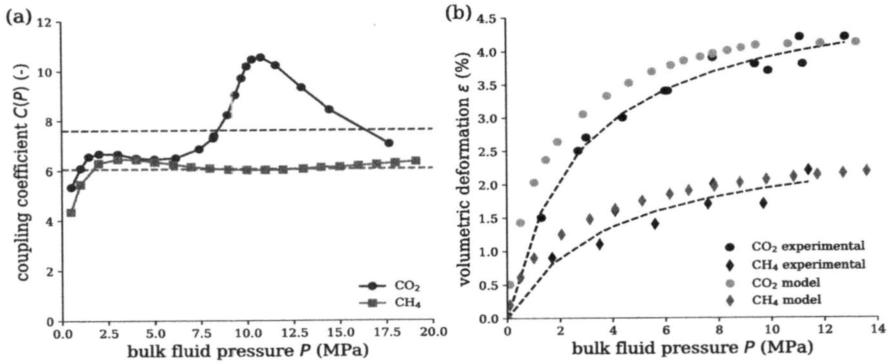

Figure 5.22. *(a) Coupling coefficient C for the adsorption of methane and carbon dioxide in coal, estimated from the adsorption and deformation measurements under immersion by Ottiger et al. (2006, 2008). The simplified poro-mechanical model assumes pressure-independent C coefficients, represented here by the dashed lines. (b) The application of the simplified poro-mechanical model using the fluid contents obtained from molecular simulations provides a reasonable estimate of coal swelling as measured by Ottiger et al. (2006, 2008)*

Another perspective on this coefficient is to consider that the adsorbed fluid behaves as a phase with density $\rho^a(\mu, T) \neq \rho^b$, which depends solely on osmotic loading and occupies a porosity ϕ (not effective!) with an incompressible solid, such that $\phi \approx \phi_0 + \varepsilon$. In this context, the fluid content takes the linearized form [5.16]: $n = \rho^a \phi = \rho^a \phi_0 \left(1 + \frac{\varepsilon}{\phi_0}\right)$, where the coefficient C is identified as the inverse of the porosity, $C = 1/\phi_0$. Here, the stress induced by adsorption simplifies to $dP_a = -d\sigma|_{\varepsilon,T} = \frac{\rho^a}{\rho^b} dP$, so the mechanical force resembles an "adsorption pressure" P_a, which is the bulk pressure scaled by the ratio between the adsorbed density and the bulk density. Assuming that the adsorbed density is independent of deformation implies treating the adsorbed phase as extensive with pore volume, leading to a form of the Gibbs–Duhem relationship for the adsorbed phase: $dP_a = \rho^a d\mu$. Of course, this is an effective perspective, and in reality, the response is heterogeneous from one pore to another.

The linearized model [5.16] can be extended to the case of a methane/carbon dioxide mixture, assuming that the coefficients C for pure fluids continue to apply to mixtures:

$$n_i(\varepsilon, \mu_j) = n_{i,0}(\mu_j)(1 + C_i \varepsilon) \quad [5.19]$$

The mechanical response induced by adsorption then becomes:

$$\sigma(\varepsilon, \mu_i) = \sigma_0(\varepsilon) - \int_{-\infty,-\infty,\dots}^{\mu_1,\mu_2,\dots} \sum_{i\in f} C_i n_{i,0} d\mu_i \quad [5.20]$$

For carbon storage and enhanced coal bed methane recovery, changes in osmotic loading correspond to changes in the fluid mixture composition while maintaining a constant bulk pressure. Adsorption isotherms can be obtained via grand canonical molecular simulations (Figure 5.9). Assuming a dry elastic linear behavior ($\sigma_0(\varepsilon) = K\varepsilon$), the differential swelling resulting from the replacement of methane with carbon dioxide can be predicted by:

$$\Delta\varepsilon = \frac{1}{K}\int_{\mu_{CH_4}^{pure},-\infty}^{\mu_{CH_4},\mu_{CO_2}} \left(C_{CH_4} n_{CH_4,0} d\mu_{CH_4} + C_{CO_2} n_{CO_2,0} d\mu_{CO_2}\right) \quad [5.21]$$

Figure 5.23 presents the predicted differential swellings for coals at various depths. It is noteworthy that this swelling depends relatively little on depth and appears almost linear with the carbon dioxide content.

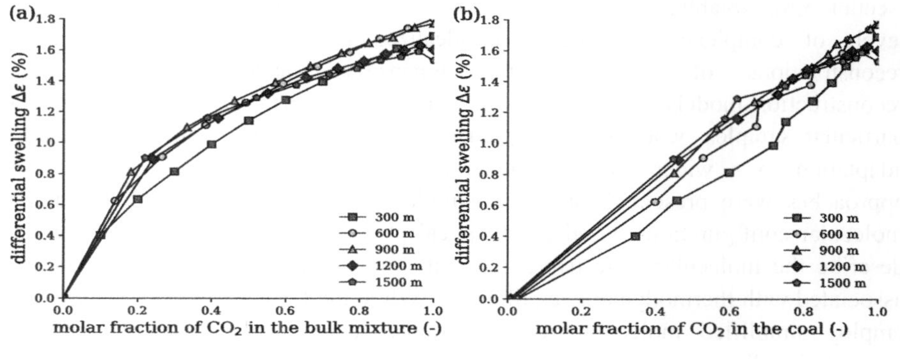

Figure 5.23. *Differential swelling when replacing coal gas with carbon dioxide at constant bulk fluid pressure, as predicted from the extended poro-mechanical model for mixtures and the mixture adsorption isotherms obtained via molecular simulation. (a) Swelling as a function of bulk composition, and (b) swelling as a function of adsorbed composition*

5.6. Conclusion

In this chapter, we have introduced molecular modeling as applied to coal geomechanics. Molecular approaches complement experimental methods, proving highly relevant for understanding and predicting physical phenomena that originate at the nanometer scale. In the case of coal, fluid adsorption within the organic matrix has driven the use of molecular methods. On the one hand, adsorption leads to fluid accumulation in micropores and mesopores, resulting in fluid contents far exceeding those in the bulk phase, making coal an unconventional hydrocarbon reservoir and a potential site for geological carbon storage. On the other hand, adsorption induces unusual poro-mechanical couplings: the organic matrix swells in response to adsorption, and this swelling depends on the nature of the fluid, strongly correlating with adsorption affinity. Therefore, any mechanical or osmotic loading of a coal reservoir leads to a response that cannot be adequately explained by conventional poro-mechanical models. Molecular approaches enable quantitative predictions of fluid adsorption, both for pure fluids and mixtures in coal, while also helping to understand and model these unusual poro-mechanical couplings. This extended poro-mechanics framework serves as the basis for more comprehensive geomechanical models at the scale of rock or reservoirs (see Chapters 4 and 6).

Throughout this chapter, we explored the molecular models of the organic matrix in coal (section 5.2), atomistic simulation methods (section 5.3), and DFT approaches (section 5.4) applied to adsorption in coal, as well as mechanics-adsorption couplings (section 5.5). Notably, we discussed various molecular models of coal with differing levels of complexity, ranging from idealized planar pore representations to reconstructions of disordered three-dimensional nanostructures. Although reconstruction models offer more realistic representations, they remain specific to particular samples, whereas using planar model pores along with PSDs enables adaptation to a wider range of materials. Two main categories of molecular approaches were presented: atomistic simulation methods, which aim to sample molecular configurations based on statistical physics, and DFT approaches, which describe the molecular state using a density functional to minimize the potential associated with thermodynamic equilibrium. While the DFT methods discussed here employ simplified molecular representations, such as planar pores, they are computationally more efficient than atomistic simulations, and recent advancements make it feasible to consider more complex geometries. Several application examples are provided, including methane, carbon dioxide and water adsorption in coal.

Lastly, understanding adsorption-mechanical couplings necessitates revisiting the conventional thermodynamic description of porous media. It emerges that the stress induced by adsorption and the fluid content are linked by a Maxwell relation, which is challenging to evaluate experimentally. Molecular simulations, using both model and realistic systems, help clarify the mechanisms underlying adsorption-mechanical couplings in coal. The molecular-scale response is highly heterogeneous and dominated by the commensurability of micropores with fluid molecules. Nonetheless, a macroscopic model can be derived to accurately capture the experimentally measured deformations induced by adsorption.

In this chapter, we have specifically focused on the contribution of molecular modeling to adsorption-mechanical coupling. This perspective is not exhaustive, as other key issues in coal geomechanics also benefit from insights gained through molecular simulation. One such issue is fluid transport in the microporous matrix, which is a major practical concern because it governs the recoverable or storable quantities of hydrocarbons and carbon, as well as the associated kinetics. For further reading, interested readers may refer to Falk et al. (2015), which demonstrates that although Darcy's law may no longer apply, there remains a linear relationship between pressure gradient and flow rate, with the proportionality coefficient depending on fluid content and directly correlating with the diffusion coefficient. This is just one aspect of the broader field, and molecular approaches prove useful for addressing many other challenges, such as chemical reactivity, coal maturation or coal-solvent interactions (see Mathews et al. (2011) for a comprehensive review of various applications).

5.7. References

Berendsen, H.J.C., Grigera, J.R., Straatsma, T.P. (1987). The missing term in effective pair potentials. *The Journal of Physical Chemistry*, 91(24), 6269–6271. doi: 10.1021/j100308a038.

Brenner, D.W. (1990). Empirical potential for hydrocarbons for use in simulating the chemical vapor deposition of diamond films. *Physical Review B*, 42(15), 9458–9471. doi: 10.1103/PhysRevB.46.1948.2.

Brenner, D.W., Shenderova, O.A., Harrison, J.A., Stuart, S.J., Ni, B., Sinnott, S.B. (2002). A second-generation reactive empirical bond order (REBO) potential energy expression for hydrocarbons. *Journal of Physics: Condensed Matter*, 14(4), 783–802. doi: 10.1088/0953-8984/14/4/312.

Brochard, L., Vandamme, M., Pellenq, R.J.M., Fen-Chong, T. (2012). Adsorption-induced deformation of microporous materials: Coal swelling induced by CO_2-CH_4 competitive adsorption. *Langmuir*, 28(5), 2659–2670. doi: 10.1021/la204072d.

Brooks, B.R., Bruccoleri, R.E., Olafson, B.D., States, D.J., Swaminathan, S., Karplus, M. (1983). CHARMM: A program for macromolecular energy, minimization, and dynamics calculations. *Journal of Computational Chemistry*, 4(2), 187–217. doi: 10.1002/jcc.540040211.

Cornell, W.D., Cieplak, P., Bayly, C.I., Gould, I.R., Merz, K.M., Ferguson, D.M., Spellmeyer, D.C., James, T.F., Caldwell, W., Kollman, P.A. (1995). A second generation force field for the simulation of proteins, nucleic acids, and organic molecules. *Journal of the American Chemical Society*, 117(19), 5179–5197. doi: 10.1021/ja00124a002.

Davis, H.T. (1996). *Statistical Mechanics of Phases, Interfaces and Thin Films*. University of Texas Press, Austin.

van Duin, A.C.T., Dasgupta, S., Lorant, F., Goddard, W.A. (2001). ReaxFF: A reactive force field for hydrocarbons. *The Journal of Physical Chemistry A*, 105(41), 9396–9409. doi: 10.1021/jp004368u.

Errington, J.R., Boulougouris, G.C., Economou, I.G., Panagiotopoulos, A.Z., Theodorou, D.N. (1998). Molecular simulation of phase equilibria for water–methane and water–ethane mixtures. *The Journal of Physical Chemistry B*, 102(44), 8865–8873. doi: 10.1021/jp981627v.

Evans, R. (1992). Density functionals in the theory of nonuniform fluids. In *Fundamentals of Inhomogeneous Fluids*, Henderson, D. (ed.). Marcel Dekker, New York. doi: 10.1201/9781003066934.

Falk, K., Coasne, B., Pellenq, R., Ulm, F.-J., Bocquet, L. (2015). Subcontinuum mass transport of condensed hydrocarbons in nanoporous media. *Nature Communications*, 6(1), 6949. doi: 10.1038/ncomms7949.

Fernandez-Alos, V., Watson, J.K., Vander Wal, R., Mathews, J.P. (2011). Soot and char molecular representations generated directly from HRTEM lattice fringe images using Fringe3D. *Combustion and Flame*, 158(9), 1807–1813. doi: 10.1016/j.combustflame.2011.01.003.

Gasteiger, J. and Marsili, M. (1980). Iterative partial equalization of orbital electronegativity – A rapid access to atomic charges. *Tetrahedron*, 36(22), 3219–3228. doi: 10.1016/0040-4020(80)80168-2.

Grégoire, D., Malheiro, C., Miqueu, C. (2018). Estimation of adsorption-induced pore pressure and confinement in a nanoscopic slit pore by a density functional theory. *Continuum Mechanics and Thermodynamics*, 30(2), 347–363. doi: 10.1007/s00161-017-0602-x.

Harris, J.G. and Yung, K.H. (1995). Carbon dioxide's liquid-vapor coexistence curve and critical properties as predicted by a simple molecular model. *The Journal of Physical Chemistry*, 99(31), 12021–12024. doi: 10.1021/j100031a034.

Jagiello, J. and Olivier, J.P. (2013). 2D-NLDFT adsorption models for carbon slit-shaped pores with surface energetical heterogeneity and geometrical corrugation. *Carbon*, 55, 70–80. doi: 10.1016/j.carbon.2012.12.011.

Jain, S.K., Pellenq, R.J.-M., Pikunic, J.P., Gubbins, K.E. (2006a). Molecular modeling of porous carbons using the hybrid reverse Monte Carlo method. *Langmuir*, 22(24), 9942–9948. doi: 10.1021/la053402z.

Jain, S.K., Gubbins, K.E., Pellenq, R.J.-M., Pikunic, J.P. (2006b). Molecular modeling and adsorption properties of porous carbons. *Carbon*, 44(12), 2445–2451. doi: 10.1016/j.carbon.2006.04.034.

Jorgensen, W.L. and Tirado-Rives, J. (1988). The OPLS [optimized potentials for liquid simulations] potential functions for proteins, energy minimizations for crystals of cyclic peptides and crambin. *Journal of the American Chemical Society*, 110(6), 1657–1666. doi: 10.1021/ja00214a001.

Jorgensen, W.L., Maxwell, D.S., Tirado-Rives, J. (1996). Development and testing of the OPLS all-atom force field on conformational energetics and properties of organic liquids. *Journal of the American Chemical Society*, 118(45), 11225–11236. doi: 10.1021/ja9621760.

Kurniawan, Y., Bhatia, S.K., Rudolph, V. (2006). Simulation of binary mixture adsorption of methane and CO_2 at supercritical conditions in carbons. *AIChE Journal*, 52(3), 957–967. doi: 10.1002/aic.10687.

Landers, J., Gor, G.Y., Neimark, A.V. (2013). Density functional theory methods for characterization of porous materials. *Colloids and Surfaces A: Physicochemical and Engineering Aspects*, 437, 3–32. doi: 10.1016/j.colsurfa.2013.01.007.

Levine, D.G., Schlosberg, R.H., Silbernagel, B.G. (1982). Understanding the chemistry and physics of coal structure A review. *Proceedings of the National Academy of Sciences*, 79(10), 3365–3370. doi: 10.1073/pnas.79.10.3365.

Lutsko, J.F. (2010). Recent developments in classical density functional theory. *Advances in Chemical Physics*, 144, 1. doi: 10.1002/9780470564318.ch1.

Malheiro, C. (2014). Adsorption de systèmes gaz/eau en milieu confiné : modélisation par une approche DFT/SAFT couplée à une étude expérimentale. PhD Thesis, Université de Pau et des Pays de l'Adour, Anglet.

Malheiro, C., Mendiboure, B , Plantier, F., Blas, F.J., Miqueu, C. (2014a). Density functional theory for the description of spherical non-associating monomers in confined media using the SAFT-VR equation of state and weighted density approximations. *The Journal of Chemical Physics*, 140(13), 134707. doi: 10.1063/1.4869996.

Malheiro, C., Mendiboure, B., Miguez, J.M., Piñeiro, M.M., Miqueu, C. (2014b). Nonlocal density functional theory and grand canonical Monte Carlo molecular simulations of water adsorption in confined media. *The Journal of Physical Chemistry C*, 118(43), 24905–24914. doi: 10.1021/jp505239e.

Malheiro, C., Mendiboure, B., Plantier, F., Guatarbes, B., Miqueu, C. (2015). An accurate model for the filling pressure of carbon slit-like micropores. *Fluid Phase Equilibria*, 404, 118–123. doi: 10.1016/j.fluid.2015.07.003.

Mathews, J.P. and Chaffee, A.L. (2012). The molecular representations of coal – A review. *Fuel*, 96, 1–14. doi: 10.1016/j.fuel.2011.11.025.

Mathews, J.P., van Duin, A.C.T., Chaffee, A.L. (2011). The utility of coal molecular models. *Fuel Processing Technology*, 92(4), 718–728. doi: 10.1016/j.fuproc.2010.05.037.

Meyers, R.A. (ed.). (1982). *Coal Structure*. Elsevier, Amsterdam. doi: 10.1016/B978-0-124-93080-3.X5001-9.

Miqueu, C. and Grégoire, D. (2020). Estimation of pore pressure and phase transitions of water confined in nanopores with non-local density functional theory. *Molecular Physics*, 118(9–10), e1742935. doi: 10.1080/00268976.2020.1742935.

Müller, E.A. and Gubbins, K.E. (2001). Molecular-based equations of state for associating fluids: A review of SAFT and related approaches. *Industrial & Engineering Chemistry Research*, 40(10), 2193–2211. doi: 10.1021/ie000773w.

Müller, E.A., Rull, L.F., Vega, L.F., Gubbins, K.E. (1996). Adsorption of water on activated carbons: A molecular simulation study. *The Journal of Physical Chemistry*, 100(4), 1189–1196. doi: 10.1021/jp952233w.

Neimark, A.V., Lin, Y., Ravikovitch, P.I., Thommes, M. (2009). Quenched solid density functional theory and pore size analysis of micro-mesoporous carbons. *Carbon*, 47(7), 1617–1628. doi: 10.1016/j.carbon.2009.01.050.

Ottiger, S., Pini, R., Storti, G., Mazzotti, M., Bencini, R., Quattrocchi, F., Sardu, G., Deriu, G. (2006). Adsorption of pure carbon dioxide and methane on dry coal from the sulcis coal province (SW Sardinia, Italy). *Environmental Progress*, 25(4), 355–364. doi: 10.1002/ep.10169.

Ottiger, S., Pini, R., Storti, G., Mazzotti, M. (2008). Competitive adsorption equilibria of CO_2 and CH_4 on a dry coal. *Adsorption*, 14(4–5), 539–556. doi: 10.1007/s10450-008-9114-0.

Puziy, A.M., Poddubnaya, O.I., Gawdzik, B., Sobiesiak, M. (2016). Comparison of heterogeneous pore models QSDFT and 2D-NLDFT and computer programs ASiQwin and SAIEUS for calculation of pore size distribution. *Adsorption*, 22(4–6), 459–464. doi: 10.1007/s10450-015-9704-6.

Rappe, A.K. and Goddard, W.A. (1991). Charge equilibration for molecular dynamics simulations. *The Journal of Physical Chemistry*, 95(8), 3358–3363. doi: 10.1021/j100161a070.

Ravikovitch, P.I., Vishnyakov, A., Russo, R., Neimark, A.V. (2000). Unified approach to pore size characterization of microporous carbonaceous materials from N_2, Ar, and CO_2 adsorption isotherms. *Langmuir*, 16(5), 2311–2320. doi: 10.1021/la991011c.

Roth, R. (2010). Fundamental measure theory for hard-sphere mixtures: A review. *Journal of Physics: Condensed Matter*, 22(6), 063102. doi: 10.1088/0953-8984/22/6/063102.

Steele, W.A. (1973). The physical interaction of gases with crystalline solids. *Surface Science*, 36(1), 317–352. doi: 10.1016/0039-6028(73)90264-1.

Tarazona, P. (1984). A density functional theory of melting. *Molecular Physics*, 52(1), 81–96. doi: 10.1080/00268978400101071.

Tenney, C.M. and Lastoskie, C.M. (2006). Molecular simulation of carbon dioxide adsorption in chemically and structurally heterogeneous porous carbons. *Environmental Progress*, 25(4), 343–354. doi: 10.1002/ep.10168.

Tsuzuki, S., Uchimaru, T., Tanabe, K. (1998). A new ab initio based model potential for methane. *Chemical Physics Letters*, 287(3–4), 327–332. doi: 10.1016/S0009-2614(98)00193-6.

Vandamme, M., Brochard, L., Lecampion, B., Coussy, O. (2010). Adsorption and strain: The CO_2-induced swelling of coal. *Journal of the Mechanics and Physics of Solids*, 58(10), 1489–1505. doi: 10.1016/j.jmps.2010.07.014.

Vassilev, S.V., Kitano, K., Vassileva, C.G. (1996). Some relationships between coal rank and chemical and mineral composition. *Fuel*, 75(13), 1537–1542. doi: 10.1016/0016-2361(96)00116-0.

Vassilev, S.V., Kitano, K., Vassileva, C.G. (1997). Relations between ash yield and chemical and mineral composition of coals. *Fuel*, 76(1), 3–8. doi: 10.1016/S0016-2361(96)00181-0.

Vishnyakov, A., Piotrovskaya, E.M., Brodskaya, E.N. (1998). Capillary condensation and melting/freezing transitions for methane in slit coal pores. *Adsorption*, 4(3/4), 207–224. doi: 10.1023/A:1008873414503.

6

Reservoir-Scale Modeling

Patrick DANGLA
Navier, ENPC, CNRS, Institut Polytechnique de Paris, École nationale des ponts et chaussées, Université Gustave Eiffel, Marne-la-Vallée, France

6.1. Introduction

Coal seams naturally contain methane (Moore 2012), and the estimated global reserves range between 3,000 and 9,000 Tscf[1] (Liu et al. 2011). Methane has long been a concern for the coal industry due to its associated safety risks. Consequently, safety concerns were the initial motivation for detecting and extracting methane from coal seams prior to mining operations (Karacan et al. 2011). Additionally, monetizing the extracted methane and reducing greenhouse gas emissions involving methane[2] are the other key reasons for methane extraction from coal deposits. Furthermore, to mitigate carbon dioxide emissions into the atmosphere, it is planned to store CO_2 in geological formations, including unmineable coal seams located at great depths; this could be combined with enhanced methane recovery techniques, which not only increase methane production but also store carbon dioxide in the process. This emerging technique has attracted significant attention from researchers in recent years (Zhang et al. 2016).

Natural coal deposits consist of two pore networks: a fracture network and another composed of much smaller pores (Cui et al. 2004). The latter pore system

1 Tscf (trillion standard cubic feet) is equivalent to 28.3 km^3, giving a global reserve estimate of between 85,000 and 255,000 km^3.

2 Methane is 80 times more potent than carbon dioxide as a greenhouse gas.

Geomechanics of Coal Seams,
coordinated by Matthieu VANDAMME and Jean-Michel PEREIRA.

includes macropores (>50 nm), mesopores (2–50 nm) and micropores (<2 nm), but it is widely accepted that most methane exists primarily as absorbed gas in the coal's micropores (Liu et al. 2011). During gas depressurization, methane molecules desorb, exerting a tensile surface stress on the pore walls, causing the coal matrix to contract, leading to the opening of fractures (at constant volume) and an increase in permeability. Conversely, the injection of carbon dioxide tends to cause the coal to swell, which closes the fractures. These hydromechanical phenomena, involving variations in permeability, have significant economic implications and are therefore critical for numerical models designed to predict behavior at the scale of coal seams and reservoirs. A critical review of permeability models can be found in Liu et al. (2011).

This chapter focuses on the description of reservoir-scale numerical modeling, based on conservation and constitutive laws, some of which were established in previous chapters. In section 6.2, we revisit the key elements of these laws, such as the behavior of pure gases and gas mixtures, their adsorption in the coal matrix, the poromechanical behavior of fractured coal and gas transfer laws. Section 6.3 provides a review of the various three-dimensional field equations that need to be considered. The presence of displacement discontinuity surfaces is also addressed, allowing for the modeling of thin joints or fractures filled with porous materials. Section 6.5 presents two case studies to illustrate reservoir-scale numerical modeling of adsorption.

6.2. Isothermal constitutive laws of coal in the presence of adsorbates

The previous chapters have detailed the mechanical behavior of coal and the methods used to measure its parameters. Here, we will recall those that will be useful for the mathematical formulation and numerical resolution of the equations governing reservoir-scale gas transport and mechanical behavior at the reservoir scale.

6.2.1. *Gas behavior*

At very low pressures, all real gases behave as ideal gases, where the pressure P and the molar volume V of the gas follow the ideal gas law $PV = RT$. As pressure increases, the behavior of real gases deviates from that of ideal gases. This deviation from ideal behavior is generally measured using the compression factor $Z = PV/RT$, which deviates from 1 as pressure rises. At very high pressures, Z is always greater than 1, while at intermediate pressures, most gases exhibit values where $Z < 1$. By analogy with the ideal gas case, the fugacity, f, is defined through the expression for the specific chemical potential (in J/kg), μ, of the gas:

$$\mu = \frac{RT}{M} \ln f \qquad [6.1]$$

where M is the molar mass of the gas. The Gibbs–Duhem relationship, expressed for one mole of pure gas, $M\mathrm{d}(\mu) = V\mathrm{d}(P) - S\mathrm{d}(T)$ (with S being the molar entropy), allows us to relate the molar volume of the gas to the fugacity by:

$$RT\left(\frac{\partial(\ln f)}{\partial P}\right)_T = V \qquad [6.2]$$

Given an expression for the compression factor, $Z(P,T)$, as a function of pressure and temperature, we can identify the fugacity $f(P,T)$ or the fugacity coefficient $\varphi = f/P$ via:

$$\ln\varphi = \int_0^P (Z-1)\frac{\mathrm{d}P'}{P'} \qquad [6.3]$$

The Redlich–Kwong model (Redlich and Kwong 1949) is a good approximation for the behavior of real gases. It improves upon the van der Waals model and takes the form:

$$P = \frac{RT}{V-b} - \frac{a/\sqrt{T}}{V(V+b)} \qquad [6.4]$$

where a and b are constants. The compression factor is then implicitly given as a function of P and T by:

$$Z = \frac{Z}{Z-\frac{b}{RT}P} - \frac{a}{bRT^{1.5}}\frac{\frac{b}{RT}P}{Z+\frac{b}{RT}P} \qquad [6.5]$$

and integrating equation [6.3] yields:

$$\ln\varphi = Z - 1 - \ln\left(Z - \tfrac{b}{RT}P\right) - \frac{a}{bRT^{1.5}}\ln\left(1 + \tfrac{b}{RT}\tfrac{P}{Z}\right) \qquad [6.6]$$

The constants a and b in the model are determined using the values of the gas' critical pressure and critical temperature, as well as the fact that the isotherm $P-V$ at the critical point exhibits a horizontal inflection point:

$$\left(\frac{\partial(P)}{\partial V}\right)_T (P_c, T_c) = 0 \qquad [6.7]$$

$$\left(\frac{\partial^2(P)}{\partial V^2}\right)_T (P_c, T_c) = 0 \qquad [6.8]$$

From this, we derive:

$$b = V_c \left(2^{1/3} - 1\right) \qquad [6.9]$$

$$a = RT_c^{1.5} \left(\frac{V_c}{V_c - b} - Z_c\right)(V_c + b) \qquad [6.10]$$

where $Z_c = 1/3$ for all gases and $V_c = Z_c RT_c/P_c$.

The Redlich–Kwong model extends to gas mixtures according to the following laws:

$$b = \sum_i y_i b_i \qquad [6.11]$$

$$\sqrt{a} = \sum_i y_i \sqrt{a_i} \qquad [6.12]$$

where y_i is the mole fraction of component i in the mixture, and a_i, b_i are the constants associated with the behavior of pure gases.

Similarly, as in equation [6.3], the partial fugacity f_i and the associated fugacity coefficient $\varphi_i = f_i/(y_i P)$ can be determined through the integration:

$$\ln \varphi_i = \int_0^P (Z_i - 1)\frac{\mathrm{d}P'}{P'} \qquad [6.13]$$

where $Z_i = PV_i/RT$ is the partial compression factor and V_i is the partial molar volume[3]. Integration of equation [6.13] gives (Redlich and Kwong 1949):

3 According to Redlich and Kwong (1949), Z_i and Z satisfy the equation

$$Z_i = Z + \left(\frac{\partial (Z)}{\partial y_i}\right)_{P,T,y_j \neq y_i} - \sum_k y_k \left(\frac{\partial (Z)}{\partial y_k}\right)_{P,T,y_j \neq y_k}$$

where each partial derivative must be taken while keeping all mole fractions constant, except for the one with respect to which the derivative is performed. This equation applies to any molar quantity Z and its corresponding partial molar quantities Z_i. It is derived by applying the usual definition of partial molar quantities, replacing the number of moles with mole fractions. However, the derivation of this equation does not assume that the mole fractions y_i are independent variables. Therefore, it is not necessary, though permitted, to eliminate one of the y_i value in Z before applying the formula.

$$\ln \varphi_i = (Z-1)\frac{b_i}{b} - \ln\left(Z - \frac{b}{RT}P\right) - \frac{a}{bRT^{1.5}}\left(2\frac{\sqrt{a_i}}{\sqrt{a}} - \frac{b_i}{b}\right)\ln\left(1 + \frac{b}{RT}\frac{P}{Z}\right) \quad [6.14]$$

As an example, Figures 6.1 and 6.2 show the density and fugacities for a carbon dioxide and methane mixture as a function of pressure and the mole fractions of the mixture. The critical pressure and temperature values used are P_c = 7.375 MPa and T_c = 304.13 K for carbon dioxide, and P_c = 4.599 MPa and T_c = 190.56 K for methane (Lide 2010).

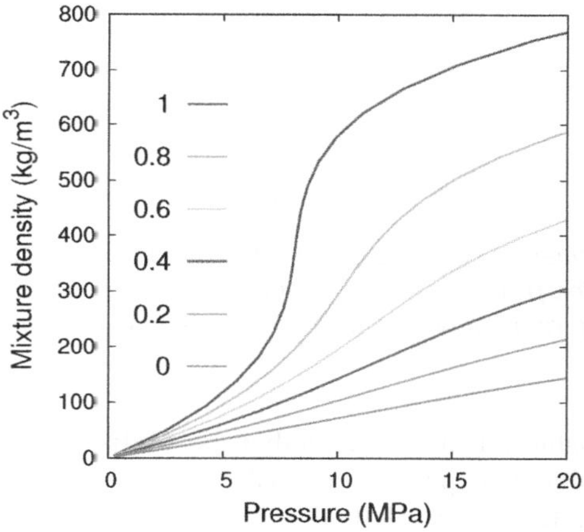

Figure 6.1. *Density of a carbon dioxide and methane mixture at 310 K. The parameter shown is the mole fraction of carbon dioxide. For a color version of this figure, see: www.iste.co.uk/vandamme/coal.zip*

6.2.2. *Adsorption in a deformable coal matrix*

Coal is a microporous material, meaning it contains a complex pore network where a significant portion of the pores are smaller than 2 nm in size (Sun et al. 2020). This gives coal its adsorption properties, which we will now recall. In what follows, temperature is considered constant.

In a microporous material, the state of the fluid within the pores depends on the nature of the material it is in, as the fluid molecules interact with the solid molecules. Therefore, the state of the fluid within the coal differs from that of a fluid in an infinite reservoir at thermodynamic equilibrium. For instance, the density of the adsorbed fluid is generally higher than that of the fluid in the reservoir.

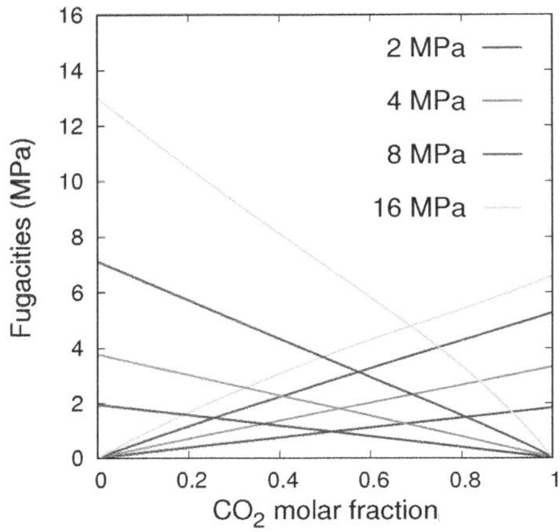

Figure 6.2. *Fugacities of carbon dioxide and methane at 310 K. The parameter shown is the pressure of the mixture. For a color version of this figure, see: www.iste.co.uk/vandamme/coal.zip*

For a non-deformable material, thermodynamics tells us that the mass of fluid adsorbed per unit volume of porous material, m^{ads}, is a function of the specific chemical potential of the adsorbed fluid, μ^{ads} : $m^{\text{ads}}(\mu^{\text{ads}})$. In the general case, however, the material is subjected to stresses and deforms. The amount adsorbed then also depends on the deformation of the material. In the case of an isotropic material, it depends only on the volumetric strain, ϵ. Since the deformation is typically small, the relation can be linearized with respect to the volumetric strain, yielding the following form:

$$m^{\text{ads}}(\epsilon, \mu) = m_0(\mu) + a(\mu)\epsilon \qquad [6.15]$$

where $m_0(\mu)$ is the quantity adsorbed in the absence of deformation, while $a(\mu)$ is a coupling term that accounts for matrix deformation. A good approximation for $a(\mu)$ is given by $c\,m_0(\mu)$, where c is a constant (Espinoza et al. 2014).

Suppose the adsorbed fluid in the microporous material is in thermodynamic equilibrium with the same fluid in a hypothetical infinite reservoir. This equilibrium requires the equality of the chemical potentials of the fluids. At equilibrium, the chemical potential of the adsorbed fluid, μ^{ads}, is equal to that of the fluid in the infinite reservoir, μ^{b}. The chemical potential of the reservoir fluid depends on the reservoir's pressure (and the temperature): $\mu^{\mathrm{b}}(p)$. We can then associate the pressure p of the infinite reservoir to the adsorbed fluid, which is in equilibrium with this reservoir, through the equality of the chemical potentials: $\mu^{\mathrm{ads}} = \mu^{\mathrm{b}}(p)$. This pressure is, by definition, the thermodynamic pressure of the adsorbed fluid. The density of the fluid in the hypothetical reservoir, ρ^{b}, is then related to the specific chemical potential by:

$$\mathrm{d}\left(\mu^{\mathrm{b}}\right) = \frac{1}{\rho^{\mathrm{b}}}\mathrm{d}\left(p\right) \qquad [6.16]$$

Note that this density is not equal to the actual density of the adsorbed fluid (note the superscript b associated to the bulk fluid of the infinite reservoir). At constant temperature, the Langmuir adsorption isotherm is commonly used to model the relationship between m_0 and p as $m_0(\mu^{\mathrm{b}}(p)) = m^{\mathrm{Lang}}(p)$ with:

$$m^{\mathrm{Lang}}(p) = m^{\mathrm{max}}\frac{p/p_0}{1+p/p_0} \qquad [6.17]$$

Table 6.1 summarizes some typical values for the parameters.

Fluid	CO_2	CH_4
m^{max} (kg.m^3)	105.6	19.2
p_0 (MPa)	1.6	1.6
$c = a(\mu)/m_0(\mu)$	11	6

Table 6.1. *Parameters of Langmuir adsorption isotherm. According to Espinoza et al. (2014)*

The above equations can be generalized to the case of fluid mixtures. For each fluid component i of the mixture, relationship [6.15] generalizes to:

$$m_i^{\mathrm{ads}}(\epsilon, \{\mu_k\}) = m_{i0}(\{\mu_k\}) + a_i(\{\mu_k\})\epsilon \qquad [6.18]$$

The limiting case for a pure fluid is recovered by taking $\mu_k = -\infty$ for $k \neq i$. The thermodynamic equilibrium with a hypothetical infinite reservoir for each component, j, of the fluid also generalizes to the form $\mu_j = \mu_j^b(p, \{c_k\})$, where p is the pressure of the mixture in the reservoir, and c_k are the mass fractions composing the mixture in the reservoir. The density of the mixture in the reservoir, ρ^{b}, is related to the specific chemical potentials by the Gibbs–Duhem relationship:

$$\sum_k c_k \,\mathrm{d}\left(\mu_k^b\right) = \frac{1}{\rho^{\mathrm{b}}}\mathrm{d}\left(p\right) \qquad [6.19]$$

The extension of the Langmuir isotherm, equation [6.17], to mixtures is also commonly employed to model the relationships $m_{i0}(\{\mu_k\})$ in the form $m_{i0} = m_i^{\mathrm{Lang}}(\{p_k\})$ (Kapoor et al. 1990):

$$m_i^{\mathrm{Lang}}(\{p_k\}) = m_i^{\max}\frac{p_i/p_{i0}}{1+\sum_k p_k/p_{k0}} \qquad [6.20]$$

where p_k are the partial (thermodynamic) pressures of the mixture.

Thermodynamics imposes the following symmetry relationships (Nikoosokhan et al. 2014):

$$\left(\frac{\partial\left(m_{i0}\right)}{\partial\mu_j}\right)_{(\mu_k\neq\mu_j)} = \left(\frac{\partial\left(m_{j0}\right)}{\partial\mu_i}\right)_{(\mu_k\neq\mu_i)} \qquad [6.21]$$

$$\left(\frac{\partial\left(a_i\right)}{\partial\mu_j}\right)_{(\mu_k\neq\mu_j)} = \left(\frac{\partial\left(a_j\right)}{\partial\mu_i}\right)_{(\mu_k\neq\mu_i)} \qquad [6.22]$$

However, these relationships are not respected by the Langmuir model (equation [6.20]), making it thermodynamically inconsistent. It is worth noting that other thermodynamically consistent approaches exist for modeling the adsorption of mixtures, such as the ideal adsorbed solution (IAS) theory (Myers and Prausnitz 1965) or the two-dimensional equation of state (2-D EOS) theory (Zhou et al. 1994).

6.2.3. *Gas content in fractured coal*

Coal naturally contains fractures, referred to as cleats. Within these fractures, the fluid or fluid mixture exists in a state similar to that which it would have if it were isolated in a large, infinite reservoir at the same pressure and temperature. In other words, the fluid does not interact with the material in which it is located, and its state is independent of the nature of the surrounding material. Specifically, the fluid's density, ρ^{b}, within the fractures of the coal, as in the infinite reservoir, depends solely on the pressure, temperature and, in the case of mixtures, on the molar or mass fractions. Throughout the text, p^{c} will denote the pressure prevailing in the fractures. The quantity of fluid, m^{c}, contained in the fractures per unit volume of the material is determined by the product of the fracture porosity, ϕ^c, and the fluid density: $m^c = \rho^b(p^c)\phi^c$.

In the presence of fractures, the total fluid quantity is the sum of the fluid present in the macroscopic fractures, m^c, and the adsorbed fluid in the isotropic coal matrix, m^m, resulting in $m = m^c + m^m$. Relative to expression [6.15], it is more relevant to express the adsorbed quantity per unit volume of fractured coal, $m^m = (1 - \phi_0^c)m^{ads}$, as:

$$m^m(p^m, \epsilon, \varphi^c) = m_0^m(p^m) + a(\mu^b(p^m))\,(\epsilon - \varphi^c) \qquad [6.23]$$

where $m_0^m(p^m) = (1 - \phi_0^c)m_0(\mu^b(p^m))$ and p^m is the thermodynamic pressure associated with the fluid adsorbed in the matrix. The volumetric strain of the material, ϵ, and the variation in fracture porosity, $\varphi^c = \phi^c - \phi_0^c$, are related to the volumetric strain of the coal matrix, ϵ^m, through:

$$\epsilon = (1 - \phi_0^c)\epsilon^m + \varphi^c \qquad [6.24]$$

Distinguishing between the pressure in the fractures, p^c, and the thermodynamic pressure, p^m, of the adsorbed fluid in the matrix introduces the possibility of a thermodynamic disequilibrium between these two fluids. This disequilibrium is characterized by the difference in the chemical potentials of the two fluids $\mu^c - \mu^m = \mu^b(p^c) - \mu^b(p^m)$.

The previous equations can be generalized to the case of fluid mixtures. For each fluid component i in the mixture, relationship [6.23] generalizes as:

$$m_i^m(p^m, \{c_k\}, \epsilon, \varphi^c) = m_{i0}^m(\{\mu_k\}) + a_i(\{\mu_k\})\,(\epsilon - \varphi^c) \qquad [6.25]$$

where each chemical potential is a function of $(p^m, \{c_k\}) : \mu_j^b = \mu_j^b(p^m, \{c_k\})$.

6.2.4. *Poroelastic behavior*

Fractured coal is a material whose fractures and matrix are saturated by either a pure fluid or a fluid mixture (e.g. carbon dioxide and methane). In both cases, the poromechanical behavior of the fractured material can be described by the following constitutive equations (Espinoza et al. 2014):

$$\boldsymbol{\sigma} + s^a\mathbf{1} = \mathbb{C} : \boldsymbol{\epsilon} - \mathbb{B}\,(p^c - s^a) \qquad [6.26]$$

$$\varphi^c = \mathbb{B} : \boldsymbol{\epsilon} + \frac{1}{N}(p^c - s^a) \qquad [6.27]$$

where $\varphi^{\mathrm{c}} = \phi^{\mathrm{c}} - \phi_0^{\mathrm{c}}$ represents the change in fracture porosity. In equation [6.26], $\mathbb{C}$ denotes the (drained) elasticity tensor of the material and $\mathbb{B}$ is the Biot tensor. The coal matrix is assumed to be isotropic, with a bulk modulus of K^{m}. The Biot tensor, $\mathbb{B}$, and the Biot modulus, N, are given by:

$$\mathbb{B} = \mathbf{1} - \frac{1}{3K^{\mathrm{m}}}\mathbb{C} : \mathbf{1} \tag{6.28}$$

$$\frac{1}{N} = \frac{1}{3K^{\mathrm{m}}}\left(\mathbb{B} : \mathbf{1} - 3\phi_0^{\mathrm{c}}\right) \tag{6.29}$$

It is important to note that any anisotropy in the fractured material arises exclusively from the anisotropic distribution of the fractures, as the matrix itself is assumed to be isotropic. The scalar s^{a} is a quantity analogous to stress (positive in compression, like pressure), depending on the nature of the solid matrix, which is assumed to be isotropic. In the case of an anisotropic matrix, the adsorption stress, s^{a}, can no longer be reduced to a scalar and must be generalized to a tensor (Espinoza et al. 2013).

For a pure fluid, $s^{\mathrm{a}}(p^{\mathrm{m}})$ is a function of the thermodynamic pressure of the fluid adsorbed in the matrix and is derived by integrating the coupling term $a(\mu)$ (Brochard et al. 2012a):

$$s^{\mathrm{a}}(p^{\mathrm{m}}) = \left(\frac{\partial\left(m^{\mathrm{ads}}\right)}{\partial\epsilon^{\mathrm{m}}}\right)_{\mu} = \int_{-\infty}^{\mu^{\mathrm{b}}(p^{\mathrm{m}})} a(\mu')\mathrm{d}\left(\mu'\right) \tag{6.30}$$

A swelling experiment through immersion, where a fractured coal sample is subjected to an isotropic confining stress, $\boldsymbol{\sigma} = -p^{\mathrm{m}}\mathbf{1}$, allows for identifying this adsorption stress through the measurement of the isotropic strain response $\boldsymbol{\epsilon} = \epsilon^{\mathrm{im}}(p^{\mathrm{m}})\frac{1}{3}\mathbf{1}$. Indeed, according to equation [6.26] and at thermodynamic equilibrium, where $p^{\mathrm{c}} = p^{\mathrm{m}}$, we obtain:

$$s^{\mathrm{a}}(p^{\mathrm{m}}) = p^{\mathrm{m}} + K^{m}\epsilon^{\mathrm{im}}(p^{\mathrm{m}}) \tag{6.31}$$

Comparing equations [6.30] and [6.31] yields:

$$a(\mu^{\mathrm{b}}(p^{\mathrm{m}})) = \rho^{\mathrm{b}}(p^{\mathrm{m}})\left(1 + K^{m}\frac{\mathrm{d}\left(\epsilon^{\mathrm{im}}\right)}{\mathrm{d}p^{\mathrm{m}}}\right) \tag{6.32}$$

This adsorption stress has also been theoretically estimated using density functional theory (Ravikovitch and Neimark 2006).

In the case of a mixture, $s^{\mathrm{a}}(\{\mu_k\})$ is a function of the chemical potentials of the fluid mixture components through the coupling terms $a_i(\{\mu_k\})$ defined in equation [6.18]. Relationship [6.30] generalizes through the integration of the following differential form Brochard et al. (2012a):

$$\mathrm{d}\left(s^{\mathrm{a}}\right) = \sum_i a_i(\{\mu_k\})\mathrm{d}\left(\mu_i\right) \qquad [6.33]$$

The symmetry relationships given by equation [6.22] ensure the integrability of the differential relationship above, allowing for the identification of the function $s^{\mathrm{a}}(\{\mu_k\})$ once the terms $a_i(\{\mu_k\})$ are determined. However, experimentally determining the terms a_i is a challenging task. For a mixture of carbon dioxide and methane, the hypothesis $a_i(\mu_1, \mu_2) = C_i m_{i0}(\mu_1, \mu_2)$, where C_i is a constant dependent on fluid i, has been successfully tested through molecular simulation (Brochard et al. 2012b). However, conditions [6.21] and [6.22] would require that $C_i = C$. In the article by Brochard et al. (2012b), the values of C_i differ by only 25%. So, assuming $C_i = C$ should not significantly affect the results. The ad hoc assumption $a_i(\{\mu_k\}) = a_i(\mu_i)$, which satisfies the condition [6.22], was adopted in the article by Nikoosokhan et al. (2014) but has not been experimentally validated.

6.2.5. *Mass transfer kinetics between coal matrix and fractures*

The permeability of fractured coal is generally dominated by the permeability of the fractures. In other words, it is reasonable, as a first approximation, to neglect fluid flow within the matrix relative to that occurring within the fractures. In this scenario, the only way for the mass of adsorbed fluid to vary is through mass exchange between the fractures and the coal matrix. One approach to modeling mass transfer between the fractures and the coal matrix is to assume, at all times, a thermodynamic equilibrium between the two types of fluids, that is, to assume that the pressures are equal at all times: $p^{\mathrm{c}} = p^{\mathrm{m}}$. In this case, there is no need to introduce an additional law to model mass transfer between the fractures and the coal matrix. This model implies that mass transfers are very rapid and that pressure equilibrium is reached almost instantaneously.

Assuming a thermodynamic imbalance between the adsorbed fluid in the matrix and the fluid in the fractures can be useful and more general. Thus, another approach to modeling mass transfer between the fractures and the matrix is to consider a discrete form of Darcy's law (Vandamme et al. 2014):

$$\frac{\partial\left(m^{\mathrm{m}}\right)}{\partial t} = \rho^{\mathrm{b}}(p^{\mathrm{c}})\frac{k_{\mathrm{m}}}{\eta_f\, l^2}\left(p^{\mathrm{c}} - p^{\mathrm{m}}\right) \qquad [6.34]$$

where k_m is the intrinsic permeability of the coal matrix, η_f is the gas viscosity and l is the characteristic distance separating two neighboring fractures at a small scale (i.e. smaller than the size of the representative elementary volume (REV)). Dimensional analysis links the factor in front of $(p^c - p^m)$ to a characteristic transfer time at this small scale:

$$\tau_d = \frac{M}{RT} \frac{\eta_f \, l^2}{\rho^b k_m} \qquad [6.35]$$

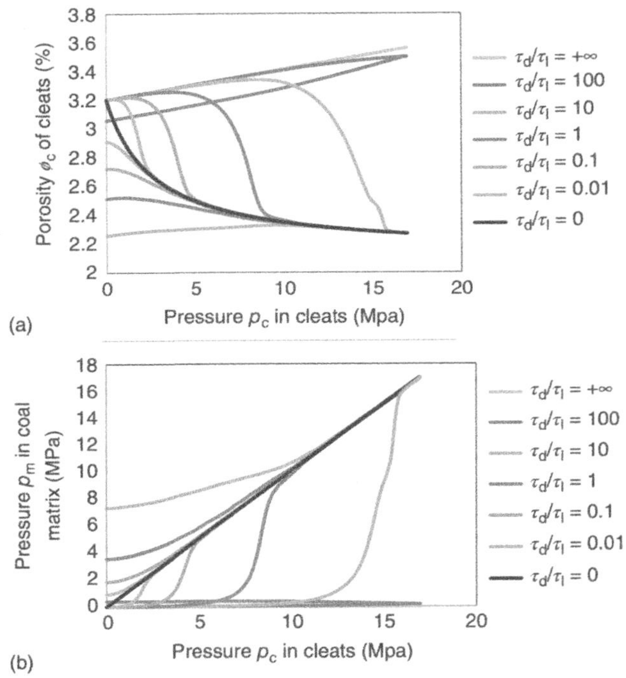

Figure 6.3. *Representative elementary volume (REV) subjected to a constant volume at pressure p^c in fractures. The pressure p^c increases to 17 MPa in a variable time defined by τ_l, then reducing to 0. REV's response depends on the ratio τ_d/τ_l. The figure is taken from the book chapter by Vandamme et al. (2014). For a color version of this figure, see: www.iste.co.uk/vandamme/coal.zip*

Figure 6.3, taken from the book chapter by Vandamme et al. (2014), illustrates the response of a REV subjected to a time-varying pressure p^c in the fractures, characterized by the time τ_l and representing the kinetics due to Darcy flow within the fractures. Two limiting cases are observed. The first case corresponds to a ratio τ_d/τ_l approaching 0, meaning a very rapid gas transfer within the coal matrix compared to gas transfer within the fractures. In this situation, the two gases are in thermodynamic equilibrium and their pressures are equal at all times: $p^{\mathrm{c}} = p^{\mathrm{m}}$. The increase in pressure p^{c} leads to swelling of the coal matrix and a reduction in fracture porosity since the REV is constrained. The second limiting case corresponds to a ratio τ_{d}/τ_l approaching infinity, meaning a very slow gas transfer within the coal matrix compared to gas transfer within the fractures. The thermodynamic pressure p^{m} of the gas adsorbed in the coal matrix does not change because the gas from the fractures does not have time to penetrate the coal matrix due to its very low permeability, implicitly defined by the large value of τ_{d}. Consequently, the REV behaves as an ordinary poroelastic material, and the fracture porosity increases linearly with pressure p^{c}. Between these two extreme cases, that is, for ratios τ_{d}/τ_l typically between 0.01 and 100, hysteresis is observed during a complete cycle of evolution of pressure p^{c}. Indeed, when pressure p^{c} in the fractures increases, the thermodynamic pressure p^{m} of the adsorbed gas increases as well, but remains less than p^{c}. Conversely, when the pressure p^{c} in the fractures decreases, the thermodynamic pressure p^{m} of the adsorbed gas also decreases but remains greater than p^{c}. These delayed variations in p^{m} naturally lead to complex variations in fracture porosities, as illustrated in Figure 6.3(a).

In the previous model, the thermodynamic pressure p^{m} of the fluid adsorbed in the matrix is, by construction, homogeneous at the scale of the REV. However, if the permeability of the coal matrix is sufficiently low, pressure p^{m} will take some time to become homogeneous. During this time, the coal matrix will experience local swelling near the fracture faces, causing temporary closure of the fracture, even if the material is subjected to constant stresses. Subsequently, the pressure will tend to homogenize, and the material will swell homogeneously, including the fractures. During this process, the fractures temporarily close and then open, leading to a non-monotonic evolution of fracture porosity. This complex phenomenon is the origin of the non-monotonic changes in permeability of fractured coals as observed in laboratory samples (Pini et al. 2009). The model developed by Abouloifa et al. (2021) accounts for the inhomogeneity of p^{m}. Quantities depending on this variable and involved in the constitutive equations [6.26], [6.27] and [6.23] must be modified to account for the inhomogeneity of p^{m} in the form:

$$\boldsymbol{\sigma} + < s^{\mathrm{a}} > \mathbf{1} = \mathbb{C} : \boldsymbol{\epsilon} - \mathbb{B}\,(p^{\mathrm{c}} - < s^{\mathrm{a}} >) \qquad [6.36]$$

$$\varphi^{\mathrm{c}} = \mathbb{B} : \boldsymbol{\epsilon} + \frac{1}{N}(p^{\mathrm{c}} - < s^{\mathrm{a}} >) \qquad [6.37]$$

$$m^{\mathrm{m}} = (1 - \phi_0^{\mathrm{c}}) < m^{\mathrm{ads}} > \qquad [6.38]$$

where the notation $< Q >$ represents the average of the quantity Q in the coal matrix. The model described by Abouloifa et al. (2021) proposes a method for calculating the average $< Q >$ based on the assumption that Q satisfies a linear diffusion equation in the coal matrix, depending on the value of Q at the edges of the fractures. The average,

$< Q(p^{\mathrm{m}}) >$, which depends linearly on $Q(p^{\mathrm{c}})$, is determined by a convolution product involving a kernel function, $J_Q(t)$, given by:

$$< Q(p^{\mathrm{m}}) >= J_Q \otimes Q(p^{\mathrm{c}}) \qquad [6.39]$$

where the Stieltjes convolution product, $a \otimes b$, is defined by[4]:

$$(a \otimes b)(t) = \int_{-\infty}^{t} a(t-\tau)\frac{\partial\,(b)}{\partial \tau}(\tau)\mathrm{d}\,(\tau) \qquad [6.40]$$

Identification of the kernels, $J_Q(t)$, through finite element simulations on elliptical fracture geometries in 2D (Abouloifa et al. 2021) leads to the following proposals for $J_Q(t)$:

$$J_{m^{\mathrm{ads}}}(t) = 1 - \exp\left(-\tfrac{t}{\tau_{\mathrm{d}}}\right) \qquad [6.41]$$

$$J_{s^{\mathrm{a}}}(t) = J_{m^{\mathrm{ads}}}(t) + C_{\mathrm{c}} \exp\left(-\tfrac{t}{\tau_{\mathrm{d}}}\right)\left(1 - \exp\left(-\tfrac{t}{\tau_{\mathrm{c}}}\right)\right) \qquad [6.42]$$

The kernel function, $J_{m^{\mathrm{ads}}}$, defined by [6.41] is a monotonic function because the adsorbed mass evolves monotonically during the diffusion process. The parameter τ_{d} is a characteristic diffusion time at the scale of the coal matrix and is associated with the half-distance separating the fractures, consistent with the characteristic time defined by [6.35]. In contrast, the kernel function, $J_{s^{\mathrm{a}}}(t)$, defined by [6.42] is the sum of two contributions: a monotonic function that is the net contribution of $< s^{\mathrm{a}} >$ to the poroelastic variation in fracture porosity, and a non-monotonic function associated with the transient evolution of fracture porosity. The parameter τ_{c} represents the characteristic time required to reach the extreme value of fracture porosity and is related to the transient process of fracture closure/opening. The constant C_{c} controls the amplitude of the transient variation in fracture porosity.

6.2.6. *Permeability of fractured coal*

The permeability of fractured coal is essentially determined by the geometry of the fractures through which all fluid flow occurs. The porosity of the coal matrix plays a

4 In the case of a discontinuity of b at certain times, the time derivative involved in defining the convolution product must be understood in the sense of distributions, that is, considering terms like: $(b(t_i^+) - b(t_i^-))\delta(t - t_i)$ where $\delta(t)$ is the Dirac distribution.

negligible role. The permeability k of fractured coal can be related to the porosity of the fractures through a cubic law (Cui and Bustin 2005):

$$k = k_0 \left(\frac{\phi^{\mathrm{c}}}{\phi_0^{\mathrm{c}}} \right)^3 \qquad [6.43]$$

This law does not account for the anisotropic evolution of stresses. To address this, some permeability models are based on dependence on variations in Terzaghi's effective stresses $\boldsymbol{\sigma}' = \boldsymbol{\sigma} + p\mathbf{1}$ (Connell et al. 2010; Wu et al. 2010; Pan and Connell 2012; Espinoza et al. 2014):

$$k = k_0 \exp\left(\alpha_h \Delta\sigma_h' + \alpha_v \Delta\sigma_v'\right) \qquad [6.44]$$

where σ_h' is the average effective horizontal stress, that is, $\frac{1}{2}(\sigma_{h1}' + \sigma_{h2}')$, and σ_v' is the effective vertical stress. α_h and α_v are two constants. For instance, for coal from the Forzando mine in South Africa, located at a depth of $\approx$ 500 m, it was found that α_h = 0.269 MPa^{-1} and α_v = 0.067 MPa^{-1} (Espinoza et al. 2014).

6.3. Reservoir scale field equations

The main field equations governing mechanical equilibrium and gas transport at every point within the domain V are outlined. The possible presence of faults, which are layers of low thickness filled with a permeable porous material, is also considered.

6.3.1. *Mechanical equilibrium*

The displacement field $\boldsymbol{u}$, and the strain field, $\boldsymbol{\epsilon}$, satisfy the compatibility equations:

$$\boldsymbol{\epsilon} = \frac{1}{2} \left(\boldsymbol{\nabla}(\boldsymbol{u}) + {}^t\boldsymbol{\nabla}(\boldsymbol{u}) \right) \qquad [6.45]$$

The stress field, $\boldsymbol{\sigma}$, and the volume forces, $\boldsymbol{f}$, satisfy the mechanical equilibrium equation at every point within the considered domain V:

$$\nabla \cdot (\boldsymbol{\sigma}) + \boldsymbol{f} = 0 \qquad [6.46]$$

The stress tensor is given by [6.26]:

$$\boldsymbol{\sigma} + s^{\mathrm{a}}\mathbf{1} = \mathbb{C} : \boldsymbol{\epsilon} - \mathbb{B}\left(p^{\mathrm{c}} - s^{\mathrm{a}}\right) \qquad [6.47]$$

Along potential displacement discontinuity surfaces, Γ, the stress vector, $\boldsymbol{\sigma} \cdot \boldsymbol{n}$ where $\boldsymbol{n}$ is the unit normal vector to Γ, remains continuous across Γ:

$$[\![\boldsymbol{\sigma} \cdot \boldsymbol{n}]\!] = 0 \tag{6.48}$$

where the notation $[\![\cdot]\!]$ represents the jump operator. In contrast, the strain tensor, $\boldsymbol{\epsilon}$, becomes singular. Its expression must then be extended in the sense of distributions as follows:

$$\boldsymbol{\epsilon} = \{\boldsymbol{\epsilon}\} + \frac{1}{2}\left([\![\boldsymbol{u}]\!] \otimes \boldsymbol{n} + \boldsymbol{n} \otimes [\![\boldsymbol{u}]\!]\right) \delta^{\Gamma} \tag{6.49}$$

where the notation $\{\cdot\}$ indicates that the operand should be understood in the usual sense of functions defined only outside of Γ, $[\![\boldsymbol{u}]\!]$ is the jump of the displacement vector and δ^{Γ} is the Dirac delta distribution supported on Γ [5].

By substituting the expression [6.49] for $\boldsymbol{\epsilon}$ into the constitutive equation [6.26], then integrating over a very small thickness h on either side of the discontinuity surface Γ and averaging over this thickness, we obtain to the first order:

$$\boldsymbol{\sigma} + s^{\mathrm{a}}\boldsymbol{1} = \frac{1}{h}\mathbb{C}^{\Gamma} : \frac{1}{2}\left([\![\boldsymbol{u}]\!] \otimes \boldsymbol{n} + \boldsymbol{n} \otimes [\![\boldsymbol{u}]\!]\right) - \mathbb{B}^{\Gamma}\left(p^{\mathrm{c}} - s^{\mathrm{a}}\right) \tag{6.50}$$

In fact, the thickness h physically represents the thickness of a thin layer composed of a material with elastic properties significantly weaker than the surrounding medium. This layer may, for example, represent a geological fault filled with soft material. The constitutive law [6.50] as formulated describes an interface without thickness, provided it is assigned an elasticity tensor equal to $\frac{1}{h}\mathbb{C}^{\Gamma}$ (thus homogeneous to surface stress) and a strain tensor equal to $\frac{1}{2}\left([\![\boldsymbol{u}]\!] \otimes \boldsymbol{n} + \boldsymbol{n} \otimes [\![\boldsymbol{u}]\!]\right)$ (homogeneous to displacement). Thus, in a finite element approach, it will not be necessary to mesh the thickness of the layer, but only to mesh the surface Γ with elements that have no thickness, consisting of pairs of nodes with the same coordinates. The nodal unknowns will be the displacements, and a pseudo-strain tensor equal $\frac{1}{2}\left([\![\boldsymbol{u}]\!] \otimes \boldsymbol{n} + \boldsymbol{n} \otimes [\![\boldsymbol{u}]\!]\right)$ will be associated at every point of the element.

From [6.50], the following expression for the stress vector is deduced:

$$\boldsymbol{\sigma} \cdot \boldsymbol{n} + s^{\mathrm{a}}\boldsymbol{n} = \boldsymbol{R}^{\Gamma} \cdot [\![\boldsymbol{u}]\!] - \mathbb{B}^{\Gamma} \cdot \boldsymbol{n}\left(p^{\mathrm{c}} - s^{\mathrm{a}}\right) \tag{6.51}$$

5 We recall that, by definition, the single-layer distribution supported by Γ, δ^{Γ}, satisfies for any infinitely differentiable test function φ with compact support in the domain V:

$$< \delta^{\Gamma}, \varphi >= \int_{\Gamma} \varphi \qquad \forall \varphi \in \mathcal{D}(V)$$

where the tensor $\boldsymbol{R}^{\Gamma}$ is given by:

$$R^{\Gamma}_{ik} = \frac{1}{4h} n_j \left(C^{\Gamma}_{ijkl} + C^{\Gamma}_{ijlk} + C^{\Gamma}_{jikl} + C^{\Gamma}_{jilk} \right) n_l \qquad [6.52]$$

6.3.2. *Gas transport*

The fluid mass flux vector, $\boldsymbol{w}$, and the mass content of the fluid, m, satisfy the equation of conservation of total fluid mass at every point in the considered domain V:

$$\frac{\partial (m)}{\partial t} + \nabla \cdot (\boldsymbol{w}) = 0 \qquad [6.53]$$

The fluid mass flux vector, $\boldsymbol{w}$, is related to the pressure gradient by Darcy's law:

$$\boldsymbol{w} = \rho^{\mathrm{b}} \frac{k}{\eta_f} \left(-\boldsymbol{\nabla} (p^{\mathrm{c}}) + \rho^{\mathrm{b}} \boldsymbol{g} \right) \qquad [6.54]$$

where $\boldsymbol{g}$ is the gravity acceleration vector.

Typically, at a regular point in the medium, fluid mass conservation requires the continuity of the fluid mass flow: $[\![\boldsymbol{w} \cdot \boldsymbol{n}]\!] = 0$. However, as with the case of surfaces of displacement field discontinuities, we consider a potential surface, Γ, where there exists a surface mass density of fluid m^{Γ} and a surface mass flux $\boldsymbol{w}^{\Gamma}$ parallel to the tangent plane of Γ. This surface physically represents a fault of small thickness, h, filled with a porous material of relatively high permeability, along which fluid preferentially flows. A fluid mass balance over a very thin volume, disk-shaped, surrounding an arbitrary part of the surface Γ leads to the fluid mass conservation law on Γ (Pouya 2012)[6]:

$$\frac{\partial (m^{\Gamma})}{\partial t} + \nabla_{\Gamma} \cdot \left(\boldsymbol{w}^{\Gamma} \right) + [\![\boldsymbol{w} \cdot \boldsymbol{n}]\!] = 0 \qquad [6.55]$$

---

6 The same result can also be obtained by extending the fields in the sense of distributions in the form:

$$m = \{m\} + m^{\Gamma} \delta^{\Gamma}$$

$$\boldsymbol{w} = \{\boldsymbol{w}\} + \boldsymbol{w}^{\Gamma} \delta^{\Gamma}$$

We recall that, by definition, the derivative of a distribution T satisfies, for any infinitely differentiable test function φ with compact support in the domain V:

$$< \frac{\partial (T)}{\partial x_i}, \varphi >= - < T, \frac{\partial (\varphi)}{\partial x_i} > \qquad \forall \varphi \in \mathcal{D}(V)$$

where $\nabla_\Gamma \cdot ()$ represents the surface divergence operator on Γ. If $\boldsymbol{t}$ is a unit vector in the tangent plane of Γ, the quantity $\boldsymbol{w}^\Gamma \cdot \boldsymbol{t}$ physically represents the fluid mass flux integrated over the thickness h of the fault. Consequently, if k^Γ denotes the permeability of the material constituting the fault, the fluid flow law of Darcy type for the surface Γ is written as (Pouya 2012):

$$\boldsymbol{w}^\Gamma = \rho^\mathrm{b} \frac{hk^\Gamma}{\eta_f} \left(-\boldsymbol{\nabla}_\Gamma \left(p^\mathrm{c} \right) + \rho^\mathrm{b} \boldsymbol{g} \right) \qquad [6.56]$$

where $\boldsymbol{\nabla}_\Gamma ()$ represents the surface gradient operator on Γ. It should be noted that in this approach, the pressure, p^c, is assumed to be continuous across Γ. The Darcy flow law [6.56] formulated in this manner describes an interface without thickness, provided it is assigned a global permeability coefficient equal to hk^Γ (homogeneous to a volume). Parallel to $\boldsymbol{w}^\Gamma$, the surface density m^Γ physically represents the total mass of fluid, per unit volume of material, integrated over the thickness h of the fault. Thus, it is a surface density. To correctly compute this quantity using the constitutive laws already introduced, that is, equations [6.27] and [6.23], it suffices to assign the interface a surface porosity equal to $h\phi_0^\mathrm{c}$, a Biot modulus equal to N/h (homogeneous to surface stress) and an amount of fluid adsorbed per unit area equal to hm_0^m. In the same manner as for the displacement field discontinuity surfaces, in a finite element approach, it will not be necessary to mesh the fault thickness but only to mesh the surface Γ with zero-thickness elements, whose nodal unknowns are the pressures. Furthermore, it will be necessary to make the pressure field continuous across Γ, meaning that the pressures at nodes with the same coordinates must be equal. This last operation can prove non-trivial.

6.3.3. *Synthesis of equations and unknowns for the problem: case of a single fluid*

To verify that the mathematical problem is well-posed, meaning there are as many unknowns as equations, we have gathered all of the equations in Table 6.2 and, in Table 6.3, all of the unknowns involved in these equations, for the case of a single fluid.

Considering that $\{\boldsymbol{w}\}$ is discontinuous across Γ, we obtain:

$$\nabla \cdot (\boldsymbol{w}) = \{\nabla \cdot (\boldsymbol{w})\} + \left([\![\boldsymbol{w} \cdot \boldsymbol{n}]\!] + \nabla_\Gamma \cdot \left(\boldsymbol{w}^\Gamma \right) \right) \delta^\Gamma - (\boldsymbol{w}^\Gamma \cdot \boldsymbol{\nu}) \delta^{(\partial\Gamma \cap V)}$$

Using the definition of the derivative of a distribution, we directly arrive at equation [6.55] by canceling out the term in factor of δ^Γ in the expression $\frac{\partial (m)}{\partial t} + \nabla \cdot (\boldsymbol{w})$. In the above equation, $\boldsymbol{\nu}$ is the unit vector tangent to Γ and outward normal to $\partial\Gamma$. The distribution $\delta^{(\partial\Gamma \cap V)}$ is the single-layer distribution supported by the part of the contour of Γ included in the domain V. This equation, in addition to equation [6.55], indicates that $\boldsymbol{w}^\Gamma \cdot \boldsymbol{\nu} = 0$ at every point on the closed boundary of Γ.

Description	Equation	Number
Conservation of fluid mass	$\frac{\partial(m)}{\partial t} + \nabla \cdot (w) = 0$	1
Total fluid mass	$m = m^{c} + m^{m}$	1
Mass in fractures	$m^{c} = \rho^{b}(p^{c})\,\phi^{c}$	1
Adsorbed mass in the matrix	$m^{m} = m_0^{m}(p^{m}) + a(p^{m})(\epsilon - \varphi^{c})$	1
Darcy's law	$w = \rho^{b}(p^{c})\,\frac{k(\sigma)}{\eta_f} \cdot (-\nabla(p^{c}) + \rho^{b}(p^{c})g)$	3
Transfer law	$\frac{\partial(m^{m})}{\partial t} = \rho^{b}(p^{c})\frac{k_m}{\eta_f\, l^2}(p^{c} - p^{m})$	1
Mechanical equilibrium	$\nabla \cdot (\sigma) + f = 0$	3
First law of poroelasticity	$\sigma + s^{a}(p^{m})\mathbf{1} = \mathbb{C} : \epsilon - \mathbb{B}(p^{c} - s^{a}(p^{m}))$	6
Second law of poroelasticity	$\varphi^{c} = \mathbb{B} : \epsilon + \frac{1}{N}(p^{c} - s^{a}(p^{m}))$	1
Compatibility equation	$\epsilon = \frac{1}{2}\left(\nabla(u) + {}^{t}\nabla(u)\right)$	6
Fracture porosity	$\phi^{c} = \phi_0^{c} + \varphi^{c}$	1
Total		25

Table 6.2. *List of equations in the case of a single fluid. Indicated are: in red, the unknown variables; in blue, given functions or constants. For a color version of this table, see: www.iste.co.uk/vandamme/coal.zip*

6.4. Numerical solution via the finite element method

The finite element method is based on two main conceptual steps: a weak formulation of the problem's equations and an approximation of the solution through the use of finite element basis functions. The latter step will not be detailed here, as it is thoroughly described in books dedicated to the finite element method (Potts and Zdravković 1999; Hughes 2012).

6.4.1. *Weak formulation of the field equations*

The numerical solution via finite elements is based on a weak formulation of the equations, which involves multiplying the equation by a test field and performing integration by parts to reduce the degree of spatial differentiation compared to the original equation (hence the term "weak formulation"). Regarding the mechanical equilibrium equation [6.46], the weak formulation is written as:

$$\mathcal{P}[\sigma, f, v] = 0 \qquad \forall v \qquad [6.57]$$

where $\mathcal{P}$ refers to a functional, that is, a function of functions (note the brackets distinguishing it from a simple function). This functional $\mathcal{P}[\boldsymbol{\sigma}, \boldsymbol{f}, \boldsymbol{v}]$ is obtained by multiplying the equilibrium equation [6.46] by a test vector field $\boldsymbol{v}$ and integrating over the domain V. After integration by parts, we finally obtain:

$$\mathcal{P}[\boldsymbol{\sigma}, \boldsymbol{f}, \boldsymbol{v}] := \int_V \boldsymbol{\sigma} : \boldsymbol{\nabla}(\boldsymbol{v}) - \int_V \boldsymbol{f} \cdot \boldsymbol{v} - \int_{\partial V} (\boldsymbol{\sigma} \cdot \boldsymbol{n}) \cdot \boldsymbol{v} \qquad [6.58]$$

Unknown Variable	Symbol	Number
Total fluid mass	m	1
Fluid mass flux	$\boldsymbol{w}$	3
Fluid mass in fractures	m^{c}	1
Fluid mass adsorbed in the matrix	m^{m}	1
Stress tensor	$\boldsymbol{\sigma}$	6
Strain tensor	$\boldsymbol{\epsilon}$	6
Pore pressure in the matrix	p^{m}	1
Pore pressure in fractures	p^{c}	1
Fracture porosity	ϕ^{c}	1
Displacement vector	$\boldsymbol{u}$	3
Variation of fracture porosity	φ^{c}	1
Total		25

Table 6.3. *List of unknown variable fields in the case of a single fluid*

In the presence of a displacement field discontinuity surface, Γ, representing a thin fault filled with a soft elastic material, the description must be enriched by considering a discontinuity in the test vector field, $[\![\boldsymbol{v}]\!]$, on this surface:

$$\mathcal{P}[\boldsymbol{\sigma}, \boldsymbol{f}, \boldsymbol{v}] := \ldots + \int_\Gamma (\boldsymbol{\sigma} \cdot \boldsymbol{n}) \cdot [\![\boldsymbol{v}]\!] \qquad [6.59]$$

Regarding the mass conservation equation [6.53], the weak formulation is written as:

$$\mathcal{M}[\tfrac{\partial(m)}{\partial t}, \boldsymbol{w}, \varphi] = 0 \qquad \forall\varphi \qquad [6.60]$$

where the functional $\mathcal{M}[\frac{\partial(m)}{\partial t}, \boldsymbol{w}, \varphi]$ is obtained by multiplying equation [6.53] by a test scalar field φ and integrating over the domain V. After integration by parts, we obtain:

$$\mathcal{M}[\tfrac{\partial(m)}{\partial t}, \boldsymbol{w}, \varphi] := \int_V \frac{\partial (m)}{\partial t} \varphi - \int_V \boldsymbol{w} \cdot \boldsymbol{\nabla} (\varphi) + \int_{\partial V} (\boldsymbol{w} \cdot \boldsymbol{n}) \varphi \qquad [6.61]$$

Similarly, in the presence of a thin but highly permeable porous fault, represented by a surface Γ through which the mass flux is discontinuous, the contribution of $[\![\boldsymbol{w} \cdot \boldsymbol{n}]\!]$ must be added:

$$\mathcal{M}[\tfrac{\partial(m)}{\partial t}, \boldsymbol{w}, \varphi] := \cdots - \int_\Gamma [\![\boldsymbol{w} \cdot \boldsymbol{n}]\!] \varphi \qquad [6.62]$$

In this case, it is also necessary to account for the weak formulation of equation [6.55] in the form:

$$\mathcal{M}^\Gamma[\tfrac{\partial(m^\Gamma)}{\partial t}, \boldsymbol{w}^\Gamma, \varphi'] = 0 \qquad \forall \varphi' \qquad [6.63]$$

where the functional $\mathcal{M}^\Gamma[\frac{\partial(m^\Gamma)}{\partial t}, \boldsymbol{w}^\Gamma, \varphi']$ is obtained in the same manner but using a scalar field φ' defined on Γ and a priori distinct from φ:

$$\mathcal{M}^\Gamma[\tfrac{\partial(m^\Gamma)}{\partial t}, \boldsymbol{w}^\Gamma, \varphi'] := \int_\Gamma \frac{\partial (m^\Gamma)}{\partial t} \varphi' - \int_\Gamma \boldsymbol{w}^\Gamma \cdot \boldsymbol{\nabla} (\varphi')$$
$$+ \int_{(\partial\Gamma \cap \partial V)} (\boldsymbol{w}^\Gamma \cdot \boldsymbol{\nu}) \varphi' + \int_\Gamma [\![\boldsymbol{w} \cdot \boldsymbol{n}]\!] \varphi' \qquad [6.64]$$

where it has been taken into account that $\boldsymbol{w}^\Gamma \cdot \boldsymbol{\nu} = 0$ on any part of $\partial\Gamma$ included in the domain V. By choosing $\varphi' = \varphi$ on Γ, it becomes easy to eliminate the unknown density $[\![\boldsymbol{w} \cdot \boldsymbol{n}]\!]$ by considering the sum of equations [6.60] and [6.63]:

$$\mathcal{M}[\tfrac{\partial(m)}{\partial t}, \boldsymbol{w}, \varphi] + \mathcal{M}^\Gamma[\tfrac{\partial(m^\Gamma)}{\partial t}, \boldsymbol{w}^\Gamma, \varphi] = 0 \qquad \forall \varphi \qquad [6.65]$$

6.4.2. *Time stepping and spatial discretization*

We employ an implicit Euler scheme for the time discretization of the mass conservation equation, that is:

$$\frac{\partial (m)}{\partial t}(t + \Delta t) = \frac{m(t + \Delta t) - m(t)}{\Delta t} = \frac{\Delta m}{\Delta t} \qquad [6.66]$$

The Darcy law is written using a semi-implicit scheme, where the transport coefficient is evaluated at time t:

$$\boldsymbol{w}(t+\Delta t) = \boldsymbol{K}^{\text{Darcy}}(t) \cdot \left(-\boldsymbol{\nabla}\,(p^{\text{c}})(t+\Delta t) + \rho^{\text{b}}(t+\Delta t)\boldsymbol{g}\right) \qquad [6.67]$$

Additionally, we consider a discretization of the domain V and possibly Γ using finite elements based cn standard normalized shape functions (i.e. expressed in terms of local coordinates):

$$\boldsymbol{x} = \sum_k \boldsymbol{x}^k N^k(\boldsymbol{\xi}) \qquad [6.68]$$

where $\boldsymbol{\xi}$ is the vector of local coordinates and $\boldsymbol{x}^k$ is the vector of the global coordinates of node k. The approximation of any field, $a(\boldsymbol{x}, t)$, by its values at the mesh nodes is expressed as:

$$a(\boldsymbol{x}, t) = \sum_k a^k(t) N^k(\boldsymbol{\xi}) \qquad [6.69]$$

where $a^k(t) = a(\boldsymbol{x}^k, t)$ is the value of a at node k. The transformation defined by equation [6.68] easily identifies zero-thickness elements belonging to Γ, as for these elements, the determinant of the Jacobian matrix, $\frac{\partial(\boldsymbol{x})}{\partial \boldsymbol{\xi}}$, is zero. For these zero-thickness elements, only the first nodes with different coordinates (i.e. eliminating nodes with the same coordinates) are used to describe the shape functions $N^k(\boldsymbol{\xi})$, which will then be used to describe surface fields. This method is applied in the following applications.

6.4.3. *Iterative Newton method for solving*

The weak formulations of the mechanical equilibrium and mass conservation equations are solved using an iterative Newton method. Suppose that at one stage of the iterative process $\mathcal{P}[\boldsymbol{\sigma}, \boldsymbol{f}, \boldsymbol{v}] \neq 0$. We then seek to correct the stress field $\boldsymbol{\sigma}(\boldsymbol{x}, t)$ to satisfy:

$$\delta\mathcal{P} = -\mathcal{P}[\boldsymbol{\sigma}, \boldsymbol{f}, \boldsymbol{v}] \qquad \forall \boldsymbol{v} \text{ kinematically admissible at zero} \qquad [6.70]$$

where

$$\delta\mathcal{P} := \int_V \delta\boldsymbol{\sigma} : \boldsymbol{\nabla}\,(\boldsymbol{v}) \qquad [6.71]$$

The variation $\delta\boldsymbol{\sigma}$ must be calculated from the constitutive law [6.26]:

$$\delta\boldsymbol{\sigma} = \left(\frac{\partial\left(\boldsymbol{\sigma}\right)}{\partial\epsilon}\right)_{p^{\mathrm{c}}} : \boldsymbol{\nabla}\left(\delta\boldsymbol{u}\right) + \left(\frac{\partial\left(\boldsymbol{\sigma}\right)}{\partial p^{\mathrm{c}}}\right)_{\epsilon}\delta p^{\mathrm{c}} \qquad [6.72]$$

The fourth-order tensor $\left(\frac{\partial(\boldsymbol{\sigma})}{\partial\epsilon}\right)_{p^{\mathrm{c}}}$ is a tangent elastic tensor:

$$\left(\frac{\partial\left(\boldsymbol{\sigma}\right)}{\partial\epsilon}\right)_{p^{\mathrm{c}}} = \mathbb{C} + \frac{\mathrm{d}\left(s^{\mathrm{a}}\right)}{\mathrm{d}p^{\mathrm{m}}}\left(\mathbb{B} - \mathbf{1}\right)\otimes\left(\frac{\partial\left(p^{\mathrm{m}}\right)}{\partial\epsilon}\right)_{p^{\mathrm{c}}} \qquad [6.73]$$

which can be approximated by $\mathbb{C}$ if we consider that the corrective term is small, or if the thermodynamic equilibrium hypothesis $p^{\mathrm{m}} = p^{\mathrm{c}}$ is assumed. Similarly, the second-order tensor $\left(\frac{\partial(\boldsymbol{\sigma})}{\partial p^{\mathrm{c}}}\right)_{\epsilon}$ is a form of tangent Biot tensor:

$$\left(\frac{\partial\left(\boldsymbol{\sigma}\right)}{\partial p^{\mathrm{c}}}\right)_{\epsilon} = -\mathbb{B} - \frac{\mathrm{d}\left(s^{\mathrm{a}}\right)}{\mathrm{d}p^{\mathrm{m}}}\left(\mathbb{B} - \mathbf{1}\right)\left(\frac{\partial\left(p^{\mathrm{m}}\right)}{\partial p^{\mathrm{c}}}\right)_{\epsilon} \qquad [6.74]$$

which must be calculated accurately.

In the same way as for the mechanical equilibrium equation, the mass conservation equation is solved by an iterative method:

$$\delta\mathcal{M} = -\mathcal{M}[\tfrac{\Delta m}{\Delta t}, \boldsymbol{w}, \varphi] \qquad \forall\varphi \text{ kinematically admissible at zero} \qquad [6.75]$$

with

$$\delta\mathcal{M} := \int_V \frac{\delta m}{\Delta t}\varphi - \int_V \delta\boldsymbol{w}\cdot\boldsymbol{\nabla}\left(\varphi\right) \qquad [6.76]$$

where the variations δm and $\delta\boldsymbol{w}$ are calculated from the constitutive and transport laws. By considering the semi-implicit scheme mentioned earlier for the transport law of $\delta\boldsymbol{w}$:

$$\delta\boldsymbol{w} = \boldsymbol{K}^{\mathrm{Darcy}}(t)\cdot\left(-\boldsymbol{\nabla}\left(\delta p^{\mathrm{c}}\right) + \frac{\mathrm{d}\left(\rho^{\mathrm{b}}\right)}{\mathrm{d}p^{\mathrm{c}}}\boldsymbol{g}\,\delta p^{\mathrm{c}}\right) \qquad [6.77]$$

From the constitutive law [6.27], we obtain for δm:

$$\delta m = \left(\frac{\partial (m)}{\partial \epsilon}\right)_{p^c} : \boldsymbol{\nabla}(\delta \boldsymbol{u}) + \left(\frac{\partial (m)}{\partial p^c}\right)_{\epsilon} \delta p^c \qquad [6.78]$$

The second-order tensor $\left(\frac{\partial (m)}{\partial \epsilon}\right)_{p^c}$ and the coefficient $\left(\frac{\partial (m)}{\partial p^c}\right)_{\epsilon}$ can theoretically be calculated from the laws [6.27] and [6.23]. However, an efficient, quick and easy technique for calculating these partial derivatives is to compute them numerically in the form:

$$\frac{\partial (m)}{\partial a} = \frac{m(a + da) - m(a)}{da} \qquad [6.79]$$

where da is a quantity as small as possible, chosen based on the nature of a and the problem at hand. This technique can obviously be applied to calculate $\delta \boldsymbol{\sigma}$, and more generally to any complex function of the main unknowns.

Needless to say, this approach is similarly applied in the presence of a discontinuity surface, where in this case, we must solve:

$$\delta \mathcal{M} + \delta \mathcal{M}^{\Gamma} = -\left(\mathcal{M}[\tfrac{\Delta m}{\Delta t}, \boldsymbol{w}, \varphi] + \mathcal{M}^{\Gamma}[\tfrac{\Delta m^{\Gamma}}{\Delta t}, \boldsymbol{w}^{\Gamma}, \varphi]\right)$$

$\forall \varphi$ kinematically admissible at zero [6.80]

with

$$\delta \mathcal{M}^{\Gamma} := \int_{\Gamma} \frac{\delta m^{\Gamma}}{\Delta t} \varphi - \int_{\Gamma} \delta \boldsymbol{w}^{\Gamma} \cdot \boldsymbol{\nabla}_{\Gamma}(\varphi) \qquad [6.81]$$

6.5. Application examples

6.5.1. *Methane production*

In this example, we simulate methane production from a vertical well with a radius of 10 cm, passing through a 2-m thick coal seam located at a depth of 500 m. The surrounding host rock, in which the coal seam lies, is assumed to be elastic, with a Young's modulus of 3 GPa and a Poisson's ratio of 0.26. Only a 10-m thick section of the rock above and below the seam is meshed (Figure 6.4). The initial stresses in place before the pumping operations are assumed to be homogeneous, isotropic and equal to -10 MPa (the weight of the overlying material). The initial methane pressure in the fractures of the coal seam is set at 10 MPa. The coal seam is assumed to be transversely isotropic with a vertical axis[7]. Its properties are gathered in Table 6.4 and are derived from the article by Espinoza et al. (2014). Moreover, the adsorption of methane in the coal matrix is defined by a Langmuir law, and the coupling coefficient $a = cm_0$ uses the parameters from Table 6.1 for methane. Thermodynamic equilibrium is assumed between the adsorbed fluid in the coal matrix and that present in the fractures: $p^{\mathrm{c}} = p^{\mathrm{m}}$. The intrinsic behavior of pure methane is assumed to be ideal. The entire system is assumed to be at a temperature of 312.15 K. To simulate methane extraction through the well, the pressure and radial compression stress applied on the well wall are reduced from 10 MPa to 1 MPa over the course of 1 h, after which they are held constant. In contrast, the pressure and radial compression stress are maintained at 10 MPa on the outer lateral boundary of the mesh located 60 m from the axis of symmetry. A vertical compression stress of 10 MPa, corresponding to the weight of the overlying material, is consistently applied on the upper part of the mesh. The calculation is performed using the Bil finite element code (Dangla 2018).

Figure 6.5 shows the daily methane production. The same figure also illustrates production assuming an adsorption stress s^{a} equal to the thermodynamic pressure p^{m}, that is, assuming the material behaves like an ordinary poroelastic material with no swelling induced by adsorption (see equation [6.31]). Under this assumption, and for the same methane depressurization, the material decompresses less, leading to

---

7 In transverse isotropy, non-zero coefficients of the elasticity tensor are of the form:

$$C_{1111} = C_{2222} = \lambda_1 + 2\mu_1; \quad C_{3333} = \lambda_3 + 2\mu_3$$

$$C_{1122} = C_{2211} = \lambda_1; \quad C_{1133} = C_{2233} = C_{3322} = C_{3311} = \lambda_2$$

$$C_{2323} = C_{2332} = C_{3223} = C_{3232} = C_{1313} = C_{1331} = C_{3113} = C_{3131} = \mu_3$$

$$C_{1212} = C_{1221} = C_{2112} = C_{2121} = \mu_1$$

and its inverse $\mathbb{L} = \mathbb{C}^{-1}$ of the form:

$$L_{1111} = L_{2222} = \tfrac{1}{E}; \quad L_{3333} = \tfrac{1}{E_3}$$

$$L_{1122} = L_{2211} = -\tfrac{\nu}{E}; \quad L_{1133} = L_{2233} = L_{3322} = L_{3311} = -\tfrac{\nu_3}{E_3}$$

$$L_{2323} = L_{2332} = L_{3223} = L_{3232} = L_{1313} = L_{1331} = L_{3113} = L_{3131} = \tfrac{1}{\mu_3}$$

$$L_{1212} = L_{1221} = L_{2112} = L_{2121} = \tfrac{2(1+\nu)}{E}$$

The relationships between the two sets of parameters are

$$2\mu_1 = \frac{E}{1+\nu}; \qquad \frac{1}{2(\lambda_1 + \mu_1)} = \frac{1-\nu}{E} - \frac{2\nu_3^2}{E_3}$$

$$\lambda_2 = 2(\lambda_1 + \mu_1)\nu_3; \quad \lambda_3 + 2\mu_3 = E_3 + 2\nu_3\lambda_2$$

a smaller increase in permeability. As a result, methane production is almost halved for the majority of the time. Figure 6.6 shows the deformed mesh near the well after one year of production, comparing the original model to the one where $s^{\mathrm{a}} = p^{\mathrm{m}}$. It is evident that the difference $s^{\mathrm{a}} - p^{\mathrm{m}}$ produces a greater vertical deformation of the coal layer. The same figures also display the magnitude of the orthoradial stress σ_{tt}. Consistent with the deformation, the orthoradial compression stress is $\sigma_{tt} = 1.6$ MPa with the adsorption stress s^{a}, while it is five times greater under the assumption $s^{\mathrm{a}} = p^{\mathrm{m}}$.

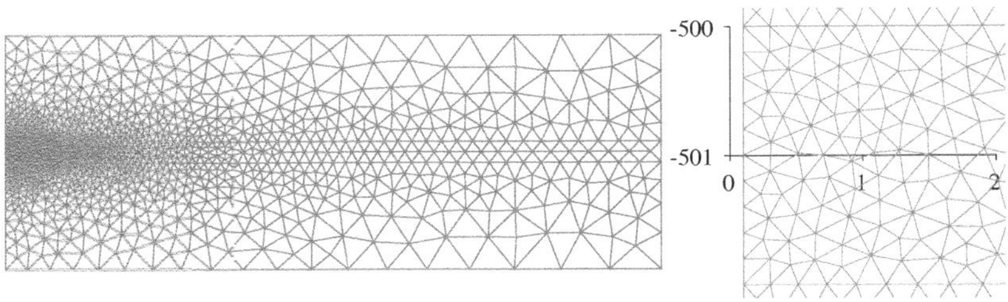

Figure 6.4. *Mesh of a cylindrical well with a radius of 10 cm passing through a 2-m thick coal seam. The axis of the well is shown in the enlargement on the right. The radius of the domain is 60 m. Its thickness, including the upper and lower parts of the host rock, is 22 m. For a color version of this figure, see: www.iste.co.uk/vandamme/coal.zip*

Description	Symbols	Value
Compression modulus of coal matrix	K^{m}	5 GPa
Fracture porosity	ϕ_0^{c}	0.08
Reference intrinsic permeability	k_0	20 nm^2
Methane viscosity	η_f	11.51 10^{-6} Pa.s
Horizontal Young's modulus	E	2.736 GPa
Vertical Young's modulus	E_3	2.551 GPa
Horizontal Poisson's ratio	ν	0.267
Vertical Poisson's ratio	ν_3	0.198
Shear modulus	μ_3	0.71 GPa
Constant in equation [6.44]	α_h	0.269 MPa^{-1}
Constant in equation [6.44]	α_v	0.067 MPa^{-1}

Table 6.4. *Properties of the coal seam*

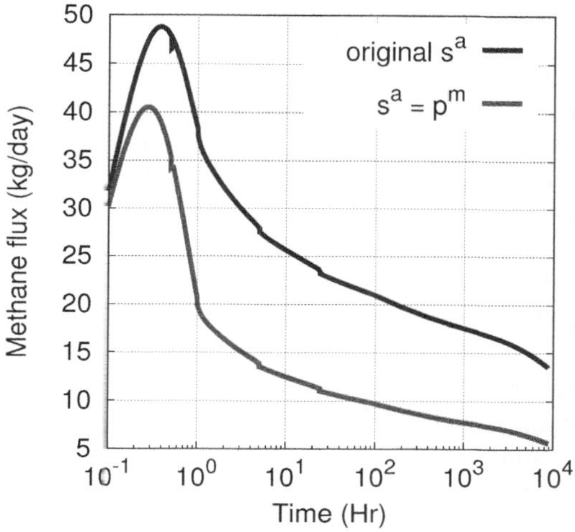

Figure 6.5. *Daily methane production calculated first with the model as described in this chapter using the adsorption isotherm parameters from Table 6.1 to calculate the adsorption stress s^a (blue curve labeled "original s^a") and then under the assumption $s^a = p^m$ (i.e. $a(\mu^b(p^m)) = \rho^b(p^m)$) where adsorption swelling effects are deliberately canceled (red curve labeled "$s^a = p^m$"). For a color version of this figure, see: www.iste.co.uk/vandamme/coal.zip*

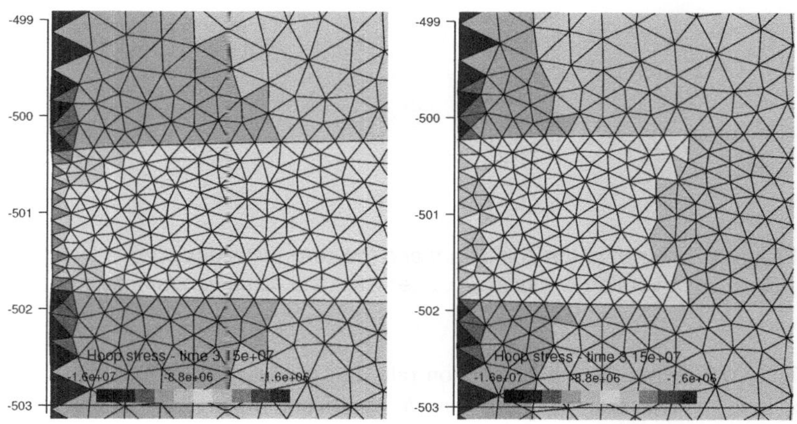

(a) (b)

Figure 6.6. *Deformed mesh near the well after one year of production. The amplification factor is 200 The colors represent the isovalues of orthoradial stress: (a) obtained with the original s^a; (b) obtained under the assumption that $s^a = p^m$. For a color version of this figure, see: www.iste.co.uk/vandamme/coal.zip*

6.5.2. *Carbon dioxide injection through a draining macroscopic fracture*

In this example, we simulate the injection of CO_2 into a 10-m-thick horizontal coal seam in two dimensions. The mesh of the seam, with a length of 40 m, is shown in Figure 6.7. The injection occurs through a porous horizontal fracture, 20 cm thick, located in the center of the seam over the first 20 m. The fracture's thickness is not meshed but is modeled as a mesh interface with zero-thickness elements. This draining fracture can represent, for example, a pre-existing fault or a crack induced by prior fracturing. The permeability of the fracture's filling material is set as a fraction of the permeability of the coal seam. The other material properties are the same as those in the previous example and are summarized in Table 6.4. The initial stresses are -10 MPa, and the initial CO_2 pressure is 1 MPa. CO_2 is injected through the open portion of the fracture located at the center and left side of the mesh (Figure 6.7) by increasing the CO_2 pressure from 1 MPa to 10 MPa over the course of 1 h. The behavior of pure CO_2 is described using the Redlich–Kwong equation, presented at the beginning of the chapter (see Figure 6.1). Figure 6.8 shows the isovalues of pressure after 500 h of CO_2 injection. Four permeability fractions are considered: $k^\Gamma/k = 100{,}10{,}1{,}0.1$. For ratios below 0.1, the pressure profile remains unchanged, and the result becomes independent of the ratio k^Γ/k because the fluid diffuses into the material from the injection point without passing through the fracture. The white horizontal line observed in the figures represents the normal displacement discontinuity of the interface due to the CO_2 pressure generated in the fracture. This opening measures 6.8 mm.

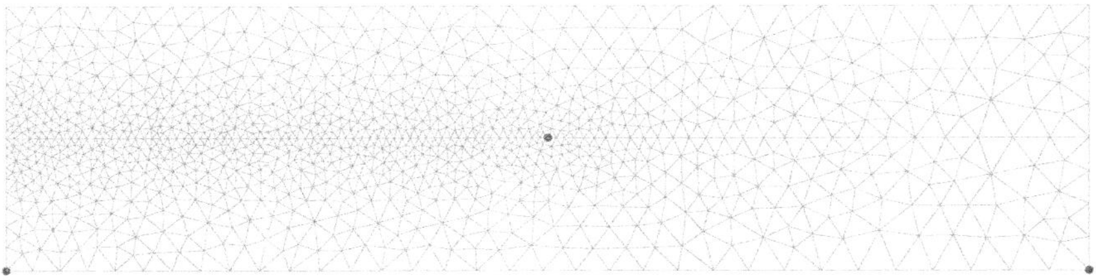

Figure 6.7. *Mesh of the coal seam and the zero-thickness porous interface. For a color version of this figure, see: www.iste.co.uk/vandamme/coal.zip*

Figure 6.9 shows the daily injection rate into the seam as a function of the ratio k^Γ/k. It is evident that this ratio has a significant impact on the injection rate. Figure 6.10 shows the fraction of CO_2 mass stored in an adsorbed state throughout the coal seam as a function of the total stored mass. It can be observed that this ratio does not depend on k^Γ/k, and that the CO_2 mass adsorbed in the coal's micropores represents more than half of the total stored mass.

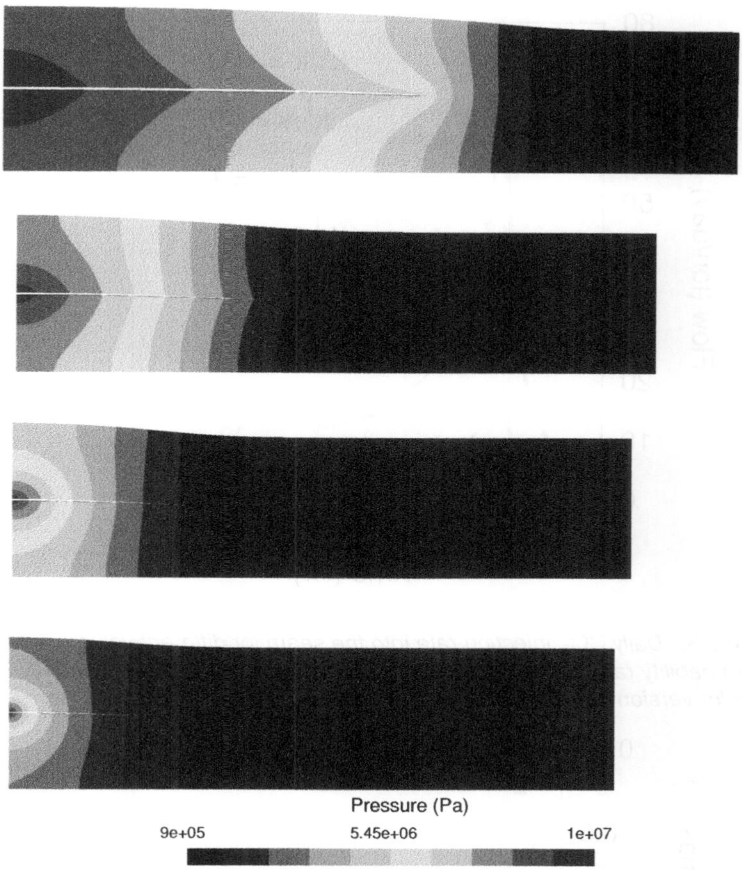

Figure 6.8. *Deformations and CO_2 isopressures after 500 h of injection for four values of the ratio k^Γ/k: from top to bottom 100, 10, 1, 0.1. Deformations are amplified by a factor of 30. For a color version of this figure, see: www.iste.co.uk/vandamme/coal.zip*

6.6. Conclusion

This chapter has focused on describing methods for solving the mechanical behavior and fluid transport at the reservoir scale. The two illustrative examples presented at the end of the chapter reflect the types of problems that can be addressed, particularly methane production or carbon dioxide injection, and the influence of stresses on the reservoir's permeability. It has also been shown that thin porous layers can be taken into account without needing to mesh their thickness. This is particularly useful for representing natural or artificial heterogeneities, such as faults or wells, through which fluid may preferentially flow.

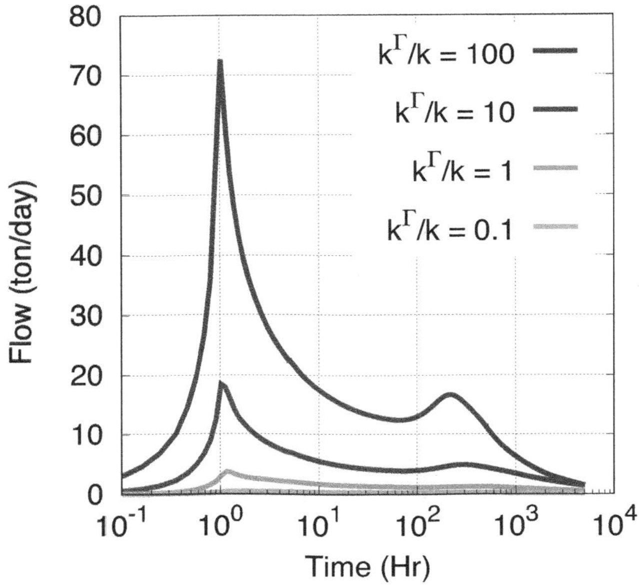

Figure 6.9. *Daily CO_2 injection rate into the seam for different values of the permeability ratio between the fracture and the coal seam: k^Γ/k. For a color version of this figure, see: www.iste.co.uk/vandamme/coal.zip*

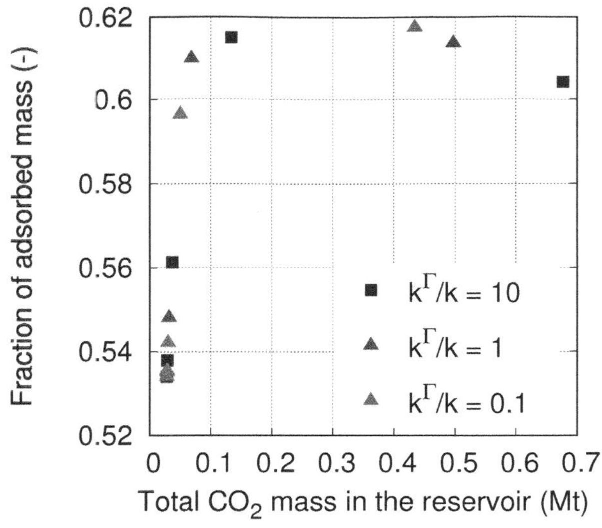

Figure 6.10. *Fraction of CO_2 mass adsorbed in the coal's micropores. For a color version of this figure, see: www.iste.co.uk/vandamme/coal.zip*

This allows for simulating issues related to the sealability or injectivity of a site. From a mechanical standpoint, the model could be further developed to account for irreversible behavior or material damage. Such behavior would be highly valuable to consider at the mentioned interfaces, as it would explicitly account for the fracturing of the material and the impact of such fracturing on fluid transport.

6.7. References

Abouloifa, N., Vandamme, M., Dangla, P. (2021). Modeling transient variations of permeability in coal seams at the reservoir scale. *Journal of Natural Gas Science and Engineering*, 88, 103796.

Brochard, L., Vandamme, M., Pellenq, R.J.-M. (2012a). Poromechanics of microporous media. *Journal of the Mechanics and Physics of Solids*, 60(4), 606–622.

Brochard, L., Vandamme, M., Pellenq, R.J.-M., Fen-Chong, T. (2012b). Adsorption-induced deformation of microporous materials: Coal swelling induced by CO_2-CH_4 competitive adsorption. *Langmuir*, 28(5), 2659–2670.

Connell, L.D., Lu, M., Pan, Z. (2010). An analytical coal permeability model for tri-axial strain and stress conditions. *International Journal of Coal Geology*, 84(2), 103–114.

Cui, X. and Bustin, R.M. (2005). Volumetric strain associated with methane desorption and its impact on coalbed gas production from deep coal seams. *AAPG Bulletin*, 89(9), 1181–1202.

Cui, X., Bustin, R.M., Dipple, G. (2004). Differential transport of CO_2 and CH_4 in coalbed aquifers: Implications for coalbed gas distribution and composition. *AAPG Bulletin*, 88(8), 1149–1161.

Dangla, P. and Bonnard, A. (2024). Universite-Gustave-Eiffel/bil: v2.9. Software, Zenodo. doi: 10.5281/zenodo.14356277.

Espinoza, D.N., Vandamme, M., Dangla, P., Pereira, J.-M., Vidal-Gilbert, S. (2013). A transverse isotropic model for microporous solids: Application to coal matrix adsorption and swelling. *Journal of Geophysical Research: Solid Earth*, 118(12), 6113–6123.

Espinoza, D.N., Vandamme, M., Pereira, J.-M., Dangla, P., Vidal-Gilbert, S. (2014). Measurement and modeling of adsorptive-poromechanical properties of bituminous coal cores exposed to CO_2: Adsorption, swelling strains, swelling stresses and impact on fracture permeability. *International Journal of Coal Geology*, 134-135, 80–95.

Hughes, T. (2012). *The Finite Element Method: Linear Static and Dynamic Finite Element Analysis*. Dover Publications, Mineola.

Kapoor, A., Ritter, J.A., Yang, R.T. (1990). An extended Langmuir model for adsorption of gas mixtures on heterogeneous surfaces. *Langmuir*, 6(3), 660–664.

Karacan, C.O., Ruiz, F.A., CotÃ¨, M., Phipps, S. (2011). Coal mine methane: A review of capture and utilization practices with benefits to mining safety and to greenhouse gas reduction. *International Journal of Coal Geology*, 86(2), 121–156.

Lide, D.R. (ed.) (2010). *CRC Handbook of Chemistry and Physics*, 90th edition (CD-ROM version). CRC Press/Taylor and Francis, Boca Raton.

Liu, J., Chen, Z., Elsworth, D., Qu, H., Chen, D. (2011). Interactions of multiple processes during CBM extraction: A critical review. *International Journal of Coal Geology*, 87(3), 175–189.

Moore, T.A. (2012). Coalbed methane: A review. *International Journal of Coal Geology*, 101, 36–81.

Myers, A. and Prausnitz, J. (1965). Thermodynamics of mixed-gas adsorption. *A.I.Ch.E. Journal*, 11(1), 121–127.

Nikoosokhan, S., Vandamme, M., Dangla, P. (2014). A poromechanical model for coal seams saturated with binary mixtures of CH_4 and CO_2. *Journal of the Mechanics and Physics of Solids*, 71, 97–111.

Pan, Z. and Connell, L.D. (2012). Modelling permeability for coal reservoirs: A review of analytical models and testing data. *International Journal of Coal Geology*, 92, 1–44.

Pini, R., Ottiger, S., Burlini, L., Storti, G., Mazzotti, M. (2009). Role of adsorption and swelling on the dynamics of gas injection in coal. *Journal of Geophysical Research: Solid Earth*, 114(B4). doi: 10.1029/2008JB005961.

Potts, D. and Zdravković, L. (1999). *Finite Element Analysis in Geotechnical Engineering: Theory*. Thomas Telford, London.

Pouya, A. (2012). Three-dimensional flow in fractured porous media: A potential solution based on singular integral equations. *Advances in Water Resources*, 35(Complete), 30–40.

Ravikovitch, P.I. and Neimark, A.V. (2006). Density functional theory model of adsorption deformation. *Langmuir*, 22, 10864–10868.

Redlich, O. and Kwong, J.N.S. (1949). On the thermodynamics of solutions. V. An equation of state. Fugacities of gaseous solutions. *Chemical Reviews*, 44(1), 233–244.

Sun, B., Yang, Q., Zhu, J., Shao, T., Yang, Y., Hou, C., Li, G. (2020). Pore size distributions and pore multifractal characteristics of medium and low-rank coals. *Scientific Reports*, 10, 22353.

Vandamme, M., Dangla, P., Nikoosokhan, S., Brochard, L. (2014). *Modeling the Poromechanical Behavior of Microporous and Mesoporous Solids: Application to Coal*. John Wiley & Sons, New York.

Wu, Y., Liu, J., Elsworth, D., Chen, Z., Connell, L., Pan, Z. (2010). Dual poroelastic response of a coal seam to CO_2 injection. *International Journal of Greenhouse Gas Control*, 4(4), 668–678.

Zhang, X.G., Ranjith, P.G., Perera, M.S.A., Ranathunga, A.S., Haque, A. (2016). Gas transportation and enhanced coalbed methane recovery processes in deep coal seams: A review. *Energy & Fuels*, 30(11), 8832–8849.

Zhou, C., Hall, F., Gasem, K.A.M., Robinson, R.L.J. (1994). Predicting gas adsorption using two-dimensional equations of state. *Industrial & Engineering Chemistry Research*, 33(5), 1280–1289.

7

Multiscale Modeling

François BERTRAND and Frédéric COLLIN
University of Liège, Belgium

7.1. Introduction

This chapter, like Chapter 6, focuses on the reservoir-scale hydromechanical modeling of coal. In particular, it aims to present specific numerical techniques that allow for integrating a model developed at the fracture scale into a more comprehensive model.

7.1.1. *Different scales*

Ideally, a numerical model should provide a detailed representation of the entire microstructure of the material, with a distinct description of each component, that is, the fractures and coal matrix blocks. However, direct modeling of the entire microstructure is not feasible due to the enormous computational costs that such an approach would entail at the reservoir scale.

Consequently, reservoir models are generally phenomenological: the behavior of all components is collectively represented by a set of macroscopic equations. While this approach is numerically efficient, it unfortunately presents certain limitations when attempting to account for an increasing number of physical phenomena occurring on the microscopic scale. This is why these macroscopic models[1] typically

1 In this chapter, the generic term "macroscopic" is used to refer to the reservoir scale, whereas the term "microscopic" refers to the scale of fractures and matrix blocks.

Geomechanics of Coal Seams,
coordinated by Matthieu VANDAMME and Jean-Michel PEREIRA.

rely on a number of simplifying assumptions. In fact, for any model, it is crucial to determine how faithfully it represents reality in relation to the objectives of the modeling. However, coal models often simplify the evolution of permeability with respect to the stress state and sorption, even though this is a critical aspect for modeling methane recovery or carbon dioxide injection into coal.

Over the past few decades, numerous permeability models for coal have been developed in Somerton et al. (1975); Schwerer and Pavone (1984); Seidle et al. (1992); Palmer and Mansoori (1996); Gilman and Beckie (2000); Lu and Connell (2007); Pan and Connell (2007); Connell et al. (2010); Gu and Chalaturnyk (2010); Pan and Connell (2012); Chen et al. (2013, 2017); Shi et al. (2014); Lu et al. (2016), to name just a few. However, these models are either analytical formulations or macroscopic models.

An alternative solution is to use a multiscale approach. The idea is to model micromechanical effects explicitly at their respective scale through direct modeling and then couple their homogenized effects to the larger scale.

7.1.2. *Multiscale approaches*

Mathematical homogenization theory is one technique that, in certain cases, can solve two-scale problems. This analytical method assumes that the microstructure is locally formed by the spatial repetition of a very small part of the entire microstructure (Fish and Wagiman 1993; Schröder 2014). The foundations of this theory are documented in Bensoussan et al. (1978); Sanchez-Palencia and Zaoui (1985). In practice, analytical homogenization is generally limited to very simple microscopic geometries.

In this chapter, we explore another method: numerical homogenization (Kouznetsova et al. 2001). The macroscopic behavior of the material results from the numerical modeling of a Representative Elementary Volume (REV), which contains the details of the microstructure. Averages are then performed on the REV to evaluate the overall response of the material to macroscopic stress. This approach replaces the use of macroscopic constitutive laws.

As the name suggests, the REV must be chosen such that it is representative of the material's microstructure. Figure 7.1 illustrates a thought experiment, demonstrating the representativity of the volume element for the concept of porosity. As the volume decreases, porosity fluctuates and the volume can no longer be considered representative of the medium (Bear 1972; Lake and Srinivasan 2004). Therefore, the volume must be sufficiently large.

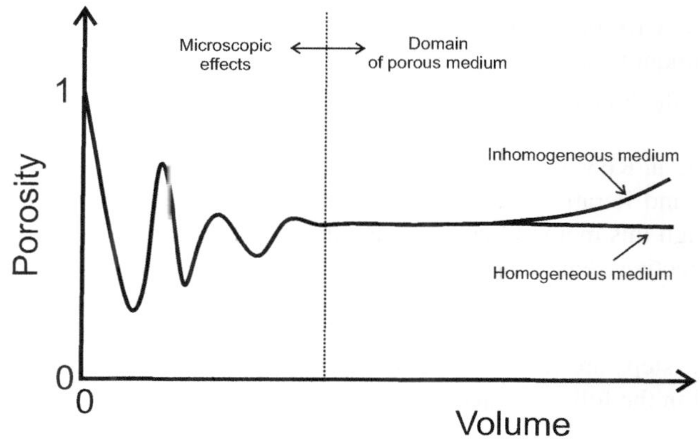

Figure 7.1. *Representativity of the REV for the concept of porosity. Adapted from Bear (1972) and Lake and Srinivasan (2004)*

Once the REV is selected, calculations on the microstructure can be performed using various methods: a finite element (FE) method (Terada and Kikuchi 1995; Smit et al. 1998; Miehe et al. 1999; Feyel and Chaboche 2000; Kouznetsova et al. 2001; Massart 2003; Özdemir et al. 2008; Larsson et al. 2010b; Su et al. 2011; Frey et al. 2013), a discrete element (DEM) method (Nitka et al. 2011; Nguyen et al. 2014), the Voronoi cell method (Ghosh et al. 1995) and a Fourier series approach (Moulinec and Suquet 1998).

When the finite element method is used at both scales, the approach is called the "finite element square method" (FE^2) (Kouznetsova et al. 2001). This is the method we applied to coal and will be developed further in this chapter.

In the context of multiphysics couplings, the FE^2 method was first applied to solve heat conduction problems (Özdemir et al. 2008). It was later used for modeling saturated porous media (Su et al. 2011; van den Eijnden 2015). In the context of coal seam gas, the method has now been extended to partially saturated media (Bertrand et al. 2020).

7.2. Principles of the FE^2 method

The numerical method employed is based on the following four consecutive steps:

1) Macro-micro transition: localization of the macroscopic load on the microstructure.

2) Solving the boundary value problem at the microscale.

3) Micro–macro transition: homogenization of microscale quantities to compute macroscopic quantities.

4) Solving the boundary value problem at the macroscale.

In practice, an REV is associated with each integration point of the FEs meshing the reservoir, and iterative methods are used until convergence is achieved at both scales. Although this method may seem computationally expensive, it is significantly more cost-effective than directly modeling the microstructure across the entire reservoir.

These four steps are reviewed below in order to introduce our hydromechanical model of coal in the following section.

7.2.1. *Macro-to-micro transition (localization)*

The two-scale approach distinguishes between a microscale and a macroscale[2]. Given a point $\hat{\mathcal{P}}$ with a displacement $u_i^M(\hat{\mathcal{P}})$, the displacement $u_i^m(\hat{\mathcal{P}})$ is identical. For a point $\mathcal{P}$ close to $\hat{\mathcal{P}}$, the displacement can be approximated by a Taylor series expansion:

$$u_i(\mathcal{P}) = u_i(\hat{\mathcal{P}}) + \frac{\partial u_i(\hat{\mathcal{P}})}{\partial x_j}(x_j - \hat{x}_j) + \frac{1}{2}\frac{\partial^2 u_i(\hat{\mathcal{P}})}{\partial x_j \partial x_k}(x_j - \hat{x}_j)(x_k - \hat{x}_k) + \cdots \quad [7.1]$$

where x_j and $\hat{x}_j$ are the *j*-th coordinates of $\mathcal{P}$ and $\hat{\mathcal{P}}$, respectively.

For the macroscopic continuum, the displacement field is approximated by:

$$u_i^M(\mathcal{P}) \approx u_i^M(\hat{\mathcal{P}}) + \frac{\partial u_i^M(\hat{\mathcal{P}})}{\partial x_j}(x_j - \hat{x}_j) \quad [7.2]$$

At the microscale, higher order terms are replaced by a fluctuation term $u_i^f(\mathcal{P})$. The microscopic displacement field $u_i^m(\mathcal{P})$ is thus decomposed into a macroscopic field $u_i^M(\mathcal{P})$ and this fluctuation $u_i^f(\mathcal{P})$:

$$u_i^m(\mathcal{P}) = u_i^M(\hat{\mathcal{P}}) + \frac{\partial u_i^M(\hat{\mathcal{P}})}{\partial x_j}(x_j - \hat{x}_j) + u_i^f(\hat{\mathcal{P}}) \quad [7.3]$$

2 The subscript m is used in the following sections to refer to the microscale, while "M" denotes macroscopic quantities.

This fluctuation is such that:

$$\int_{\Gamma_0} u_i^f \, N_j \, d\Gamma_0 = 0 \qquad [7.4]$$

where N_j is the unit normal vector to Γ_0 the outer surface of the REV.

This condition can be satisfied by applying various types of boundary conditions to the REV (Kouznetsova et al. 2001; Özdemir et al. 2008). It is through these boundary conditions that the macroscopic load is transferred to the microstructure (localization) (Geers et al. 2010).

In fact, with Dirichlet or Neumann boundary conditions, we tend to overestimate or underestimate the material's strength when the REV is not fully representative (Suquet 1987). Periodic boundary conditions, however, lead to results that fall between these two boundaries (van der Sluis et al. 2000; Terada et al. 2000).

In theory, the multiscale method presented here is limited to cases where the scale of variation of the macroscopic field is large compared to the microstructure and the size of its REV. This is the concept of scale separation (Kouznetsova et al. 2002). Indeed, since $u_i^m(\hat{\mathcal{P}}) = u_i^M(\hat{\mathcal{P}})$ must hold for any point in the macrostructure, it follows that:

$$\frac{\partial u_i^M(\hat{\mathcal{P}})}{\partial x_j}(x_j - \hat{x}_j) + u_i^f(\hat{\mathcal{P}}) \ll u_i^M(\hat{\mathcal{P}}) \qquad [7.5]$$

If this assumption holds, then the boundary conditions applied to the REV boundary value problem can adequately represent the macroscopic deformation gradient. However, if the macroscopic deformation field varies significantly in certain areas (e.g. in the case of strain localization bands), a higher order homogenization technique should be applied (Schröder 2014) to avoid truncating the deformation field after the second term in the Taylor series expansion (Kouznetsova et al. 2002). In this work, we limit ourselves to first-order homogenization, which assumes a separation of scales.

7.2.2. *Solving the microscale problem*

The resolution of the microscale problem begins with selecting the REV. This REV must be large enough to be representative of the microstructure, yet small enough to maintain the separation between scales. Once the boundary conditions are applied to the REV, the task is to solve a boundary value problem.

The REV is assumed to be in equilibrium. Since it is small, gravitational forces can be neglected, and the equilibrium equation for the first Piola–Kirchhoff stress tensor[3] P_{ij}^m is simply written as:

$$\frac{\partial P_{ij}^m}{\partial X_j} = 0 \qquad [7.6]$$

where X_j represents the coordinates in the reference configuration.

Each component of the microstructure is characterized by its constitutive law, specifying the relationship between stress and strain. A priori, any microscopic constitutive law can be introduced into the FE^2 method. The REV is discretized using FEs to solve the boundary value problem.

7.2.3. *Micro–macro transition (homogenization)*

Macroscopic stresses are derived from a volume-averaged stress within the REV:

$$P_{ij}^M = \frac{1}{\Omega_0}\int_{\Omega_0} P_{ij}^m \, d\Omega_0 = \frac{1}{\Omega_0}\int_{\Gamma_0} \bar{t}_i \, X_j \, d\Gamma_0 \qquad [7.7]$$

where Ω_0 refers to the reference volume of the REV, and its outer surface is denoted as Γ_0. $\bar{t}_i$ is the traction vector on the surface.

This homogenization technique must satisfy the macro-homogeneity condition, also known as the Hill–Mandel condition (Hill 1965; Mandel 1972), which requires that the variation of local work at the macroscale equals the average variation of work over the REV:

$$\delta W^M = \frac{1}{\Omega_0}\int_{\Omega_0} \delta W^m \, d\Omega_0 \qquad [7.8]$$

Formulated using the pair P_{ij} and F_{ij}, with $F_{ij} = \dfrac{\partial x_i}{\partial X_j}$ as the deformation gradient, this principle is expressed as:

$$P_{ij}^M \, \delta F_{ij}^M = \frac{1}{\Omega_0}\int_{\Omega_0} P_{ij}^m \, \delta F_{ij}^m \, d\Omega_0 = \frac{1}{\Omega_0}\int_{\Gamma_0} \bar{t}_i \, \delta x_i \, d\Gamma_0 \qquad [7.9]$$

We will verify that the boundary conditions chosen later ensure that this condition is satisfied when macroscopic quantities are homogenized using equation [7.7].

3 The Piola–Kirchhoff stress relates forces acting in the current configuration to the surface of the element in the reference configuration.

7.2.4. *Solving the macroscale problem*

At the macroscale, the structure is assumed to be homogeneous. It is discretized using FEs. However, no specific constitutive behavior is defined at the macro level. The constitutive relationship at the macroscale is obtained by assigning an REV to each Gauss point in the macroscopic mesh. Indeed, the macroscopic strain tensor is transmitted to the REV through boundary conditions, and the local macroscopic stress is returned to the macroscopic integration point after homogenization. Therefore, the stiffness matrix required at each macroscopic integration point for the Newton–Raphson scheme must be computed numerically from the relationship between variations in macroscopic stress and strain at that point. This can be obtained via numerical perturbation (Miehe 1996) or static condensation (Kouznetsova et al. 2001).

7.2.5. FE^2 *method algorithm*

To conclude this section on the general principles of the FE^2 method, it can be summarized in the following algorithm (Kouznetsova et al. 2001):

1) The macrostructure is discretized using FEs. The load is applied incrementally, following discrete time steps.

2) The macroscopic strain gradient is calculated at each macroscopic integration point based on the estimated macroscopic nodal displacements. During initialization, it is assumed that the deformation is zero throughout the macrostructure.

3) A REV representing the microstructure morphology of the material is assigned to each macroscopic integration point. Appropriate displacements are applied to the REV boundaries based on the macroscopic strain tensor.

4) The stress and strain distributions at the microscale are computed by solving boundary value problems after discretizing each REV with FEs.

5) The average stress in the REV is calculated and returned as the local macroscopic stress at the macroscopic integration point.

6) The macroscopic internal nodal forces are computed based on the stresses obtained for each integration point. If these forces are balanced with the external load, convergence is achieved, and the next time increment can be evaluated. Otherwise, the procedure is continued by updating the estimation of the macroscopic nodal displacements.

7.3. Hydro-mechanical model

Based on the principles presented in the previous section, the FE^2 method is used to develop a two-scale hydro-mechanical model for partially saturated conditions in coal seams.

To this end, new variables need to be introduced: water pressure p_w and gas pressure p_g. The number of degrees of freedom increases by two compared to the purely mechanical problem. As with displacements, the water and gas pressure fields at the microscale can be decomposed into macroscopic components and fluctuation terms:

$$p_w^m(\mathcal{P}) = p_w^M(\hat{\mathcal{P}}) + \frac{\partial p_w^M(\hat{\mathcal{P}})}{\partial x_j}(x_j - \hat{x}_j) + p_w^f(\mathcal{P}) \quad [7.10]$$

and

$$p_g^m(\mathcal{P}) = p_g^M(\hat{\mathcal{P}}) + \frac{\partial p_g^M(\hat{\mathcal{P}})}{\partial x_j}(x_j - \hat{x}_j) + p_g^f(\mathcal{P}) \quad [7.11]$$

7.3.1. *Representative volume element*

Coal, being generally naturally fractured, consists of small matrix blocks separated by fractures. An example of an REV is shown in Figure 7.2. In reality, this corresponds more to a unit cell. For simplicity, a fracture is not represented along the boundary of this representative cell. This cell consists of three parts: the solid matrix, the hydraulic network formed by the fractures and the mechanical interfaces between the solid blocks.

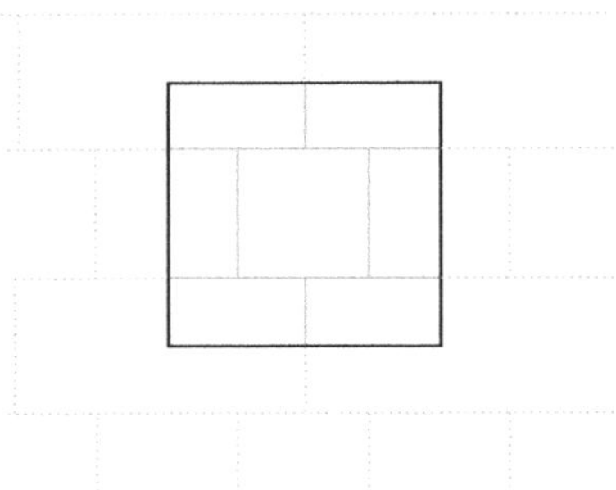

Figure 7.2. *Example of the microstructure used. For a color version of this figure, see: www.iste.co.uk/vandamme/coal.zip*

Note that from a mechanical point of view, defining the REV size is not strictly necessary. However, the fracture openings must have a physical dimension to obtain a meaningful permeability tensor. Practically, the microstructure is defined with a size of one by one, and with the macroscale expressed in meters, a factor defining the REV length (unit or fraction of a unit) is used to convert all quantities involved in the localization and homogenization steps.

7.3.2. Boundary conditions

Periodic boundary conditions are adopted for our microstructure. Besides their effectiveness in transferring the macroscopic strain gradient to the microstructure, this is a natural choice for coal, given its relatively periodic structure (although this is not a requirement for applying such boundary conditions).

The boundary Γ surrounding the REV is periodic, meaning it can be divided into two parts (Figure 7.3): lead Γ^L and follow Γ^F. The kinematics of any point x^F on the following boundary depend directly on its counterpart x^L on the leading boundary.

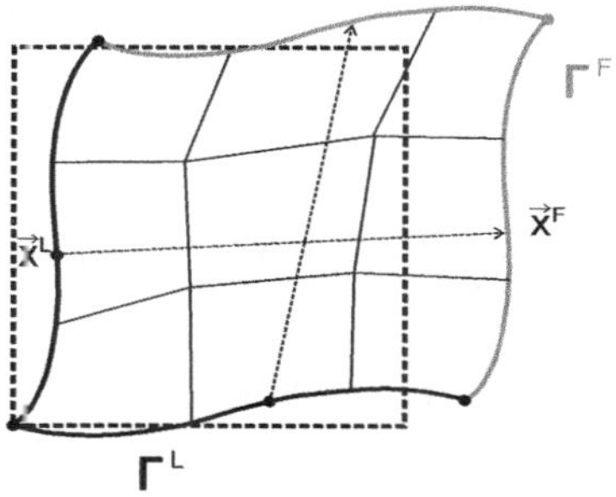

Figure 7.3. *Periodic boundary conditions on the REV. For a color version of this figure, see: www.iste.co.uk/vandamme/coal.zip*

The periodicity conditions for the mechanics can be written as:

$$\begin{aligned} x_i^F &= x_i^L + \frac{\partial x_i^M}{\partial X_j} ({X_j}^F - {X_j}^L) \\ &= x_i^L + F_{ij}^M ({X_j}^F - {X_j}^L) \end{aligned} \quad [7.12]$$

It is important to note that one point on the leading boundary is fixed to prevent rigid body motion. Regarding displacements, it can also be written that:

$$u_i^F = u_i^L + \frac{\partial u_i^M}{\partial X_j} ({X_j}^F - {X_j}^L) \quad [7.13]$$

The fluctuation difference between the two boundary parts is zero:

$${u_i^f}^F - {u_i^f}^L = 0 \quad [7.14]$$

Moreover, periodic conditions lead to the anti-symmetry of traction vectors on the boundaries:

$$\bar{t}_i^F + \bar{t}_i^L = 0 \tag{7.15}$$

We can verify that the adopted boundary conditions satisfy the Hill–Mandel condition:

$$\begin{aligned}
\frac{1}{\Omega_0}\int_{\Omega_0} \delta W^m \, d\Omega_0 &= \frac{1}{\Omega_0}\left(\int_{\Gamma_0^F} \bar{t}_i^F \, \delta x_i^F \, d\Gamma_0 + \int_{\Gamma_0^L} \bar{t}_i^L \, \delta x_i^L \, d\Gamma_0\right) \\
&= \frac{1}{\Omega_0}\int_{\Gamma_0^F} \bar{t}_i^F \, (\delta x_i^F - \delta x_i^L) \, d\Gamma_0 \\
&= \frac{1}{\Omega_0}\int_{\Gamma_0} \bar{t}_i^F \, (X_j^F - X_j^L) \, d\Gamma_0 \quad \delta F_{ji}^M \\
&= \frac{1}{\Omega_0}\int_{\Gamma_0} \bar{t}_i \, X_j \, d\Gamma_0 \quad \delta F_{ij}^M \\
&= P_{ij}^M \, \delta F_{ij}^M \\
&= \delta W^M
\end{aligned} \tag{7.16}$$

Similarly, for hydraulics, the boundary conditions for water and gas are given by:

$$p_w^F = p_w^L + \frac{\partial p_w^M}{\partial x_j} \left({x_j}^F - {x_j}^L\right) \tag{7.17}$$

and

$$p_g^F = p_g^L + \frac{\partial p_g^M}{\partial x_j} \left({x_j}^F - {x_j}^L\right) \tag{7.18}$$

We also observe an anti-symmetry of fluxes on the boundaries:

$$\bar{q}_w^F + \bar{q}_w^L = 0 \tag{7.19}$$

and

$$\bar{q}_g^F + \bar{q}_g^L = 0 \tag{7.20}$$

7.3.3. *Equilibrium equations*

The first Piola–Kirchhoff stress tensor is recommended by Hill to coherently account for the macro-homogeneity condition. However, in this case, the problem at the microscopic scale is formulated under the assumption of small deformations. Therefore, the Cauchy stress tensor σ_{ij}^m can be directly used.

Neglecting gravity, the equilibrium equation for the REV means that:

$$\frac{\partial \sigma_{ij}^m}{\partial x_j} = 0 \qquad [7.21]$$

The weak form of this equilibrium equation is obtained by applying the principle of virtual power, considering an admissible virtual velocity field v_i^{*m}, leading to:

$$\int_\Omega \sigma_{ij}^m \frac{\partial v_i^{*m}}{\partial x_j} \, d\Omega = \int_\Gamma \bar{t}_i \, v_i^{*m} \, d\Gamma \qquad [7.22]$$

Since the REV consists of blocks and interfaces, the boundary Γ in equation [7.22] can be divided into external boundaries Γ^{ext} (periodic boundaries of the REV) and internal boundaries Γ^{int} (the interfaces). The integral over the external boundary is zero due to periodicity. Denoting $T_i^{+/-}$ as the forces acting on the different faces $\Gamma^{int,+}$ and $\Gamma^{int,-}$, the following is expressed:

$$\int_\Omega \sigma_{ij}^m \frac{\partial v_i^{*m}}{\partial x_j} \, d\Omega = \int_{\Gamma^{int,+}} T_i^+ \, v_i^{*m} \, d\Gamma + \int_{\Gamma^{int,-}} T_i^- \, v_i^{*m} \, d\Gamma \qquad [7.23]$$

For the hydraulic part, the equilibrium equations for water and gas are written as:

$$\dot{M}_w^m + \frac{\partial f_{w_i}^m}{\partial x_i} = 0 \qquad [7.24]$$

and

$$\dot{M}_g^m + \frac{\partial f_{g_i}^m}{\partial x_i} = 0 \qquad [7.25]$$

where $f_{w_i}^m$ and $f_{g_i}^m$ are the total internal fluxes for water and gas, and $\dot{M}_w^m$ and $\dot{M}_g^m$ represent the fluid mass variations. Assuming a steady-state condition at the microscopic scale, these latter terms can be eliminated. Considering the virtual pressure fields p_w^{*m} and p_g^{*m}, the equilibrium equations in their weak form are:

$$\int_{\Omega} \left[\cancel{\dot{M}_w^m p_w^{*m}} - f_{w_i}^m \frac{\partial p_w^{*m}}{\partial x_i} \right] d\Omega = - \int_{\Gamma} \bar{q}_w^m \, p_w^{*m} \, d\Gamma \qquad [7.26]$$

$$\int_{\Omega} \left[\cancel{\dot{M}_g^m p_g^{*m}} - f_{g_i}^m \frac{\partial p_g^{*m}}{\partial x_i} \right] d\Omega = - \int_{\Gamma} \bar{q}_g^m \, p_g^{*m} \, d\Gamma \qquad [7.27]$$

7.3.4. *Microscopic constitutive laws*

The multi-scale method presented in this chapter is not limited to a specific microscopic constitutive model. A model at the scale of fractures and matrix blocks is briefly introduced here to demonstrate a numerical application for coal at the end of this chapter.

This model consists of mechanical laws for the solid matrix and fractures, the evolution of adsorbed gas pressure, and finally a multiphase flow law within the fractures.

7.3.4.1. *Mechanical behavior of the solid matrix*

For the mechanical aspect, it is assumed that the material's complex behavior (orthotropy, nonlinearity and even plasticity) arises from the assembly of its various constituents. Therefore, a simple constitutive law is employed for the solid matrix: a linear elastic, isotropic law:

$$\Delta\sigma_{ij} = 2G_m \Delta\varepsilon_{ij} + \lambda_m \Delta\bar{\varepsilon} \, \delta_{ij} + K_m \Delta\varepsilon_{v_s} \delta_{ij} \qquad [7.28]$$

where ε_{ij} is the small strain tensor, given by $\frac{1}{2}\left(\frac{\partial u_i}{\partial x_j} + \frac{\partial u_j}{\partial x_i}\right)$, $\bar{\varepsilon} = \varepsilon_{ij}\delta_{ij}$ with the Kronecker delta δ_{ij}, G_m the shear modulus and λ_m the first Lamé parameter. The last term represents the swelling/shrinkage related to sorption/desorption, where K_m is the isostatic elasticity modulus and $\Delta\varepsilon_{v_s}$ is the volumetric strain induced by sorption. This strain depends on the adsorbed gas content $V_g^{Ad}[m^3/kg]$ and a linear coefficient $\beta_\varepsilon[kg/m^3]$:

$$\Delta\varepsilon_{v_s} = \beta_\varepsilon \, \Delta V_g^{Ad} \qquad [7.29]$$

In the presence of initial stresses σ_{ij_0}, the stress tensor at a given time is expressed as:

$$\sigma_{ij} = \sigma_{ij_0} + \Delta\sigma_{ij} \qquad [7.30]$$

where $\Delta\sigma_{ij}$ represents the stress increment between the current step and the initial state. It is important to note that the effective and total stresses are equivalent in the matrix.

7.3.4.2. *Mechanical law of interfaces*

In this chapter, only elastic behavior is considered for the fractures. The mechanical law of the interfaces is defined by two parameters: normal stiffness K_n and tangential stiffness K_s, which describe the relationship between stresses and the displacements of the fracture surfaces. The tangential stiffness is considered constant, $K_s = K_s^0$, while a Bandis-type law is used for normal stiffness (Goodman 1976; Bandis et al. 1983):

$$K_n = \frac{K_n^0}{\left(1 + \frac{h - h_0}{h_0}\right)^2} \quad [7.31]$$

where K_n^0 is the stiffness corresponding to the initial fracture opening h_0. The initial opening is defined for zero stress. When an initial stress is considered, an initial closure is calculated as:

$$\Delta h_0 = \frac{-\sigma_0' \, h_0}{K_n^0 \, h_0 + \sigma_0'} \quad [7.32]$$

where σ_0' is the initial effective normal stress. This way, the fracture opening can evolve with fluid pressure variations without changing the total stress. It should be noted that the definition of this effective stress does not involve a Biot coefficient, as might appear in a macroscopic formulation.

The effective stress is the difference between the total stress and fluid pressures, weighted by the saturation degree of each phase:

$$\sigma_{ij}' = \sigma_{ij} - \left[S_r \, p_w^M + (1 - S_r) \, p_g^M\right] \delta_{ij} \quad [7.33]$$

The degree of saturation $S_r(p_c)$ is obtained through the following retention curve:

$$p_c = p_w^M - p_g^M = p_e \cdot (S_r^*)^{-\frac{1}{\lambda}} \quad [7.34]$$

where p_e is the air entry pressure and λ is an index representing the distribution of the sizes of capillary tubes opened within the fracture. S_r^* is the normalized saturation:

$$S_r^* = \frac{S_r - S_{r,res}}{1 - S_{r,res} - S_{r_g,res}} \quad [7.35]$$

with $S_{r,res}$ and $S_{r_g,res}$ being the residual saturations for water and gas.

Since the tangential stiffness is constant, the initial tangential closure is directly given by:

$$\Delta s_0 = \frac{\tau_0}{K_s^0} \tag{7.36}$$

where τ_0 is the initial tangential stress. Given the initial stresses σ_{xx_0} and σ_{yy_0}, and the orientation θ of the fracture relative to the horizontal, the normal and tangential stresses on the fracture are given by:

$$\sigma_0 = \sigma_{xx_0} \cos^2(\theta) + \sigma_{yy_0} \sin^2(\theta) \tag{7.37}$$

$$\tau_0 = \frac{1}{2}(\sigma_{yy_0} - \sigma_{xx_0}) \sin(2\theta) \tag{7.38}$$

Finally, the relationships between the stresses and the displacements of the fracture lips are:

$$\sigma_0' + \Delta\sigma' = \frac{K_n^0 \, h_0}{h} \left(\Delta h_0 + \Delta h\right) \tag{7.39}$$

$$\tau_0 + \Delta\tau = K_s^0 \left(\Delta s_0 + \Delta s\right) \tag{7.40}$$

7.3.4.3. *Evolution of adsorbed gas pressure in the matrix*

At equilibrium, the pressure of adsorbed gas is determined using the Langmuir isotherm. If the assumption of instantaneous equilibrium between the pressure in the fractures and the adsorbed content in the matrix is too restrictive, a sorption time $\mathcal{T}$ can be introduced to account for the diffusion process within the matrix.

The average pressure of the adsorbed gas is given by:

$$p_g^{Ad}(t) = p_g^{Ad}(t - \Delta t) + \frac{1}{\mathcal{T}} \left(p_g^{Ad,lim} - p_g^{Ad}(t - \Delta t)\right) \Delta t \tag{7.41}$$

with $p_g^{Ad,lim}$ evaluated for an instantaneous equilibrium. The value of p_g^{Ad} at the end of the time step is more accurately computed by performing sub-time steps δt to reach Δt.

7.3.4.4. *Flow law in the fractures*

The water flow in the interface is given by:

$$\omega_w = -\rho_w \, h_b \, k_{rw} \, \frac{h_b^2}{12} \, \frac{1}{\mu_w} \, \frac{dp_w^m}{ds} = -\rho_w \, \frac{k_{rw}}{\mu_w} \, \kappa(s) \frac{dp_w^m}{ds} \quad [7.42]$$

where $\kappa(s)$ is the geometric transmissivity function, with s as the coordinate along the flow channel. The steady-state condition implies that this flux ω_w remains constant along the channel. Furthermore, due to the scale separation, the variation in fluid pressure is small compared to the size of the REV, rendering the effect on water density ρ_w negligible, and this density can be considered constant through the REV. Under isothermal conditions, the water density depends solely on the macroscopic water pressure p_w^M:

$$\rho_w = \rho_{wo} \left(1 + \frac{p_u^M - p_{wo}^M}{\chi_w}\right) \quad [7.43]$$

with ρ_{wo} the water density at pressure p_{wo}^M and $1/\chi_w$ the compressibility of water. At 20°C, $1/\chi_w = 5 \cdot 10^{-10} Pa^{-1}$.

Similarly, the relative permeability k_{rw} depends only on the macroscopic pressures.

Finally, the water flux between s_1 and s_2 is:

$$\begin{aligned} \omega_w^l &= [p_w^m(s_2) - p_w^m(s_1)] \; \frac{\rho_w(p_w^M) \, k_{rw}(p_w^M, p_{g,f}^M)}{\mu_w} \left(\int_{s_1}^{s_2} \frac{1}{\kappa(s)} ds \right)^{-1} \\ &= [p_w^m(s_2) - p_w^m(s_1)] \; \frac{\rho_w(p_w^M) \, k_{rw}(p_w^M, p_{g,f}^M)}{\mu_w} \, \Phi^l \end{aligned} \quad [7.44]$$

where Φ^l is the geometric transmissivity factor of the channel.

For gas, the flux is:

$$\omega_g^l = [p_g^m(s_2) - p_g^m(s_1)] \; \frac{\rho_g \, k_{rg}}{\mu_g} \Phi^l + [p_w^m(s_2) - p_w^m(s_1)] \; H_g \, \rho_g \, \frac{k_{rw}}{\mu_w} \Phi^l \quad [7.45]$$

where the gas density is determined based on the macroscopic pressure using the ideal gas law. Diffusive fluxes are not considered in the fractures. Moreover, no direct flux is considered in the matrix[4]; the fluxes $\overline{q}_w$ and $\overline{q}_g$ appearing in the equilibrium equations and boundary conditions correspond only to the fluxes in the fractures, ω_w^l and ω_g^l.

7.3.5. *Microscopic numerical solution*

The microscopic-scale problem is solved iteratively in two parts: first, the mechanical problem, followed by the hydraulic problem. This sequence is repeated until convergence is achieved.

7.3.5.1. *Mechanical part*

Equation [7.23], which governs the boundary value problem, is solved numerically using a Newton–Raphson method. The volume is discretized using FEs, including interface elements for the fractures. Starting from a configuration at time t already in equilibrium with the boundary conditions, the goal is to determine a new configuration in equilibrium at the end of the time step. The first estimated configuration is likely to show non-zero residuals, indicating that it is not in equilibrium. The objective is to reduce these residuals to zero in order to find the correct updated configuration.

7.3.5.2. *Hydraulic part*

For the hydraulic part, the goal is to find the water and gas pressures that satisfy the boundary conditions for a given mechanical configuration. Simultaneously, the average of the pressure fields must correspond to the macroscopic pressures.

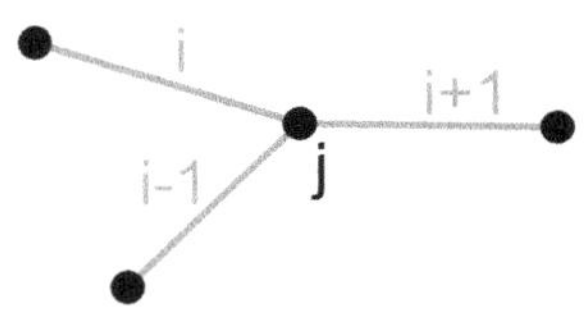

$$\omega^{i-1} + \omega^i + \omega^{i+1} = 0 \qquad [7.46]$$

with $\omega^i \propto \Phi^i(p_j - p_{j-1})$

Figure 7.4. *Mass balance at node j. For a color version of this figure, see: www.iste.co.uk/vandamme/coal.zip*

The conservation of mass along the channel implies that, for each node of the hydraulic network and for each fluid, the sum of incoming flows must equal the sum

4 Diffusion fluxes within the matrix are indirectly accounted for using the sorption time parameter in equation [7.41].

of outgoing flows (Figure 7.4, equation [7.46]). This leads to a system of equations with as many equations as there are nodes in the network. The system is well-posed when boundary conditions are considered, and the value of pressure (the macroscopic pressure) is imposed at one of the nodes. The system of equations for gas is solved after the system for water. As such, the water pressures appearing in the terms for dissolved gas in water are no longer unknowns by the time the second system of equations is solved.

The Φ^i values do not depend on the pressures, so the resulting systems of equations are linear.

7.3.6. *Homogenized response*

7.3.6.1. *Macroscopic quantities*

The micro–macro transition is derived from the Hill–Mandel condition, which requires that the average microscopic work equals the macroscopic work. Due to the presence of discontinuities in the REV, calculating the microscopic virtual work is not straightforward. Nevertheless, it can be shown that the average expression:

$$\sigma_{ij}^M = \frac{1}{\Omega} \int_\Omega \sigma_{ij}^m \, d\Omega \qquad [7.47]$$

satisfies the Hill–Mandel condition (when combined with periodic boundary conditions) for our REV.

The same approach can be applied to the homogenization of fluxes (Özdemir 2009; Massart and Selvadurai 2014). It follows that the macroscopic fluxes are the integrals of the microscopic fluxes at the boundaries:

$$f_{w_i}^M = \frac{1}{\Omega} \int_\Gamma \overline{q}_w^m \, x_i \, d\Gamma \qquad [7.48]$$

$$f_{g_i}^M = \frac{1}{\Omega} \int_\Gamma \overline{q}_g^m \, x_i \, d\Gamma \qquad [7.49]$$

Thus, the macroscopic fluxes are computed as the sum of the microscopic fluxes across the “follow” boundary.

Finally, the masses of the fluids M_w^M and M_g^M need to be homogenized. Since water is assumed to be present only in the fractures of volume Ω^{int}, we have:

$$M_w^M = \frac{1}{\Omega} \int_{\Omega_w^{int}} \rho_w \, d\Omega = \rho_w \, S_r \, \phi_f \qquad [7.50]$$

where ρ_w is constant over the REV, and S_r is the saturation defined by $\frac{\Omega_w^{int}}{\Omega^{int}}$ and ϕ_f by $\frac{\Omega^{int}}{\Omega}$.

The gas content M_g^M is the amount of gas in the interfaces (both the gaseous phase and gas dissolved in water) and the amount of gas adsorbed in the matrix:

$$\begin{aligned} M_g^M &= M_{g,f}^g + M_{g,f}^d + M_{g,m}^{Ad} \\ &= \frac{1}{\Omega}\left(\int_{\Omega_g^{int}} \rho_g \, d\Omega + \int_{\Omega_w^{int}} \rho_g^d \, d\Omega + \int_{\Omega} \rho_g^{Ad} \, d\Omega\right) \\ &= \rho_g \,(1 - S_r)\, \phi_f + \rho_g^d \, S_r \, \phi_f + \rho_g^{Ad} \cdot (1 - \phi_f) \end{aligned} \quad [7.51]$$

with ρ_g is constant over the REV, and ρ_g^{Ad} is related to the pressure p_g^{Ad}, which evolves according to equation [7.41].

Finally, the storage terms $\dot{M}_w^{M,t}$ and $\dot{M}_g^{M,t}$ of interest are obtained by applying a time difference over the interval Δt:

$$\dot{M}_w^{M,t} \approx \frac{M_w^{M,t} - M_w^{M,t-\Delta t}}{\Delta t} \quad [7.52]$$

$$\dot{M}_g^{M,t} \approx \frac{M_g^{M,t} - M_g^{M,t-\Delta t}}{\Delta t} \quad [7.53]$$

7.3.6.2. *Macroscopic tangent matrix*

Finally, the macroscopic-scale calculations are governed by:

$$\begin{bmatrix} [K_{mm}^M]_{(4x4)} & [K_{mw}^M]_{(4x3)} & [K_{mg}^M]_{(4x3)} \\ [K_{wm}^M]_{(3x4)} & [K_{ww}^M]_{(3x3)} & [K_{wg}^M]_{(3x3)} \\ [K_{gm}^M]_{(3x4)} & [K_{gw}^M]_{(3x3)} & [K_{gg}^M]_{(3x3)} \end{bmatrix} \left\{ \begin{matrix} \{\delta\varepsilon^M\}_{(4)} \\ \left\{ \begin{matrix} \delta\nabla p_w^M \\ \delta p_w^M \end{matrix} \right\}_{(3)} \\ \left\{ \begin{matrix} \delta\nabla p_g^M \\ \delta p_g^M \end{matrix} \right\}_{(3)} \end{matrix} \right\} = \left\{ \begin{matrix} \{\delta\sigma^M\}_{(4)} \\ \left\{ \begin{matrix} \delta f_w^M \\ \delta\dot{M}_w^M \end{matrix} \right\}_{(3)} \\ \left\{ \begin{matrix} \delta f_g^M \\ \delta\dot{M}_g^M \end{matrix} \right\}_{(3)} \end{matrix} \right\} \quad [7.54]$$

which can be expressed in a condensed form as

$$[A^M]_{(10x10)}\{\delta U^M\}_{(10)} = \{\delta\Sigma^M\}_{(10)} \quad [7.55]$$

where $[A^M]$ is the stiffness matrix, $\{\delta U^M\}$ contains the infinitesimal variations of the macroscopic variables and $\{\delta \Sigma^M\}$ represents their responses. The stiffness matrix $[A^M]$ is obtained through numerical perturbations, implying that the response of the microstructure to a given load must be computed multiple times.

7.4. Application

The multiscale model presented in the previous sections is now applied to model seam gas production from production wells.

7.4.1. *Case study*

The macroscopic mesh and the geometry of the microstructure used in this model are shown in Figure 7.5. Axisymmetric conditions around the well are employed, with the coal seam being 2 m thick and the microstructure's size being 1 cm by 1 cm. All parameters used in the simulation are summarized in Table 7.1.

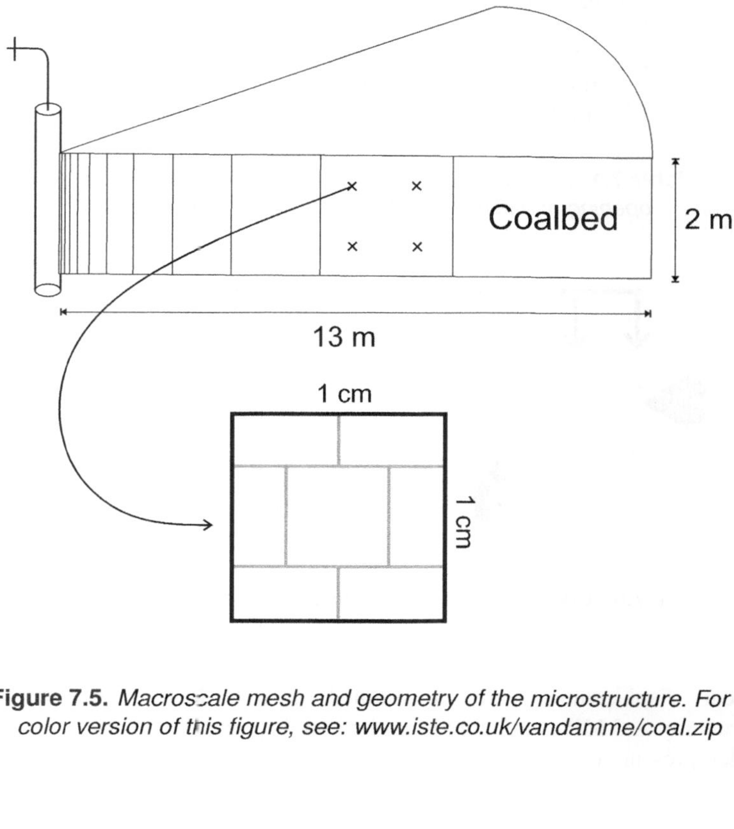

Figure 7.5. *Macroscale mesh and geometry of the microstructure. For a color version of this figure, see: www.iste.co.uk/vandamme/coal.zip*

Parameters	Values
Seam thickness (m)	2
Drainage radius (m)	13
Temperature (K)	293.15
Water pressure (Pa)	3E6
Gas pressure (Pa)	3E6
Surcharge (Pa)	5E6
Well transmissibility T (m^3)	1E-10
Coal density ρ_c (kg/m^3)	1500
Young's modulus of matrix E_m (Pa)	1.21E9
Poisson's ratio of matrix ν_m	0.16
Matrix block width w (m)	$0.005 - 0.01$
Fracture opening* h (m)	20E-6
Fracture normal stiffness* K_n (Pa/m)	100E9
Fracture tangential stiffness K_s (Pa/m)	100E9
Minimum fracture opening (m)	1E-6
Sorption time $\mathcal{T}$ (hours)	10
Langmuir volume V_L (m^3/kg)	0.02
Langmuir pressure P_L (Pa)	1.5E6
Matrix shrinkage coefficient β_ε (kg/m^3)	0.4
Air entry pressure p_e (Pa)	1E5
Fracture distribution index λ	0.25
Residual water saturation $S_{r,res}$	0.1
Residual gas saturation $S_{r_g,res}$	0.0

Table 7.1. *Parameters used for model application. *Fracture opening and stiffness values are given for zero stress*

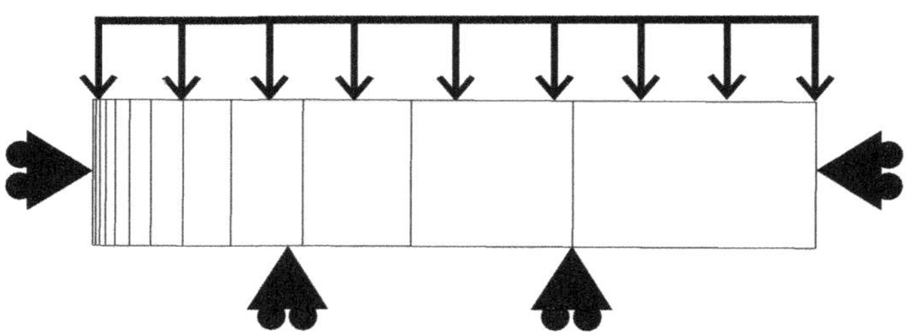

Figure 7.6. *Mechanical boundary conditions on the macrostructure*

Vertical displacements are constrained at the lower boundary of the macrostructure. A 5 MPa surcharge is applied on the upper boundary to represent the overburden pressure exerted on the coal seam. The smallest FE near the well is 5 cm wide. The element size increases by a geometric factor of 1.5, reaching 13 m after 12 elements. The drained volume[5] is thus approximately 1,000 m^3.

The initial water and gas pressures are set at 3 MPa, and the reservoir is water-saturated. Due to the 5 MPa surcharge, the effective stress in the fractures is 2 MPa. Under this stress, the fracture opening[6] is 10 μm and the normal stiffness is $225 \cdot 10^9$ Pa/m. The modeling carried out involves applying a pressure drop at the well from 3 Mpa to 1 MPa over the course of one day. This pressure drop induces water and gas flow toward the well.

7.4.2. *Results*

Figure 7.7 shows the water pressure profile following the pumping process. The pressure profile extends to the reservoir boundaries, allowing the model to replicate the bell-shaped production curve observed in a closed reservoir (with impermeable boundaries).

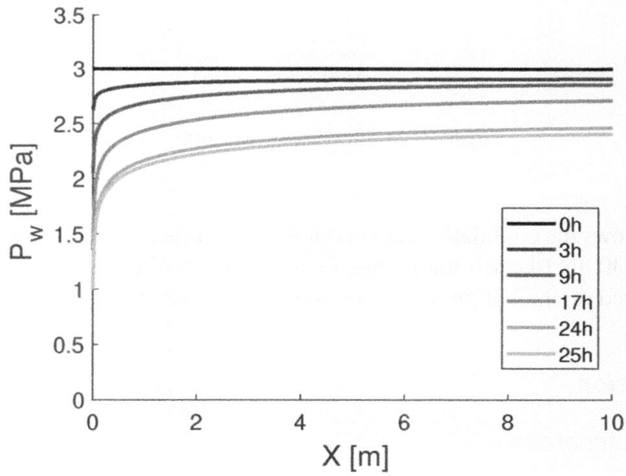

Figure 7.7. *Water pressure decline profile at different times*

Water and gas production curves are illustrated in Figure 7.8 over a period of 1.5 years. Water production peaks rapidly at the beginning, with a rate of less than

5 A real reservoir would be considerably larger. This case study uses a synthetic setup to highlight the dynamics during production.

6 Based on equation [7.32].

2 m^3/day. As the reservoir desaturates, gas permeability increases, causing the gas production curve to peak a few days after water production, with a maximum rate exceeding 400 m^3/day (volume calculated under standard conditions). The extent of the bell-shaped curve corresponds to the reservoir size.

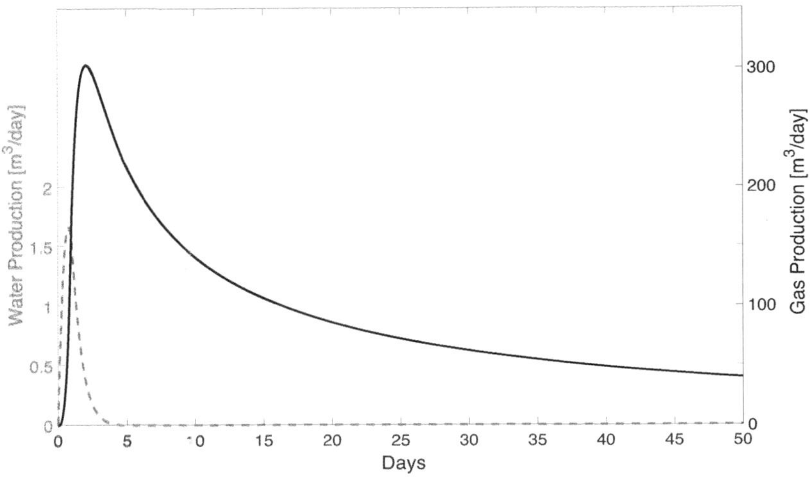

Figure 7.8. *Production curves*

Figure 7.9 shows the cumulative gas production over time. The total gas production converges to 8,330 m^3 (dashed line), which matches the initial gas volume present in the reservoir based on the Langmuir parameters used for this model (see Table 7.1).

7.4.3. *Discussion*

7.4.3.1. *Scale separation*

As explained earlier, the multiscale method employed assumes a separation of scales, meaning that the characteristic microscopic length is much smaller than the characteristic length of the macroscopic deformation field. The microscopic length is logically the size of the REV l_{REV}, since the macroscopic deformation field is applied to the REV boundaries, so $l_{REV} = {x_j}^F - {x_j}^L$. It is then natural to think that the macroscopic length is inversely proportional to the macroscopic pressure gradient $\dfrac{\partial p^M}{\partial x_j}$ (or ∇p^M) for the hydraulic part. This must be verified by the following equation:

$$l_{REV} \ll \frac{p^M}{\nabla p^M}. \qquad [7.56]$$

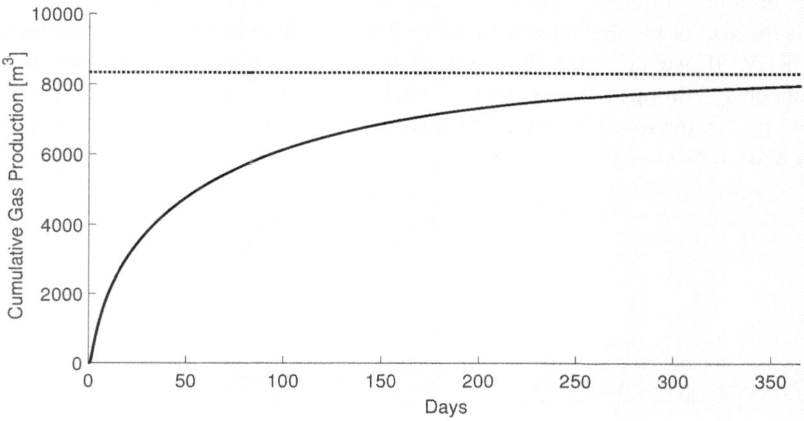

Figure 7.9. *Cumulative gas production curve*

The most critical water pressure gradient is observed near the well at the last moment before maintaining the pressure at 1 MPa, that is, at 24 h into the simulation. At this moment, the 0.05-m wide element closest to the well presents a water pressure difference of 0.44 MPa between the left edge ($p_{well} = 1$ MPa) and its center ($p_{cent} = 1.44$ MPa)[7]. This corresponds to a pressure gradient of 17.6 MPa/m. This result obviously depends on the reservoir's permeability (about 10^{-14} m^2 (10 mD) for this simulation, a value that corresponds to an equivalent macroscopic permeability calculated based on the fracture opening). Consequently, this pressure profile gives a macroscopic length of 6.9 cm in the first half of the first element. In the second half, the characteristic length is already 22.6 cm, which is much larger than the 1 cm of the REV.

$$L_{c_{crit}} = \bar{p} \cdot \frac{\Delta x}{\Delta p} = \frac{p_{cent} + p_{well}}{2} \frac{\Delta x}{p_{cent} - p_{well}} = 0.069 \text{ m} \qquad [7.57]$$

For the same pressure drop of 2 MPa, different well stimulation rates can be tested. Figure 7.10 shows that the slower the stimulation, the larger the characteristic macroscopic length (increasing the separation between scales). The minimum length is obtained for instantaneous stimulation (horizontal asymptote). This minimum value can be determined for a given mesh before the simulation. Indeed, the well pressure instantly reaches 1 MPa, while the pressure at the center of the first element remains at the initial value of 3 MPa. In this particular case, where both the average

7 The analysis was performed on a half-element because eight-node finite elements with four integration points were used.

and the pressure difference equal 2 MPa, the critical macroscopic characteristic length is the size of the first half-element, or 2.5 cm, which is still larger than the size of the REV. If we consider the even more critical case where the pressure is instantaneously dropped to a value much smaller than the initial pressure ($p_{well} \ll p_{cent}$), the macroscopic characteristic length tends toward half the length of the first half-element, $\frac{5\text{ cm}}{4} > 1$ cm.

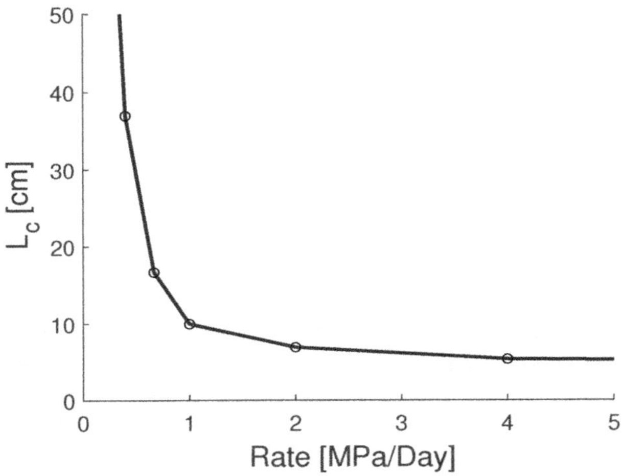

Figure 7.10. *Evolution of the critical characteristic length with loading rate*

The macroscopic mesh must be chosen so that the smallest element is sufficiently large to ensure scale separation (and sufficiently small to avoid numerical oscillations depending on the material's permeability). We can ultimately say that, a priori, selecting the smallest element to be at least four times the size of the REV ensures that the macroscopic characteristic length remains larger than the microscopic characteristic length for any pressure drop, at any time, and in any area of the domain. Afterward, depending on the applied load, we can verify a posteriori how large the scale difference is.

It should be noted that scale separation can also be temporal. Since the microstructural size is small compared to the macro-scale, it is generally consistent to consider a steady-state regime at the lower scale. Instantaneous equilibrium between the fluxes and the boundary conditions applied to the REV should only be assumed if:

$$\mathcal{T} \ll \frac{p^M}{\dot{p}^M} \qquad [7.58]$$

where $\mathcal{T}$ is the characteristic time at the microscale, which in this case is the sorption time. This depends on the diffusion coefficient and the size of the matrix blocks. The term on the right-hand side of the inequality corresponds to the macroscopic characteristic time.

For the simulation carried out, the pressure drop at the well is 2 MPa/day. By the time the well reaches the minimum pressure of 1 MPa, the macroscopic characteristic time is 0.5 day. Thus, if a sorption time of one day is considered, the double-porosity effect related to diffusion in the matrix cannot be neglected.

When condition [7.58] is not met, numerical homogenization must account for the inertia effects of the microstructure. For example, Larsson et al. (2010a, 2010b) did not assume a steady-state regime when studying heat flux problems, using higher order homogenization. A suitable scale transition is also presented by Pham et al. (2013) for modeling heterogeneous media subjected to nearly instantaneous loading. This approach was not followed here because such loading is rather extreme for reservoir simulation. However, it is evident that introducing the sorption time into the calculation of adsorbed gas pressure was relevant. This pressure directly influences the macroscopic storage term.

7.4.3.2. *Evolution of permeability*

The production curves presented in the results section inevitably conceal changes in permeability over time. However, since no macroscopic permeability is explicitly used (as there are no macroscopic constitutive laws), this information must be derived through other means.

Traditionally, permeability connects flow to hydraulic gradients. In the multiscale model, the stiffness matrix still effectively relates these two quantities, while also capturing the evolution of relative permeability[8] as desaturation occurs. The intrinsic permeability is then given by:

$$k = \frac{\mu_w}{k_{rw}\rho_w} \cdot K^M_{wwx} \qquad [7.59]$$

where K^M_{wwx} is the term from the macroscopic tangent matrix that links the water flux toward the well and the corresponding pressure gradient. It is determined at each integration point of the macroscopic FEs.

Figure 7.11 illustrates the permeability profiles at different times along the coal seam. The permeability decreases by nearly a factor of 2 after two days for the

8 The relative permeability can be isolated by using the same retention curve for all fractures (where the air entry pressure is independent of the fracture opening). The multiscale approach is not restricted to this case, of course.

parameters used. This indicates that the effect of effective stress outweighs the matrix shrinkage due to desorption.

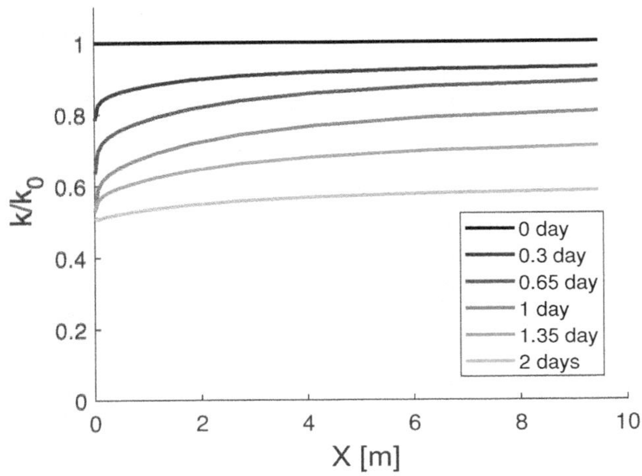

Figure 7.11. *Permeability profiles at different times*

7.4.3.3. *Macroscopic boundary conditions and couplings*

The simulation presented above considers an overburden applied to the coal seam. It is worth investigating the influence of this boundary condition on the production curve. Figure 7.12 compares this curve to the case where vertical displacements are constrained at the boundary. It is observed that this latter boundary condition increases the production peak.

Indeed, the effect of matrix shrinkage on fracture aperture is amplified under this new boundary condition. Figure 7.13 compares the evolution of permeability under the two types of boundary conditions (reservoir conditions versus constant volume).

The impact of matrix shrinkage on the production curve is minimal under reservoir boundary conditions. Indeed, Figure 7.14 shows that the difference between the case without shrinkage ($\beta = 0$) and the cases with shrinkage ($\beta \neq 0$) is minor. In fact, the volume variation due to shrinkage is primarily converted into subsidence rather than fracture opening.

Figure 7.15 shows that the subsidence of the coal seam is proportional to the shrinkage/swelling coefficient β. The subsidence observed for $\beta = 0$ is due to the increase in effective stress caused by reservoir depletion.

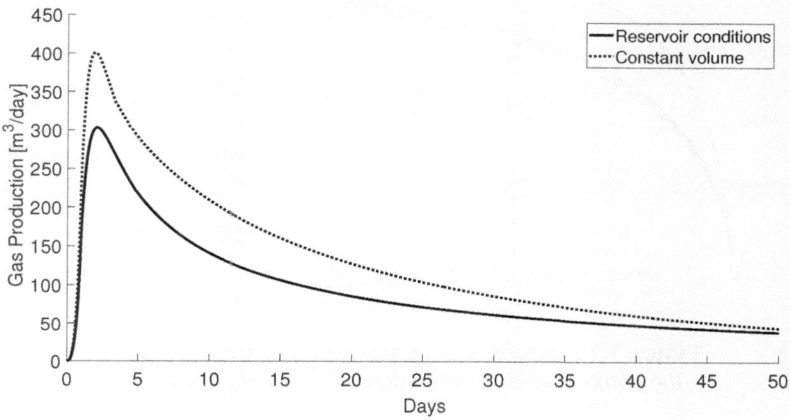

Figure 7.12. *Influence of macroscopic boundary conditions on production*

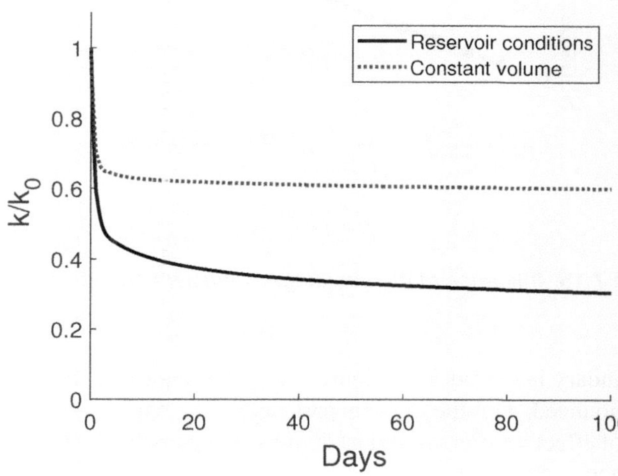

Figure 7.13. *Evolution of intrinsic permeability near the well (x = 6.6 cm)*

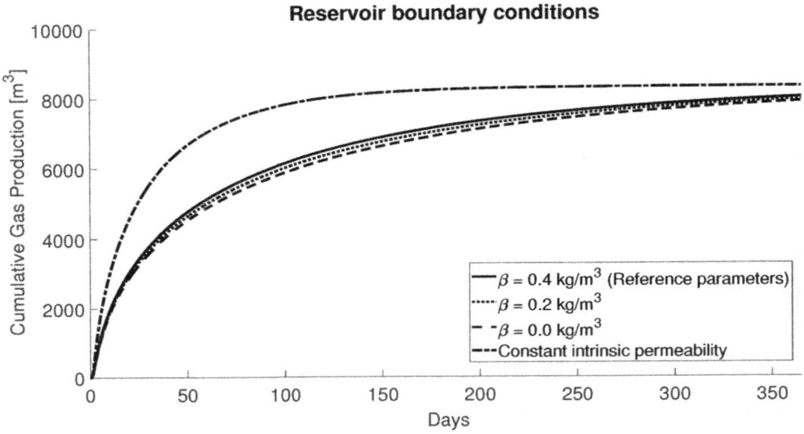

Figure 7.14. *Influence of hydro-mechanical couplings on the cumulative production curve (reservoir conditions)*

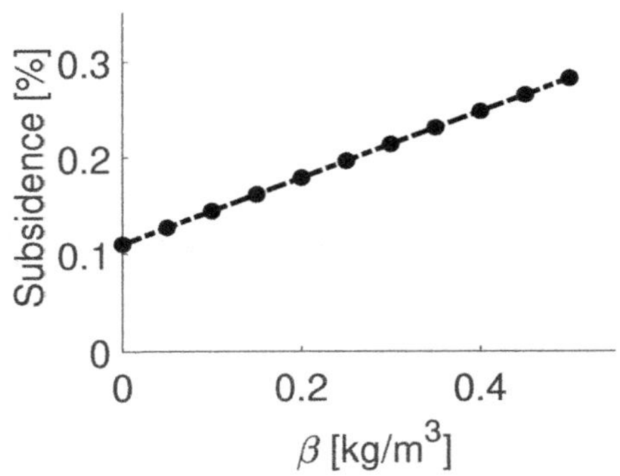

Figure 7.15. *Influence of the shrinkage coefficient on subsidence*

If a rigid boundary is considered (Figure 7.16), the impact of shrinkage becomes much more pronounced. For the chosen parameter set, with $\beta = 0.4$ kg/m^3, the negative impact of effective stress is almost entirely compensated by the positive effect of matrix shrinkage.

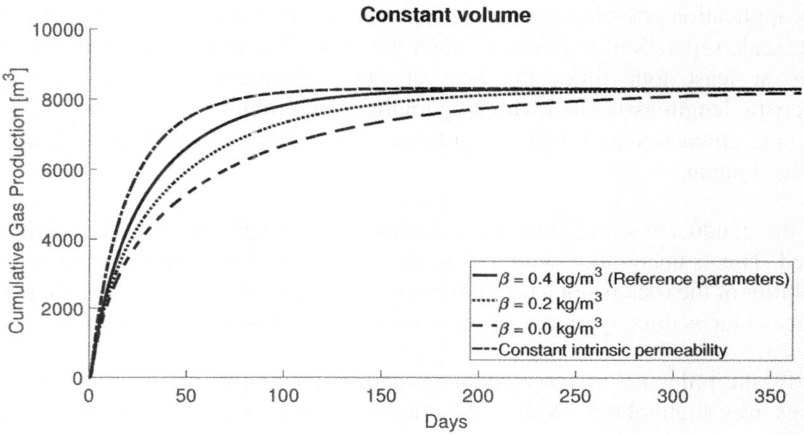

Figure 7.16. *Influence of hydro-mechanical couplings on the cumulative production curve (constant volume)*

The question now arises as to which boundary conditions should be used. In fact, these two types of boundary conditions can be considered as lower and upper bounds for the production curve of a real reservoir. Indeed, with a 2D microstructure, the fractures are necessarily continuous; otherwise, the permeability would be zero. This type of structure is minimally affected by matrix shrinkage if subsidence is allowed. In reality, 3D matrix blocks have attachment points between them that locally represent constant volume conditions, meaning that fractures can be more affected by shrinkage than what is modeled under reservoir conditions with a 2D structure. Depending on the application, whether gas recovery or injection, the most critical boundary conditions will vary.

7.5. Conclusion

In this chapter, the finite element square method was chosen to develop a multiscale hydro-mechanical model for coal. The general principles of the method were presented: localization of the loading on the microstructure, solving the problem at the fracture scale, homogenization of the microstructure's response to this loading and then solving the problem at the reservoir scale. The advantage of this method is that it eliminates the need to develop constitutive laws at the reservoir scale. The hydro-mechanical couplings observed in coal, such as the effects of effective stress and matrix shrinkage on fracture aperture, are directly accounted for at the fracture scale. The developed model was then applied to the simulation of a production well.

This application provided an opportunity to revisit an important assumption of the method: scale separation. In practice, when the smallest element of the macroscopic mesh is at least four times the size of the microstructure, the macroscopic characteristic length associated with a hydraulic load will always be greater than the microscopic characteristic length, regardless of the pressure drop, time or location within the domain.

For the conducted simulation, the evolution of the rock's permeability could be evaluated. This is uncommon with multiscale models since the intrinsic macroscopic permeability of the rock is not a useful parameter in the model. To achieve this, it was necessary to focus directly on the stiffness matrix of the macroscopic problem.

Lastly, the influence of macroscopic boundary conditions on hydro-mechanical couplings was highlighted. With a 2D microstructure, it is not possible to fully represent the reality of attachment points between matrix blocks. Considering different boundary conditions allows for determining extreme bounds for the real case.

Compared to macroscopic models, the presented model allows for the consideration of highly varied microstructure geometries and accounts for hydro-mechanical couplings at the fracture scale. Compared to microscopic models, some simplifications were made, as the problem is solved in steady-state at the lower scale. However, the sorption time is directly incorporated into the macroscopic storage to account for a potential double-porosity effect. In the end, the multiscale model represents a good compromise between the necessary computational time and model sophistication.

7.6. References

Bandis, S., Lumsden, A., Barton, N. (1983). Fundamentals of rock joint deformation. *International Journal of Rock Mechanics and Mining Sciences & Geomechanics Abstracts*, 20(6), 249–268.

Bear, J. (1972). *Dynamics of Fluids in Porous Media*. Courier Corporation, Chelmsford.

Bensoussan, A., Lions, J.-L., Papanicolaou, G. (1978). *Asymptotic Analysis for Periodic Structures*. North-Holland Publishing Company, Amsterdam.

Bertrand, F., Buzzi, O., Bésuelle, P., Collin, F. (2020). Hydro-mechanical modelling of multiphase flow in naturally fractured coalbed using a multi-scale approach. *Journal of Natural Gas Science and Engineering*, 78, 103303.

Chen, D., Pan, Z., Liu, J., Connell, L.D. (2013). An improved relative permeability model for coal reservoirs. *International Journal of Coal Geology*, 109, 45–57.

Chen, S., Yang, T., Ranjith, P., Wei, C. (2017). Mechanism of the two-phase flow model for water and gas based on adsorption and desorption in fractured coal and rock. *Rock Mechanics and Rock Engineering*, 50(3), 571–586.

Connell, L.D., Lu, M., Pan, Z. (2010). An analytical coal permeability model for tri-axial strain and stress conditions. *International Journal of Coal Geology*, 84(2), 103–114.

van den Eijnden, B. (2015) Multi-scale modelling of the hydro-mechanical behaviour of argillaceuous rocks. PhD Thesis, Université Grenoble Alpes.

Feyel, F. and Chaboche, J.-L. (2000). FE^2 multiscale approach for modelling the elastoviscoplastic behaviour of long fibre SiC/Ti composite materials. *Computer Methods in Applied Mechanics and Engineering*, 183(3), 309–330.

Fish, J. and Wagiman, A. (1993). Multiscale finite element method for a locally nonperiodic heterogeneous medium. *Computational Mechanics*, 12(3), 164–180.

Frey, J., Chambon, R., Dascalu, C. (2013). A two-scale poromechanical model for cohesive rocks. *Acta Geotechnica*, 8(2), 107–124.

Geers, M.G., Kouznetsova V.G., Brekelmans, W. (2010). Multi-scale computational homogenization: Trends and challenges. *Journal of Computational and Applied Mathematics*, 234(7), 2175–2182.

Ghosh, S., Lee, K., Moorthy, S. (1995). Multiple scale analysis of heterogeneous elastic structures using homogenization theory and Voronoi cell finite element method. *International Journal of Solids and Structures*, 32(1), 27–62.

Gilman, A. and Beckie, R. (2000). Flow of coal-bed methane to a gallery. *Transport in Porous Media*, 41(1), 1–16.

Goodman, R.E. (1976). *Methods of Geological Engineering in Discontinuous Rocks*. West Publishing Company, Eagan.

Gu, F. and Chalaturnyk, R. (2010). Permeability and porosity models considering anisotropy and discontinuity of coalbeds and application in coupled simulation. *Journal of Petroleum Science and Engineering*, 74(3), 113–131.

Hill, R. (1965). A self-consistent mechanics of composite materials. *Journal of the Mechanics and Physics of Solids*, 13(4), 213–222.

Kouznetsova, V., Brekelmans, W., Baaijens, F. (2001). An approach to micro-macro modeling of heterogeneous materials. *Computational Mechanics*, 27(1), 37–48.

Kouznetsova, V., Geers, M.G., Brekelmans, W.M. (2002). Multi-scale constitutive modelling of heterogeneous materials with a gradient-enhanced computational homogenization scheme. *International Journal for Numerical Methods in Engineering*, 54(8), 1235–1260.

Lake, L.W. and Srinivasan, S. (2004). Statistical scale-up of reservoir properties: Concepts and applications. *Journal of Petroleum Science and Engineering*, 44(1–2), 27–39.

Larsson, F., Runesson, K., Su, F. (2010a). Computational homogenization of uncoupled consolidation in micro-heterogeneous porous media. *International Journal for Numerical and Analytical Methods in Geomechanics*, 34(14), 1431–1458.

Larsson, F., Runesson, K., Su, F. (2010b). Variationally consistent computational homogenization of transient heat flow. *International Journal for Numerical Methods in Engineering*, 81(13), 1659–1686.

Lu, M. and Connell, L.D. (2007). A dual-porosity model for gas reservoir flow incorporating adsorption behaviour – Part I. Theoretical development and asymptotic analyses. *Transport in Porous Media*, 68(2), 153–173.

Lu, S., Cheng, Y., Li, W. (2016). Model development and analysis of the evolution of coal permeability under different boundary conditions. *Journal of Natural Gas Science and Engineering*, 31, 129–138.

Mandel, J. (1972). *Plasticité classique et viscoplasticité*. Springer, Vienna.

Massart, T.J. (2003). Multi-scale modeling of damage in masonry structures. PhD Thesis, Technische Universiteit Eindhoven.

Massart, T.J. and Selvadurai, A. (2014). Computational modelling of crack-induced permeability evolution in granite with dilatant cracks. *International Journal of Rock Mechanics and Mining Sciences*, 70, 593–604.

Miehe, C. (1996). Numerical computation of algorithmic (consistent) tangent moduli in large-strain computational inelasticity. *Computer Methods in Applied Mechanics and Engineering*, 134(3–4), 223–240.

Miehe, C., Schröder, J., Schotte, J. (1999). Computational homogenization analysis in finite plasticity simulation of texture development in polycrystalline materials. *Computer Methods in Applied Mechanics and Engineering*, 171(3–4), 387–418.

Moulinec, H. and Suquet, P. (1998). A numerical method for computing the overall response of nonlinear composites with complex microstructure. *Computer Methods in Applied Mechanics and Engineering*, 157(1–2), 69–94.

Nguyen, T.K., Combe, G., Caillerie, D., Desrues, J. (2014). FEM × DEM modelling of cohesive granular materials: Numerical homogenisation and multi-scale simulations. *Acta Geophysica*, 62(5), 1109–1126.

Nitka, M., Combe, G., Dascalu, C., Desrues, J. (2011). Two-scale modeling of granular materials: A DEM-FEM approach. *Granular Matter*, 13(3), 277–281.

Özdemir, I.I. (2009). Multi-scale modelling of thermal shock damage in refractory materials. PhD Thesis, Technische Universiteit Eindhoven.

Özdemir, I.I., Brekelmans, W., Geers, M. (2008). Computational homogenization for heat conduction in heterogeneous solids. *International Journal for Numerical Methods in Engineering*, 73(2), 185–204.

Palmer, I. and Mansoori, J. (1996). How permeability depends on stress and pore pressure in coalbeds: A new model. *SPE Annual Technical Conference and Exhibition*, 1(6), 539–544.

Pan, Z. and Connell, L.D. (2007). A theoretical model for gas adsorption-induced coal swelling. *International Journal of Coal Geology*, 69(4), 243–252.

Pan, Z. and Connell, L.D. (2012). Modelling permeability for coal reservoirs: A review of analytical models and testing data. *International Journal of Coal Geology*, 92, 1–44.

Pham, K., Kouznetsova, V.G., Geers, M.G. (2013). Transient computational homogenization for heterogeneous materials under dynamic excitation. *Journal of the Mechanics and Physics of Solids*, 61(11), 2125–2146.

Sanchez-Palencia, E. and Zaoui, A. (1985). *Homogenization Techniques for Composite Media*. Springer, Berlin, Heidelberg.

Schröder, J. (2014). A numerical two-scale homogenization scheme: The FE2-method. In *Plasticity and Beyond*, Schröder, J. and Hackl, K. (eds). Springer, Vienna.

Schwerer, F. and Pavone, A. (1984). Effect of pressure-dependent permeability on well-test analyses and long-term production of methane from coal seams. In *SPE Unconventional Gas Recovery Symposium*. Society of Petroleum Engineers, Pittsburgh.

Seidle, J., Jeansonne, M., Erickson, D. (1992). Application of matchstick geometry to stress dependent permeability in coals. In *SPE Rocky Mountain Regional Meeting*. Society of Petroleum Engineers, Casper.

Shi, J.-Q., Pan, Z., Durucan, S. (2014). Analytical models for coal permeability changes during coalbed methane recovery Model comparison and performance evaluation. *International Journal of Coal Geology*, 136, 17–24.

van der Sluis, O., Schreurs, P., Brekelmans, W., Meijer, H. (2000). Overall behaviour of heterogeneous elastoviscoplastic materials: Effect of microstructural modelling. *Mechanics of Materials*, 32(8), 449–462.

Smit, R., Brekelmans, W., Meijer, H. (1998). Prediction of the mechanical behavior of nonlinear heterogeneous systems by multi-level finite element modeling. *Computer Methods in Applied Mechanics and Engineering*, 155(1), 181–192.

Somerton, W.H., Söylemezoglu, I., Dudley, R. (1975). Effect of stress on permeability of coal. *International Journal of Rock Mechanics and Mining Sciences & Geomechanics Abstracts*, 12(5–6), 129–145.

Su, F., Larsson, F., Runesson, K. (2011). Computational homogenization of coupled consolidation problems in micro-heterogeneous porous media. *International Journal for Numerical Methods in Engineering*, 88(11), 1198–1218.

Suquet, P. (1987). Elements of homogenization for inelastic solid mechanics. *Homogenization Techniques for Composite Media*, 272, 193–278.

Terada, K. and Kikuchi, N. (1995). Nonlinear homogenization method for practical applications. *ASME Applied Mechanics Division AMD*, 212, 1–16.

Terada, K., Hori, M., Kyoya, T., Kikuchi, N. (2000). Simulation of the multi-scale convergence in computational homogenization approaches. *International Journal of Solids and Structures*, 37(16), 2285–2311.

List of Authors

François BERTRAND
University of Liège
Belgium

Didier BONIJOLY
Geology, mineral resources and geothermal energy expert
Retired from BRGM
Orléans
France

Laurent BROCHARD
Navier
ENPC, CNRS
Institut Polytechnique de Paris
École nationale des ponts et chaussées
Université Gustave Eiffel
Marne-la-Vallée
France

Frédéric COLLIN
University of Liège
Belgium

Patrick DANGLA
Navier
ENPC, CNRS
Institut Polytechnique de Paris
École nationale des ponts et chaussées
Université Gustave Eiffel
Marne-la-Vallée
France

Derek ELSWORTH
Department of Energy and Mineral Engineering
EMS Energy Institute and G3 Center
and
Department of Geosciences
Pennsylvania State University
University Park
USA

D. Nicolas ESPINOZA
Hildebrand Department of Petroleum and Geosystems Engineering
University of Texas at Austin
USA

David GRÉGOIRE
LFCR
E2S UPPA, CNRS
Université de Pau et des Pays de l'Adour
Anglet
and
Institut Universitaire de France
Paris
France

Jishan LIU
Department of Civil, Environmental and Mining Engineering
School of Engineering
University of Western Australia
Nedlands
Australia

Shimin LIU
Department of Energy and Mineral Engineering
EMS Energy Institute and G3 Center
Pennsylvania State University
University Park
USA

Christelle MIQUEU
LFCR
E2S UPPA, CNRS
Université de Pau et des Pays de l'Adour
Anglet
France

Jean-Michel PEREIRA
Navier
ENPC, CNRS
Institut Polytechnique de Paris
École nationale des ponts et chaussées
Université Gustave Eiffel
Marne-la-Vallée
France

Brandon SCHWARTZ
Terraforma
State College
Pennsylvania
USA

Matthieu VANDAMME
Navier
ENPC, CNRS
Institut Polytechnique de Paris
École nationale des ponts et chaussées
Université Gustave Eiffel
Marne-la-Vallée
France

Shugang WANG
Department of Energy and Mineral Engineering
EMS Energy Institute and G3 Center
Pennsylvania State University
University Park
USA

Index

A, B

C, D

E, F

G, H

I, M

O, P

R, S

T, V

nted and bound by CPI Group (UK) Ltd, Croydon, CR0 4YY
13/09/2026
14961072-0004